Integrating the Subsistence Economy

edited by
Martin Jones

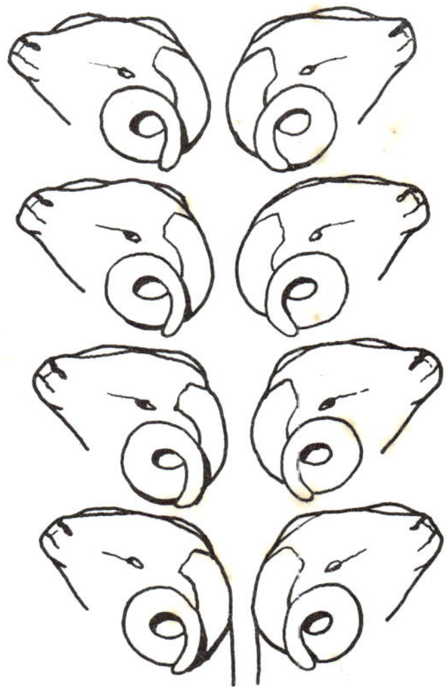

Symposia of the
Association for Environmental Archaeology No. 4
BAR International Series 181
1983

B.A.R.

122 Banbury Road, Oxford OX2 7BP, England

B.A.R.-S181, 1983 : 'Integrating the Subsistence Economy'

Price £ 14.00 post free throughout the world. Payments made in currency other than sterling must be calculated at the current rate of exchange. Cheques should be made payable to B.A.R. and sent to the above address.

ISBN 0 86054 232 7

For details of all BAR publications in print please write to the above address. Information on new titles is sent regularly on request, with no obligation to purchase.

Volumes are distributed direct from the publisher. All BAR prices are inclusive of postage by surface mail anywhere in the world.

Printed in Great Britain

CONTENTS

Page

LIST OF CONTRIBUTORS

Colleen Batey,
Department of Archaeology, 46 Saddler Street, Durham. DH1 3NU

Frank Chambers,
Department of Geography, University of Keele, Keele, Staffordshire. ST5 5BG

Andrew Fleming,
Department of Prehistory & Archaeology, University of Sheffield, Sheffield. S10 2TN

Willy Groenman-van-Waateringe,
Universiteit van Amsterdam, IPP, Singel 453, 1012 WP Amsterdam.

Andrew Jones,
Environmental Archaeology Unit, University of York, Heslington, York. YO1 5DD

Martin Jones,
Department of Archaeology, 46 Saddler Street, Durham. DH1 3NU

David Maguire,
Department of Geography, University of Lancaster, Bailrigg, Lancaster. LA1 4YN

Chris Morris,
Department of Archaeology, 46 Saddler Street, Durham. DH1 3NU

Peter Murphy,
University of East Anglia, Earlham Hall, Norwich. NR4 7TJ

Barbara Noddle,
Department of Anatomy, University College, P.O. Box 78, Cardiff. CF1 1XL

Terry O'Connor,
Environmental Archaeology Unit, University of York, Heslington, York. YO1 5DD

Norbert Paap,
Universiteit van Amsterdam, IPP, Singel 453, 1012 WP Amsterdam.

Jan-Peter Pals,
Universiteit van Amsterdam, IPP, Singel 453, 1012 WP Amsterdam.

James Rackham,
Department of Archaeology, 46 Saddler Street, Durham. DH1 3NU

Nic Ralph,
Department of Prehistory & Archaeology, University of Sheffield, Sheffield. S10 2TN

David Robinson,
Department of Botany, The University of Glasgow. G12 8QQ

Mark, Robinson,
University Museum, Parks Road, Oxford. OX1 3PW

Michael Ryder,
A.R.C. Animal Breeding Research Organisation, Field Laboratory, Roslin, Midlothian. EH25 9PS

Ken Thomas,
University of London, Institute of Archaeology, 31-34 Gordon Square, London. WC1H 0PY

The following conventions have been observed in this volume:

(i) 'bc', 'bp' and 'ad' are used to denote dates measured in radiocarbon years, calculated using the conventional Libby value of 5568 years for the half-life of radiocarbon.

(ii) 'BC' and 'AD' are used to denote dates measured in calendar years.

(iii) 'Meadow' has been used to denote a piece of grassland which is mown for hay, and 'pasture' a piece of grassland used for grazing.

1. INTRODUCTION

Martin Jones

It is one of the ironies of modern archaeology that our most detailed knowledge of past human activity is so often stranded in a sea of ignorance. While our increasingly refined techniques probe ever deeper into some aspects of the private lives of people in the past discovering things that they would not themselves know or even wish to know, they barely scratch the surface of other aspects. Thus when looking at subsistence patterns we may be able to establish the structure of flock or herd, and the way in which a farmer manipulated and utilised his animals. We may be able to estimate the mesh-size of the sieve he used to clean his grain, and observe from his bones the symptoms of leprosy he would contract after consuming the saponin-rich seeds of *Agrostemma githago* that the sieve had failed to remove. Yet with all this we may remain ignorant of the relative importance of plants and animals to his community.

In part this illustration reflects the limitations inherent in our data base, but there are possibilities within the data that have not fully been explored. In part it reflects the divisions that are artificially imposed by the circumstances of research between different forms of environmental analysis, and there are certainly possibilities here for development and change. It was to the various possibilities of integration between different forms of environmental evidence, between different lines of research, and ultimately between different strands of subsistence economy that the 1982 conference of the Association for Environmental Archaeology was addressed.

As the papers reflect, there are a number of aspects of the problem of integration, and a number of themes which recur through the volume. I shall attempt to comment on some of those themes.

Multidisciplinary studies

In advocating the multidisciplinary approach adopted by many of the other contributors, David Maguire stresses the need for examining pollen and other types of environmental evidence 'which occur in close juxtaposition' such that one may be 'calibrated' against the other. The collection of a variety of forms of evidence from close juxtaposition is no straightforward matter. The traditional sources of; for example, pollen and snails respectively, may be segregated into distinct and separate geographical areas.

The papers in section two pursue the course pioneered by Geoffrey Dimbleby of taking pollen analysis out of the remote peat bogs and late sediments, and into the very heart of the settled landscape. In this way it is possible to place pollen evidence alongside evidence from seeds, agricultural sediments, and human habitation, and relate it directly to the cultural stratigraphy. The potential is illustrated from the level of the landscape (Maguire et al; Chambers; Groenman-van Waateringe and Pals) to the level of the individual feature (David Robinson). A complementary repertoire of environmental evidence comprising fish, bird and mammal bones, land and marine molluscs and charred plant remains, demonstrates the possibility of a multidisciplinary study on a site in which alkaline sediments predominate (Jones et al.). The contrast between this latter paper and the four in section two does demonstrate how regional variations in the depositional environment may still constrain the juxtaposition of different forms of evidence. In some areas this may be overcome by operating at the scale of the landscape rather than the site, thereby exploiting the diversity of depositional environments within a region. Peter Murphy's study of the Brecklands provides an elegant example. By studying a range of sites from valley floors, valley terraces, the adjacent slopes and upland areas, he is able to complement peat deposits from the valley floor with buried alkaline soils in the upland areas and settlement refuse bearing bones and charred seeds in the intervening zone.

Taphonomy

If these different forms of evidence are to be juxtaposed and compared, and especially if they are to be quantified, then problems of taphonomy emerge. In David Maguire's critique of the current state of play in the art of arable/pastoral indices from pollen, and in Mark Robinson's examination of the possibility of deriving such indices from insect remains, each author emphasises the importance of understanding how pollen and insects respectively are dispersed and enter the archaeological record. In the absence of such an understanding, any derived quanta may be difficult to interpret. Pollen taphonomy is of course central to the papers in section two that examine pollen from the complex depositional environment of a settled landscape. These papers nevertheless demonstrate that such complexity, if handled in the right way, may be a source of evidence rather than error. It is interesting to observe that both papers that attempt a subdivision of cereal pollen into genus-type find a contrast between the pollen and charred cereals (Maguire et al.; Groenman-van Waatering and Pals). Maguire et al. make the interesting and reasonable suggestion that the pollen evidence in this case is more likely to reflect the local situation than the charred grains. Taphonomic constraints will operate on each form of evidence. Their implications for the quantification of bone data are rigourously explored in this volume by James Rackham.

Ecology and Population Biology

In the same way that the death assemblages that constitute our data base cannot be treated as simple unbiased samples of living populations, neither can those living populations as unswerving 'land-use markers' flocking religiously to the appropriate anthropologically defined land-unit. As pointed out by Mark Robinson in the case of insects, each species is constrained by a different set of environmental factors, some of which may not be even perceived, let alone considered important, by the human observer. A dung beetle may be far more concerned with the microtopography of a cow-pat than with the macrotopography of an agrarian landscape. If plant and animal species are to be used as indicators of past human landscapes, then it must be done through, and not in spite of, the ecology and population biology of each species concerned. Such an approach may enhance powers of discrimination in some aspects of landscape reconstruction, while tempering others with caution. Thus while Robinson argues that insects may not be ideally suited to reflecting arable activity, the nuances of their ecology and population biology may allow grasslands to be usefully subdivided and characterised. On another aspect of biological diversity, Barbara Noddle demonstrates that some of the variations in domestic animal bones may be attributed to differences in breed.

The advantages of framing our influence with a clear understanding of ecology and population biology is not limited to the various plants and animals examined. Our approach to the human populations at the centre of such inference may also be framed in ecological terms. In handling the various elements of palaeo-economic evidence we may tend to think in rather static terms, of some past landscape on which the various kinds of land-use are mapped out. Ken Thomas argues that we should instead 'apply some dynamic to them, involving the movement of energy, materials and people'. In moving away from a map approach to an energy-flow approach, Thomas prefers to identify potential routes through the energy pathway, rather than quantify areas of land use. Such an approach is echoed by the point made by Maguire et al. that the ratio between pastoral and arable land is a different thing from the ratio between pastoral and arable products, which bears a closer relation to energy pathways. The pathways so constructed may form intricate food-chains, and the diversity of routes is well illustrated by Michael Ryder's discussion of a single component, milk.

Such an approach leads Thomas to define more precisely the 'subsistence economy' we are trying to integrate, the various scales at which it operates, and how these relate to the archaeological landscapes from which our data is derived. Taking Thomas's definition of a subsistence economy as one in which 'the units of production are the same as the units of consumption', the question becomes not so much whether something may be treated as a subsistence economy, but what is the scale at which it is appropriate to do so.

Terry O'Connor can in this way treat Lincoln as a subsistence economy, simply be expanding the unit of analysis to encompass the whole county. By doing this, some important points emerge about rural settlements, that could not be discerned from a consideration of such settlements in isolation. The need to frame our palaeo-economic inferences within a hierarchy of nested systems is made clear by the various papers that draw attention to the intimate links between agricultural production and the wider economy. Norbert Paap's useful attempt to follow these links through to social divisions within an urban context is a prime example.

In summary, some valuable points may be drawn from the conference proceedings. It is within a framework of multidisciplinary studies and sound taphonomic guidelines that our miscellany of evidence can be translated into living systems. We may wish to perceive those systems as some kind of land-use map, but may be constrained in so doing by the very nature of the evidence. An alternative approach is to perceive those systems in terms of their own inherent structure. The various biological components of the system may be understood in terms of their own biology, and the various components integrated, not into a static landscape reconstruction, but into a dynamic food chain, with energy flowing along the various pathways that lead to humans, and in the case of the unfortunate leper who generated this discussion, to the higher trophic levels that lie beyond.

The contributors have demonstrated their intention not to be constrained by the presumed limitations of the data base. Neither need they be constrained by the few personal comments offered above, which are intended not as a summary of the proceedings, but merely as a series of markers that may hopefully guide the reader through the many valuable contributions to this volume.

Acknowledgements

Many colleagues in the AEA were involved in some aspect of the organisation of the conference, and thanks are particularly due to Neil Lang, James Rackham, Ian Simmons, Judy Turner and Maryke Van der Veen. We also benefited from the extensive co-operation of the Departments of Archaeology, Botany and Geography, the Administrative Office, and St. Aidan's College at Durham University. The cover logo and various incidental graphics were executed by Yvonne Brown and the entire manuscript was typed by Deborah Clough to whom our gratitude is extended.

2. THE IDENTIFICATION OF AGRICULTURAL ACTIVITY USING POLLEN ANALYSIS

David J. Maguire

INTRODUCTION

The objective of much palaeoenvironmental research is the identification of agricultural activity from the pollen incorporated within terrestrial sediments (see Edwards 1979; Simmons and Tooley 1981). Following pollen analytical investigations, statements about whether or not agricultural activity was practised are frequently made. It is statements about the amount of agricultural activity in a given area, during a particular time period, which prove most interesting and useful. Before such remarks can be made ecological relationships, taphonomic processes and data collection must be examined critically (Fig 2:1).

This paper considers in the first section, the taphonomic processes which lead to the formation of the fossil assemblage in the second section the procedures involved in the collection of pollen data and in the third section the relationship between agricultural activity and its associated vegetation. Also considered in the third section, are the three traditionally adopted approaches to the problem of identifying agriculture: the indicator, comparative and statistical approaches. Some additional approaches are proposed. The discussion in this paper is concerned with terrestrial sites, although some of the arguments may be relevant to lacustrine and marine sites.

TAPHONOMIC PROCESSES

Production

The anthers of plants produce varying amounts of pollen. It is well known that entomophilous (insect pollinated) plants, such as _Trifolium_, _Calluna_ and _Tilia_, and cleistogamous (closed flower) plants, such as _Viola_, are low pollen producers. In general, anemophilous (wind pollinated) plants, such as _Quercus_, _Pinus_ and _Alnus_, produce more pollen per anther. Production of pollen per unit area of land per unit of time (measured in grains per square metre per year), is however of more relevance to the reconstruction of past plant communities in space and time, as the work of Andersen (1973), Bradshaw (1981) and others has indicated. Certain low pollen producers, such as _Tilia_, actually produce more grains per square metre per year than some grasses, for example _Secale_, largely because of their greater vertical extension.

Cereals, with the exception of the genus Secale, are self-pollinated and tend to produce low amounts of pollen (Ralska-Jascewiczowa 1968). In very general terms the production of pollen in grains per square metre per year will tend to be in the order trees > pastoral agriculture > arable agriculture (see for example Pohl 1937).

Dispersal

Dispersal of pollen grains may be by wind, water or biological agencies. Pollen grains liberated into the atmosphere are dispersed according to the laws which govern all small particles. Generally speaking, large heavy grains and those with low volume: surface area ratios tend to move shorter distances. A number of general models which consider pollen movement have been developed. Tauber (1965, 1967) considered situations in which pollen is moved to sites in wooded areas, whilst Janssen (1966) investigated the proportion of pollen derived from various parts of the pollen source area of a site. These models are discussed and extended by Edwards (1979, 1982) and Jacobson and Bradshaw (1981) and are embodied in Turner's (1970, 1975) three-dimensional approach to vegetation reconstruction. The importance of these studies to the problem considered here is that they allow quantitative rather than qualitative estimates of the amount of agriculture in space and time.

In woodlands pollen dispersal is influenced by the time of year of flowering, because wind speeds decrease after tree leaves open (Tauber 1965). Therefore, the pollen of plants associated with agriculture in wooded environments will tend to move further if liberated before the tree leaves are fully open.

Water functions as a dispersal mechanism at terrestrial sites in a number of different ways. Certain taxa may be concentrated by the wash-off effect of water removing pollen grains deposited on vegetation (Krzywinksi 1977). Overland flow, interflow and channel flow may also be important dispersal mechanisms, particularly on sloping sites.

A number of biological agencies are important in the dispersal of pollen grains. Some grains are expelled forcibly by active plant processes, others may be transported by man and other animals.

Cereal pollen grains, because of their large size and the tendency of Triticum, Hordeum and Avena to remain in their hulls, (Vuorela 1973, Behre 1981) are normally poorly transported by natural agencies. Heim (1962) has established that modern cereal pollen will attain percentages of 3-4% when arable agriculture is being practised within about 2 km of the sampling site. He also mentions that percentages only rise above 4-5% when crops are grown in the immediate vicinity of the site.

The anthropogenic transport of cereal pollen grains, especially around archaeological sites, may be more important than transport by natural agencies. Vuorela (1973) in her investigations into recent pollen deposition has demonstrated that the natural dispersal of Hordeum and Avena is very limited, but increases considerably during the harvesting and threshing period. Also, she was able to show that cereal pollen grains were concentrated along the routes which harvesters used. Robinson and Hubbard (1977) have suggested that cereal pollen grains may be transported to archaeological sites on cereal fragments. Anthropogenically induced transport mechanisms may be particularly important on sites with plaggen (man-made) soils which have been manured. As Krzywinksi pointed out at the 1980 A.E.A. meeting, cereal pollen grains may survive transport through alimentary tracts and so could be transported by animals and in manure.

When dealing with low pollen values, as is often the case in the identification of agricultural activity, the 'chance' movement of grains caught in animal, or deposited in bird, faeces must not be forgotten. Birks and Birks (1980) point out there is a small probability of finding a pollen grain of any given taxon anywhere in the world. Similarly long distance transport by wind must not be overlooked as, for example, Tyldesley's (1973) work shows.

Sedimentation

The ability of a sediment surface to collect pollen depends not only on the position and orientation of the sediment relative to the pollen source and dispersal mechanism, but also upon the character of the surface. Different vegetation types will obviously vary in their pollen collection characteristics. Rough wet surfaces, *ceteris paribus*, might be expected to collect more pollen per metre squared per year than dry smooth ones. Similarly the windward side of moss polsters could collect more pollen grains than the downwind side. Salmi (1962) has shown that there may be differential horizontal deposition in bog pools; grains and bladders of for example, *Pinus* and *Picea* tend to float and may be blown to the downwind side of pools.

It is often assumed that pollen landing upon a bog surface is simply preserved as the peat grows around it. Although Rowley and Rowley (1956) demonstrated differential downward movement of grains, Clymo's (1973) studies indicate that a compact layer often forms by a depth of 5 to 10 cm. Movement may still be important when considering a series of short term agricultural episodes or taking thin samples for analysis.

Pollen grains tend to move slowly and continuously through soils (Dimbleby 1961), possibly with differential sorting and so sedimentation may be further complicated. This produces a certain amount of blurring in the pollen record with grains of different ages present at any given level (see Evans 1975 Fig 35). On archaeological sites this simplistic model may be complicated by ploughing, fertilizing, manuring or burial of the soil.

Pollen grains are very resistant to decay because of their sporopollenin content. However, they are still susceptible to breakdown by oxidation both in aerobic sediments and by biological agencies. Havinga (summarised in Havinga 1974) has investigated the susceptibility of different types of grains to corrosion in various sediments. The decay of pollen grains tends to cause more problems in soil rather than peat samples.

The fossil assemblage of a site may, in certain circumstances, contain pollen which has been redeposited due to sediment erosion. Again this process may be especially important in archaeological contexts because arable and pastoral farming frequently institute erosion. There is a suggestion that some of the pollen in the bog at Cefn Graeanog may have been redeposited (Chambers this volume). Similarly the medieval ditch from Holne Moor probably contains probably contains reworked sediment and pollen from earlier periods (Maguire, Fleming and Ralph this volume).

COLLECTION OF POLLEN DATA

Sampling and counting

Unfortunately pollen analysts have little control over the taphonomic processes which combine to produce fossil assemblages. However, in the collection of pollen data (Fig 2:1), pollen analysts may have a good deal of control over the events which could lead to erroneous results. Consideration of the methods of sampling and counting which are used to estimate the fossil assemblage is therefore of importance.

The selection of sites for palaeoenvironmental reconstruction has recently been reviewed by Jacobson and Bradshaw (1981). They suggest that different sizes of sampling sites will have different sizes of pollen source areas. Thus the majority of the pollen rain at a small site, such as a forest clearing 25 m in diameter, will be derived from the local area, that is within about 20-30 m of the site. By the same reasoning a large sampling site, such as a blanket bog, will derive most of its pollen rain from several hundred metres to several km away. Clearly sites such as acid soils and small forest clearings will provide the most useful records for reconstructing local agriculture, small peat bogs will be best suited for reconstructing extralocal agriculture and blanket peat sites should be used when trying to reconstruct regional agriculture. It must not be forgotten that human influences upon vegetation, for example deforestation and agriculture, could change the size of a sampling site and consequently may affect the pollen rain.

Sediment resolution (accumulation per year) could be important when attempting to identify the amount and type of agriculture. In marginal upland sites arable activity tends to occur for relatively short periods of time, intercalated between much longer periods of pastoral activity. In sediments where the depth of sample (traditionally 1 cm thick) represents 10-20 years, arable agriculture may be completely missed. If the sample represents 100-200 years several phases of agriculture could be present.

Traditional methods of extracting and concentrating fossil pollen generally involve boiling in potassium hydroxide, acetolysis, treatment with hydrofluoric acid or oxidation and mounting in glycerol jelly (see Faegri and Iversen 1975 and Moore and Webb 1978). One major objection to the use of glycerol jelly is the fact that pollen grains tend to swell when mounted in it (Faegri and Deuse 1960), making it difficult to employ size statistics in identification. Andersen (1960) therefore proposes the use of

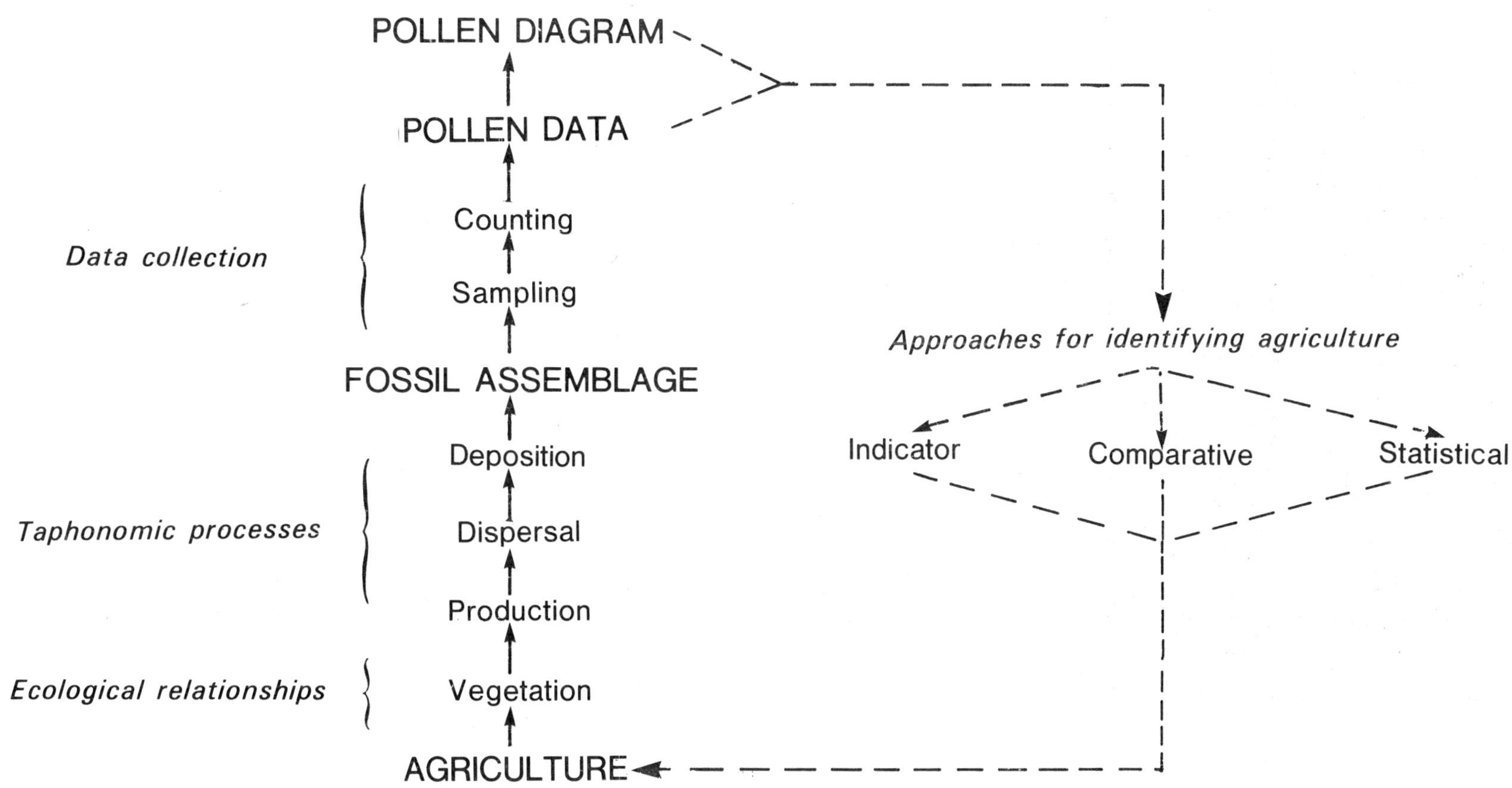

Fig. 2:1 The relationship between pollen data and agriculture and the approaches which may be adopted for identifying agriculture from pollen data.

silicone fluid as a mounting medium because it does not cause swelling. Other stages of the preparation procedure have also been shown to influence pollen size (Reitsma 1969). This led Faegri and Iversen (1975) to develop the concept of 'standard size' which they define as the size of a grain under conditions in which *Corylus avellana* has a mean size of 25 microns. This technique was used successfully by Maguire *et al.* (this volume). Correction factors of this nature should be used with care since there are several other influences on pollen grain size. According to Birks and Birks (1980) pollen size may vary not only within species, due to clinal variation or different geographical range, but even within individual plants.

All samples of pollen which are examined contain grains which are either indeterminable or determinable. If a large proportion of the grains are indeterminable the inferences which can be drawn from the data may be restricted. Silicone fluid is preferred as a mounting medium because it allows grains to be manipulated about slides and therefore identified more easily.

The splitting of Gramineae into characteristic taxa is of great interpretative value in the identification of agriculture, 'a collective curve is of little value because grains may be important in many different plant communities' (Faegri 1964). Unfortunately, due to the homogeneous nature of grass pollen, specific type identifications are very difficult. Investigations into the size distributions of modern grass pollen indicate that cereal pollen is characterised by a large grain size and annulus diameter. Analytical measurements of grain diameters in excess of 37 microns (Beug 1961) or 40 microns (Faegri and Iversen 1975) are taken to be indicative of cereal type grains. Even on this basis a few wild-type grasses remain indistinguishable. With a view of clearer differentiation, studies of surface sculpturing have been undertaken on both the light microscope, using transmitted and phase contrast microscopy (Grohne 1957, Rowley 1960), and the scanning electron microscope (Andersen and Bertelsen 1972). The author has found that using a key adapted from the size measurements of Beug (1961) and key of Faegri and Iversen (1975) it is possible to assign Gramineae grains to several general categories. Faegri (1964) distinguishes between *Dactylis* and *Festuca* (pastoral) and cereal (arable). Care must be taken when inferring arable or pastoral farming from such data since, for example, in the key of Faegri and Iversen (1975) '*Hordeum*-type' includes the cereal grains *Hordeum vulgare* L.emend and *Triticum monococcum* L. and also the wild-type grass *Agropyrum junceum*. In conclusion it must be said that when using the light microscope, only when a reasonable number of Gramineae grains are available for identification should analysts say with any certainty that cereal pollen is present.

The problem of differentiation is also found in the assignation of agricultural indicators to taxa. For example, although *Plantago lanceolata* can normally be identified with some confidence, it is difficult to differentiate between *Humulus* and *Cannabis* and consequently they are often grouped as *Cannabis*-type. In the case of Chenopodiaceae the situation is even worse since they can rarely be split at all (Faegri and Iversen 1975, Moore and Webb 1978).

Pollen counting is very time consuming and therefore a pollen count should be of the minimum size (normally 300-500 grains) required to give a reproducable sample. It is possible (Mosimann 1965; Maher 1972, 1981) to place confidence limits about taxa depending on the size of certain parameters. Confidence limits are useful when interpreting values of both individual and sequences of agricultural indicators, as well as arable: pastoral ratios. Counting statistics are especially important in studies which attempt to identify agriculture because agricultural indicators normally occur in relatively small numbers. This problem may be circumvented on some occasions by increasing the size of the pollen count (for example to 1000 grains).

APPROACHES FOR IDENTIFYING AGRICULTURE/ECOLOGICAL RELATIONSHIPS

In the palaeoenvironmental literature, the three most commonly adopted approaches for identifying agriculture using pollen analysis are the indicator, comparative and statistical approaches. These approaches all use data in the form of pollen counts.

Indicator approach

This approach involves the extension backwards in time of known sociological and ecological preferences of individual taxa. It explicitly uses the principle of uniformitarianism assuming that the preferences of the taxa in question and their response to competition have not changed through time.

It will be clear from the previous section that, because of low production and poor dispersal, the absence of cereal pollen in a sample may not be significant; likewise its presence does not necessarily indicate cultivation. Small numbers of grains could be deposited by chance or man, or could be introduced as contamination during preparation, for example. Larger numbers may be transported to archaeological sites attached to animals, man or as part of the harvest. The largest source of pollen could perhaps be from cereal plants growing as weeds on open ground around sites. For example, Godwin (1975) suggests that *Secale cereale* L. originally occurred as a weed before it rose to the status of crop-plant, possible towards the end of the Iron Age. *Avena fatua* also has a tendency to grow as a weed.

Apart from crops themselves several other species have been used as agricultural indicators. These are plants which grow as weeds when crops fail to exploit all the available resources. Bunting (1960) suggests that the characteristic conditions in which weeds occur are a modified soil surface, an absence of shading and a large amount of nutrients. Such requirements can be fulfilled by both anthropogenic and natural processes. Indeed Godwin (1960, 1975) has demonstrated that many so called 'agricultural indicators', for example, *Centaurea cyanus*, *Chenopodium album*, *Cirsium arvense*, *Plantago lanceolata*, *P. major* and *P. media* were present during previous glacial and inter-glacial times, long before agriculture reached this country. Edwards has already made the point elsewhere (1979) that supposedly pastoral indicator pollen types may represent the growth of field edge weeds and so their appearance in pollen counts could erroneously invoke an influence of mixed farming. Also, Behre (1981) is of the opinion that the weed communities of cultivation as known today, would not have developed by the early stages of agriculture.

The major problems of the indicator approach are that few plants have narrow ecological preferences or tolerances. Those which do, especially the cereals tend to have poor pollen production and dispersal and so their records are also difficult to interpret. This approach also makes unprovable assumptions about the ecological tolerances and sociological preferences of some plants. However it does offer a relatively simple solution to a complex problem.

Comparative approach

This approach involves characterisation of pollen spectra of modern agricultural communities (usually from surface samples in pollen traps or moss polsters) and then their comparison with fossil spectra.

The lack of detailed analyses of the pollen spectra of modern agricultural communities provides a major problem with this approach. Characterisation of modern agricultural communities consists of work by Potter and Rowley (1960) in New Mexico, Heim (1962) in Belgium, Turner (1964) in England, Vuorela (1973) in Finland, Caseldine (1981) in Scotland and incidental observations by various writers.

When comparing fossil and present day spectra the, not always tenable, assumption is made that the pollen source area of the collecting sites is similar. Like the previous approach, this one utilises the uniformitarianism principle. Changes in agricultural techniques can therefore complicate interpretation. The coming of the mouldboard plough, for example, resulted in a new weed spectra in cultivated fields when many perennial plants were destroyed, to be replaced by annual species (Behre 1981). The curtailment of the length of time land lay fallow would also have had similar consequences. Some of these problems may be circumvented by examining pollen spectra from sites such as the Butzer experimental farm (Reynolds 1979) or the Draved Forest experiment (Iversen 1956) which attempt to model ancient agriculture.

The advantages of the comparative approach are that the problems of sampling, identification, preparation and counting are likely to be similar for both fossil and modern samples. The multivariate nature of the data will facilitate consideration of interaction between pollen taxa and plant species. 'The comparative approach is probably the soundest method currently available for the reconstruction of past communities' (Birks and Birks 1980: 255).

Statistical approach

This approach is really an extension of the comparative approach, worthy of special treatment because of its popularity amongst some palaeoenvironmentalists. The technique attempts, by means of a ratio of arable:pastoral pollen indicators, to determine the levels of arable and pastoral farming in an area. The ratios are created and calibrated using present day pollen spectra.

Amongst the first attempts at an arable:pastoral ratio was that of Steckhan (1961) who used the ratio of cereal:Plantago lanceolata as the criterion, a ratio also favoured by Lange (for example see Lange 1975). The ratio of cereal:non-cultivated Gramineae has been employed by Kramm (1978). The most commonly used ratio in this country is the arable/pastoral index of Turner, which in its original form was 'the number of Plantago grains expressed as a percentage of the total number of Plantago, Compositae, cereal, Cruciferae, Artemisia and Chenopodiaceae pollen grains' and 'with one or two exceptions it is below 15% in the arable region and above 50% in the pastoral region' (Turner 1964:81). This index has subsequently been modified several times (Roberts, Turner and Ward 1973, Donaldson and Turner 1977, Brown 1977). Riezebos and Slotboom (1978) also preferred a sophisticated version of the ratio selecting cereal, Fagopyrum, Linum, Rumex, Artemisia and Centaurea as arable indicators and Gramineae, Papilionaceae and Plantago lanceolata as pastoral indicators.

The major criticism of this work is the relatively small amount of conceptual and empirical research which has been conducted into the relationships between agriculture and its palynological indicators. Turner (1964) states that her ratio was created on the basis of the examination of only 21 modern samples. She has since been careful to point out that the ratio was devised to identify only general trends. What would happen to such ratios when pollen sites are close to the coast where Agropyrum junceum, Rumex crispus, Aster trifolium and Plantago maritima flourish? As was pointed out above agricultural practices may have changed dramatically over the last few decades. Mechanisation and the use of herbicides has increased, both of which will tend to reduce weed populations, perhaps selectively. In view of the very different taphonomies of some arable and pastoral farming indicators comparison of modern pollen rain and fossil spectra at different sites is further complicated. The uncertainty associated with such ratios is highlighted by the use of Rumex. Donaldson and Turner (1977) used Rumex acetosa/acetosella - type as a pastoral indicator, whilst Riezebos and Slotboom (1978) preferred to use Rumex as an arable indicator!

Several different arable:pastoral ratios have been calculated from the Holne Moor pollen data (Maguire et al. this volume). The ratios have all be calculated in a similar fashion (Maguire et al. Table 4:5) and are therefore comparable. It must be stressed that only the direction of the changes and not the absolute values should be used to draw inference. The ratios are generally in agreement, each curve exhibiting an approximately similar pattern. Brown's ratio diverges most frequently from the general trend. In the horizon of the Moorland diagram (Maguire et al. Figure 4:11), for example, the values for Brown's ratio decrease whilst the other ratios increase; they actually cross each other.

In Holne Moor data set there were no occurrences of Fagopyrum, Linum, Centaurea or Papilionaceae grains which form an important component of Riezebos and Slotboom's ration. Relatively few cereal grains were present and this may influence Steckhan, Kramm and Lange's ratios, which rely heavily upon this taxon. In such situations these ratios should be interpreted with caution.

In Figs 4:11–4:13 in Maguire et al. (this volume) the number of cereal pollen grains present in the samples has also been indicated. There seems to be a good correlation between the number of cereal pollen grains and low values (that is arable) for the ratios. The inclusion of cereals in some of the ratios must contribute to this relationship.

Other approaches

Two other approaches to the problem of identifying agriculture are proposed. A true statistical approach, that is one which attempts to characterise pollen data simply upon its numeric content, appears an attractive alternative. However, because farming indicators usually occur in small quantities it is unlikely to become widely applicable. A multidisciplinary approach may prove more useful. This approach involves the calibration of pollen data, using other types of palaeoenvironmental evidence for agriculture which occur in close juxtaposition. Such evidence may include snails, beetles, plant macrofossils, soils and archaeology.

CONCLUSIONS

Research into past agriculture should concentrate on the identification and quantification of agriculture in space and time. Pollen analytical studies directed towards such problems should observe the strictures espoused in this article.

The three commonly adopted palynological pathways for the identification and quantification of agriculture are the indicator, comparative and so-called statistical approaches. The major problem common to all three is lack of detailed conceptual and methodological study. The comparative approach is probably the soundest of the the three. The less common multidisciplinary approach holds promise for the future.

The tenor of this article is critical but should not be construed as negative. It should be remembered that without the experimentation and innovation of new concepts and methods, palaeoenvironmentalists cannot solve the complex problems the subject poses. At the same time specialists have a responsibility to explain their observations and interpretations to the uninitiated in order that mis-use be avoided.

ACKNOWLEDGEMENTS

An early draft of this paper was commented upon by Dr. Christopher Caseldine, Department of Geography, University of Exeter.

Bibliography

Andersen, S. Th. (1960) Silicone oil as a mounting medium for pollen grains. Danmarks Geologiske Undersøgelse IV. 4(1) 1-24

Andersen, S. Th. (1973) The diffential pollen productivity of trees and its significance for the interpretation of a pollen diagram from a forested region. In: Quaternary Plant Ecology. (ed Birks, H.J.B. and West, R.G.) Oxford: Blackwell pp. 109-115

Andersen, S. Th. and Bertelsen, F. (1972) Scanning Electron Microscope studies of pollen of cereals and other grasses. Grana 12 78-86

Behre, K.E. (1981) The interpretation of anthropogenic indicators in pollen diagrams. Pollen et Spores 23 225-245

Beug, H.J. (1961) Leifaden der Pollenbestimmung. Lieferung 1.: Fischer

Birks, H.J.B. and Birks, H.H. (1980) Quaternary palaeocology London: Edward Arnold

Bradshaw, R.H.W. (1981) Modern pollen-representation factors for woods in south-east England. Journal of Ecology 69 45-70

Brown, A.P. (1977) Late-Devensian and Flandrian vegetation history of Bodmin Moor, Cornwall. Philosophical Transactions of the Royal Society 276 B 251-320

Bunting, A.H. (1960) Some reflections on the ecology of weeds. In: The biology of weeds (ed Harper, J.L.) Oxford: Blackwell pp. 11-26

Caseldine, C.J. (1981) Surface pollen studies across Bankhead Moss, Fife, Scotland. Journal of Biogeography 8 7-25

Clymo, R.S. (1973) The growth of Sphagnum: some effects of environment. Journal of Ecology 61 849-869

Dimbleby, G.W. (1961) Transported material in the soil profile. Journal Soil Science 12 12-22

Donaldson, A.M. and Turner, J. (1977) A pollen diagram from Hallowell Moss, near Durham City, U.K. Journal of Biogeography 4 25-33

Edwards, K.J. (1979) Palynological and temporal interference in the context of prehistory, with special reference to the evidence from lake and peat deposits. Journal of Archaeological Science 6 255-270

Edwards, K.J. (1982) Man, space and the woodland edge - speculations on the detection and interpretation of human impact in pollen profiles. In: Archaeological aspects of woodland ecology. British Archaeological Review International Series 146 (ed Bell, M.G. and Limbrey, S.): Oxford pp. 5-22

Evans, J.G. (1975) The Environment of Early Man in the British Isles. London: Elek

Faegri, K. (1964) In: Kirk, S.M. (1974) High altitude cereal growing in Co. Down, Northern Ireland?: A note Ulster Journal of Archaeology 37 99-100

Faegri, K. and Deuse, P. (1960) Size variations in pollen grains with different treatments. Pollen et Spores 2 293-298

Faegri, K. and Iversen, J. (1975) Textbook of Pollen Analysis. 3rd Edn. Oxford: Blackwell

Godwin, H. (1960) The history of weeds in Britain. In: The biology of weeds (ed Harper, J.L.) Oxford: Blackwell pp. 1-10

Godwin, H. (1975) The history of the British flora. 2nd Edn. Cambridge: University Press

Grohne, U. (1957) The importance of phase-contrast microscopy in pollen analysis, demonstrated on corn-type Gramineae pollen. Photographie Und Forschung. 7 237-248

Havinga, A.J. (1974) Problems in the interpretation of pollen diagrams from mineral soils. Geologie en Mijnbouw 53 449-453

Heim, J. (1962) Recherches sur les relations entre la végétation actuelle et le spectre pollinique récent dans les Ardennes Belges. Bulletin de la Societe Royale de Botanique de Belgique 96 5-92

Iversen, J. (1956) Forest clearance in the Stone Age. Scientific American 194 36-41

Jacobson, G.L. and Bradshaw, R.H.W. (1981) The selection of sites for palaeovegetational studies. Quaternary Research 16 80-96

Janssen, C.R. (1966) Recent pollen spectra from the deciduous and coniferous deciduous forests of Northeastern Minnesota: a study in pollen dispersion. Ecology 47 804-825

Kramm, E. (1978) Pollenanalytische hochmooruntersuchungen zur floren- und siedlungsgeschichte zwischen ems und hase. Abhandlungen Landesmuseum fur Naturkunde Munster 40(4) 445

Krzywinski, K. (1977) Different pollen deposition mechanism in forest: a simple model. Grana 16 199-202

Lange, E. (1975) The development of agriculture during the first millenium A.D. Geologiska Foreningens i Stockholm Forhandlingar 97 115-124

Maher, L.J. (1972) Nomograms for computing 0.95 confidence limits of pollen data. Revue of Palaeobotany and Palynology 13 85-93

Maher, L.J. (1981) Statistics for microfossil concentration measurements employing samples spiked with marker grains. Revue of Palaeobotany and Palynology 32 153-191

Moore, P.D. and Webb, J.A. (1978) An illustrated guide to pollen analysis. London: Hodden and Stoughton

Mosimann, J.E. (1965) Statistical methods for the pollen analyst: multinomial and negative multinominal techniques. Handbook of Palaeontological Techniques III. (ed Kummel, B.G. and Raup, D.M.). San Francisco: Freeman pp. 636-673

Pohl, F. (1937) Die Pollenerzeugung der Windbluter. Botanisch. Centralblatt 56A 365-470

Potter, L.D. and Rowley, J. (1960) Pollen rain and vegetation, San Augustin Plains, New Mexico. Botanical Gazette 122 1-25

Ralska-Jascewiczowa, M. (1968) Traces of prehistoric settlement in pollen diagrams from the Polish territory (English summary). Folia Quaternaria 29 163-182

Reitsma, T. J. (1969) Size modification of recent pollen grains under different treatments. Revue of Palaeobotany and Palynology 9 175-202

Riezebos, P.A. and Slotboom, R.T. (1978) Pollen analysis of the Husterbaach peat (Luxemborg): its significance for the study of subrecent geomorphological events. Boreas 7 75-82

Reynolds, P.J. (1979) Iron-age farm: The Butser experiment. London: British Museum

Roberts, B., Turner, J. and Ward, P. (1973) Recent forest history and land use in Weardale, N. England. In: (ed Birks, H.J.B. and West, R.G.) Quaternary Plant Ecology Oxford: Blackwell pp. 207-221

Robinson, M. and Hubbard, R.N.L.B. (1977) The transport of pollen in the bracts of hulled cereals. Journal of Archaeological Science 4 197-199

Rowley, J.R. (1960) The exine structure of 'cereal' and 'wild' type grass pollen. Grana Palynologica 2:2 9-15

Rowley, J.R. and Rowley, J. (1956) Vertical migration of spherical and aspherical pollen in a Sphagnum bog. Proceedings of the Minnesota Academy of Science 24 29-30

Salmi, M. (1962) Investigations on the distribution of pollens in an extensive raised bog. Compte Rendu Societe Geologique Finland 34 159-193

Simmons, I. G. and Tooley, M.J. (1981) The environment in British prehistory. Liverpool: Duckworth

Steckhan, H.U. (1961) Pollenanalytisch-vegetationsgeschichtliche undersuchungen zur früten siedlungsgeschichte im Vogelsberg, Knüll und Solling. Flora 150 514-551

Tauber, H. (1965) Differential pollen dispersion and the interpretation of pollen diagrams. Danmarks Geologiske Undersøgelse II 89 1-69

Tauber, H. (1967) Investigations of the mode of pollen transfer in forested areas. Review of Palaeobotany and Palynology 3 277-86

Turner, J. (1964) The anthropogenic factor in vegetational history. I. Tregaron and Whixall Mosses. New Phytologist 63 73-90

Turner, J. (1970) Post-Neolithic disturbances of British vegetation. In: (ed Walker, D. and West, R.G.) Studies in the Vegetational history of the British Isles Cambridge: University Press pp 97-116

Turner, J. (1975) The evidence for land use by prehistoric communities: the use of 3-dimensional pollen diagrams. In: (ed Evans, J.G., Limbrey, S. and Cleere, H.) The effect of man on the landscape: the Highland Zone CBA: Research Report pp 86-95

Tyldesley, J.B. (1973) Long-range transmission of tree pollen to Shetland. I, II and III. New Phytologist 72 175-181, 183-190, 691-697

Vuorela, I. (1973) Relative pollen rain around cultivated fields. Acta Botanica Fennica 102 1-27

3. ARABLE/PASTORAL RATIOS FROM INSECTS?

Mark Robinson

INTRODUCTION

The arable/pastoral ratio is taken to refer to the proportion of the landscape surrounding a site which is cultivated and the proportion which is grassland. Obviously, it is an important factor in understanding the agricultural economy of an area, whether subsistence or not. There seem to be three approaches which can be used in attempting to establish past arable/pastoral ratios from non-documentary evidence.

It is possible to speculate how an archaeological site might have functioned and on the likely uses of different soils around it. Site catchment analysis can be useful, but it is giving details of potential rather than actual use. Attempts have been made to reconstruct the agricultural economy of some Roman villa estates on the basis of the size and presumed function of excavated agricultural buildings, but the conclusions have tended to be contentious (Applebaum 1972).

The second, and most reliable approach, is to examine the ancient soils themselves and so establish their use (cf. Maguire et al. this volume). Pasture can usually be recognised by the presence of some or all of the following features: a turf line, stone-free earthworm-sorted soil and a characteristic molluscan fauna. Evidence for cultivation can be in the form of soil moved by ploughing to create lynchets, plough marks on the underlying bedrock, and a general mixing of the soil which is visible with the microscopic examination of thin soil sections if not by the incorporation of small stone fragments throughout the profile. Phosphate analysis combined with archaeological field work on ancient boundaries can also give an indication of land usage. There are three problems with these direct methods of soil investigation. Firstly the ancient soil must have been protected from subsequent disturbance. For some of the lines of evidence, even later earthworm sorting can prove disruptive and only examinations of soils sealed beneath earthworks are practicable. Secondly, it is often difficult to date ancient soils: again it is useful if they are stratified beneath archaeological earthworks. Finally, the information is about conditions at a particular point, and it is not always possible to extrapolate results with confidence.

Despite the problems, these direct methods can prove very informative. The mapping of surviving cultivation ridges and lynchets is a standard technique of field archaeology. Much important work has been done to establish prehistoric land use on the chalk of southern England using molluscan analyses combined with soil observations (Evans 1972). Soil phosphate analysis is also proving its value (Maguire et al. this volume).

The third approach to the arable/pastoral problem is to use evidence from the surrounding landscape which has accumulated at a particular collecting point. Some information can be derived from the remains of crop plants and animal bones which have been brought to an archaeological site by man. However, pollen analysis is the most informative method for this approach. Interpretation of pollen results can be refined by the calculation of arable/pastoral indices, the best known being Turner's, in which the number of Plantago pollen grains is expressed as a percentage of the total Plantago + Cruciferae + Chenopodiaceae + Artemisia + cereal type (Turner 1964: 81-2). The value for this index tends to be below 15% in arable regions and above 50% in pastoral regions. There are various other indices, which differ slightly in the species used (Maguire, this volume). It is not argued that, for example, every grain of Compositae pollen came from weeds growing in arable fields, the plant families and genera are used to show general trends. Such investigations are limited to very acid soils and waterlogged deposits in which pollen is preserved. Cereal pollen presents a problem because, apart from Secale, it is not carried by the wind in quantity but remains enclosed by the bracts. The size of the pollen catchment area of a particular deposit is often uncertain. These problems are outweighed by the advantages and, in suitable localities, pollen analysis is probably the most efficient means of establishing an arable/pastoral ratio. Many polleniferous deposits contain sufficient organic material for radiocarbon dating and often a deposit will have accumulated over a long period, enabling changes with time to be established.

Can insect remains be used to give an arable/pastoral index in the same way as pollen? Insects show great species diversity in both grassland and amongst crops, many insects have good dispersive powers by flight and, although not as resistant to decay as pollen, insect remains are preserved under anaerobic waterlogged conditions. Not all insect families are suitable. The remains of most species of Hymenoptera are at present unidentifiable. Diptera might seem promising because they include many pests of grassland and arable but the potentially identifiable phase of their life cycle, the puparium, remains at the place where the larva developed. The dispersive phase is the adult, remains of which tend to be badly preserved and hard to identify. Hemiptera are not usually abundant in archaeological deposits and species preservation is somewhat patchy. For example, aphids only seem to survive when they have been attacked by the minute hymenopteran parasite Aphidius. Thysanoptera might prove a promising group for future research. Many feed on a single plant genus or species (Pitkin 1976). There are major differences in the thrip (thysanopteran) populations of grassland from cereal fields. For example Chirothrips manicatus Hal. flies in vast numbers from some types of grassland while Limothrips cerealium Hal. has been recorded in great abundance flying from wheat and oats (Lewis 1961: 90). However, while thrips have been recorded from archaeological deposits (Kenward 1979), they are not generally found at present.

This leaves Coleoptera. Coleoptera are sufficiently mobile, while their remains are abundant, well preserved and fragments are readily identifiable from archaeological deposits.

Archaeological work has been undertaken on Coleoptera for about the past 20 years and it has long been recognised that open landscapes can be differentiated from forested environments simply by considering the habitats of the various species identified (Girling 1982; Osborne 1978). Sometimes mention has been made of arable and pastoral aspects indicated by a beetle assemblage from an archaeological context (Coope and Osborne 1968). Dung beetles have often been used to indicate grazing. Remains of scarabid dung beetles were so abundant in the Bronze Age shaft at Wilsford, Wilts., that they provided overwhelming evidence for the importance of pasture around the site (Osborne 1969). Indicators of arable are more elusive. It is worth mentioning here that grain beetles, for example Sitophilus granarius (L.) which has been recorded in large numbers from some Roman sites, infest stored grain and do not occur amongst the crops in the arable fields.

Various attempts have been made to adopt a quantitative approach to the interpretation of Coleoptera from archaeological deposits. The methods of Kenward (1978) are particularly suited to deriving information from beetles which lived in the deposit being investigated or in its immediate vicinity. They are concerned with the recognition of species associations, the comparison of species diversity and a consideration of "superabundant" species. Not all of these methods are useful for the interpretation of the more distant component in the death assemblage and this is the component which will provide details of general land use. Robinson (1981: 279-82) has suggested various species groupings which tend to be associated with particular habitats and calculated the percentage of individuals within each of the groupings for sites. This method will be elaborated upon later.

APPROACHES

Before ancient deposits can be considered, it is important to characterise the differences between the coleopterous faunas of arable and grassland. Most recently published British general ecological investigations of beetles in these habitats have had one of three aims: the study of the relationship between predators and arable pests, the effect of modern agriculture on the availability of arthropods for game birds to feed on, and whether grassland on nature reserves should be cut to improve insect species diversity. The faunal lists from these surveys are usually incomplete, with a tendency to concentrate on the Carabidae. Some beetle families have been very neglected. This work alone cannot form the basis for interpreting archaeological results. It can, however, be supplemented by the detailed ecological investigations which have been done for some beetles of grassland and arable, habitat information from identification keys and the European literature on coleopteran ecology.

If it is difficult to obtain a complete picture of the Coleoptera of grassland and arable, the ecological investigation of some present-day faunas might seem an alternative approach. Unlike plants, where only a few visits are needed to establish the flora of a locality, it is not easy to assess the total beetle fauna of a locality. Beetles are mobile, many of them are cryptic and some species have a very short active adult life with a synchronisation of emergence. With plants, one knows where they live because they are rooted to the site whereas stray beetles (Kenward 1975; 1976) and beetles which only use the habitat for hibernation (Atty 1967; Luff 1965) may be present. It would have taken too long to have surveyed the Coleoptera of sufficient types of grassland and arable to provide the sole foundation for the interpretation of archaeological results.

It was decided to investigate the coleopterous fauna of a limited number of sites in 1982. These results have been used to link the information from the literature and to form a background against which the faunas of the various habitats have been considered.

THE SITES

The coleopterous faunas of four localities on the Thames floodplain near Oxford were investigated. The reason that the floodplain was chosen is that unimproved grassland remains present and it was thought that some possible indicators of arable are only associated with sandy soil. The soils on these sites are alluvial clay loams, and all suffered flooding in the winter of 1981-82, the last floods receeding on 2nd March, although pools of water remained on one of the sites, Port Meadow, for several weeks.

Pixey Mead (SP479104)

Pixey Mead has probably been a hay meadow for the past 900 years and is still managed along traditional lines. Hay is cut in late July/early August and then the aftermath is grazed by cattle and sheep. The meadow is unimproved and has a very diverse flora. It is hard to establish which plants are dominant.

Port Meadow (SP500075)

Port Meadow seems to have been cattle and horse pasture (sheep are rigorously excluded) for at least 900 years. It is unimproved and suffers from overgrazing. The part of Port Meadow investigated was wetter than the other three sites and some plants that are usually thought of as aquatic grow there including *Glyceria fluitans* and *Rorippa nasturtium-aquaticum*. This waterlogging is in part due to soil poaching resulting from overstocking, so it is still thought reasonable to contrast the fauna of Port Meadow with the other sites. The flora, especially the grasses, is much less diverse than Pixey Mead. The turf is short, tending to be about 5cm or less high.

Binsey Lane (SP4940640)

It is noticeable just how weed free most cereal fields have become with modern methods of management over the past ten years. Therefore a plot of 10m x 10m in a field of mostly rather neglected allotments off Binsey Lane was cultivated and sown with rows of the following crops on 5th April: Triticum spelta L., T. dicoccum Schübl., spring hulled Hordeum vulgare L.emend., Papaver somniferum L. and Vicia faba L. v. minor. A few Agrostemma githago seeds were scattered at the time of sowing but the many other weeds established themselves (including Centaurea cyanus and pink, as opposed to pale lilac, flowered P. somniferum, both escapes from other allotments). Weeding was limited to preventing the crops from being completely overwhelmed. Both crops and weeds flourished, Chenopodium album being the dominant weed.

North Hinksey (SP498054)

A large field of winter wheat was investigated near North Hinksey. Elsewhere in the field, the crop had been damaged by flooding and had been re-seeded with spring barley, but here the crop was in good condition. It had been subjected to the full range of agricultural chemicals and insecticide was sprayed towards the end of the first week of pitfall trapping, eliminating a developing aphid infestation. Herbicides kept the crop almost weed-free and there were four species of weed, all rare. Of these, one, Hordeum vulgare, was a volunteer and another, Anisantha sterilis, had only invaded along the field edge from the boundary.

The Flora of the Sites

The flora of Port Meadow and Pixey Mead was given by Baker (1937), but the localities sampled have been re-surveyed because some changes have taken place over the past 45 years. The flora (excluding crops) of the sites sampled is given in Tables 3:1-3:4. Plants growing elsewhere in these fields have not been listed.

METHODS

A series of 5 matched pitfall traps (2 coffee jars and 3 steep-sided plastic beakers) was set out in each site on 10th July 1982, to catch beetles active on the ground surface. Each trap was buried to its rim and filled with about 4cm of water containing detergent to sink and drown the beetles which fell in. Formalin was not added as a preservative because it attracts certain beetles (Luff 1968). In retrospect, this was probably a mistake with traps only emptied at weekly intervals, because towards the end of the week they became very foul. The traps were emptied on 17th and 24th July. Sweep netting was undertaken in the vegetation around the traps on 10th and 17th July in the early afternoon to catch the beetles which were on it.

RESULTS

The Coleoptera caught by each method from the various localities are listed in Tables 3:5 and 3:6. In order of total coleopteran biomass from the pitfall traps, Pixey Mead had the largest, followed by North Hinksey, then Binsey Lane and finally Port Meadow. Only one of the species captured is of particular interest in its own right, a single individual of Ptomaphagus varicornis from a pitfall trap at North Hinksey. This distinctive beetle is very rare in Britain and there are no previous records of it from Oxfordshire.

The results by no means give a complete coleopteran fauna of the respective habitats. There is a marked succession of adult activity for various species throughout the year, for instance the carabid *Pterostichus caerulescens* tends to be active early in the season while *Calathus fuscipes* (which certainly occurs on some of the sites investigated but was not captured) has its period of activity later (Greenslade 1965: 168). Likewise, species of *Aphodius* show seasonal succession (White 1960). Some beetles only climb vegetation at night, so would not have been caught by the sweep-netting. For example *Tachyporus* spp. ascend cereal stems at night to feed on aphids and fungal material (Sunderland 1975: 511; Vickerman and Sunderland 1975). Different species of Carabidae may show differential susceptibility to pitfall trapping according to size, behaviour and the strata in which they are active in the ground vegetation. Catches of a single species may vary with different types of ground cover depending upon the resistance they present to movement. Therefore, quantitative comparisons between results of pitfall trapping must be made cautiously (Greenslade 1964: 309). The sweep-netting was not sufficiently matched for each site for percentage comparisons between the results to be made.

The trapping and netting did however, probably catch a major proportion of the species present on the various sites and the results certainly show great differences from site to site. With care, the results can be used to characterise meadowland, pasture, weedy arable and weed-free arable.

THE COLEOPTEROUS FAUNA OF GRASSLAND AND ARABLE

In attempting to explain the differences between the coleopterous faunas of various types of grassland and arable in Britain, the characteristics of the various habitats must be considered along with the ecology of the beetles.

Ploughing

The most obvious difference between arable and ancient grassland is that only the soil of the former is disturbed by cultivation. Ploughing imposes severe limitations on Coleoptera. Wireworms, the larvae of various species of *Agriotes* and *Athous*, are sometimes thought of as agricultural pests, but serious infestations only occur in newly ploughed old grassland. They persist in arable for three or four years (Gair *et al*. 1972: 42-4; Potts and Vickerman 1974: 156). The larvae, which feed on roots, take a long time to complete their life cycles, for example four or more years in the case of *Agriotes obscurus* (Evans 1944: 235) and they are very vulnerable to ploughing. Also, it has been observed that *A. obscurus* only lay eggs in turf covered with grass (Edwards and Evans 1950: 251). *A. obscurus* was recovered from the pitfalls on both Pixey Mead and Port Meadow but not from the arable sites. *Selatosomus incanus* and *Fleutiauxellus quadripustulatus* are grassland Elateridae with similar life cycles. They were only caught on Pixey Mead.

Another group of Coleoptera with a somewhat similar life cycle, the larvae spending a few years feeding on roots in permanent grassland and susceptible to ploughing, is the chafer group of

Scarabidae (Gair et al. 1972: 44). Phyllopertha horticola (L.) is a common example (Raw 1951-52: 644-5). None was recovered in this survey but I have excavated chafer larvae elsewhere on Port Meadow. Soil waterlogging is perhaps a limiting factor for grassland elaterids and chafers on the part of Port Meadow investigated.

Some of the large Carabidae burrow below plough depth, for example Harpalus rufipes (Luff 1978: 61), the most abundant carabid from the two arable sites. Various other Carabidae seem to be able to survive under conditions of ploughing (Thiele 1977: 158-60) and the same must be true for some of the Staphylinidae.

Migration to the crop in spring and away from it in the autumn is another way in which the ravages of ploughing can be avoided. It has been shown for a few species, for example Agonum dorsale (Pollard 1968: 130-7; Speight 1976: 97-8), which tended to overwinter in hedge-bottoms, dispersed in May to June and after breeding, returned to the hedges where the beetles are subsequently joined by a new generation. It is possible that many small Carabidae and Staphylinidae show this pattern of behaviour.

Annual ploughing followed by a crop also means that arable experiences considerably greater variation in the physical environment over the year, bare earth followed by plant cover. This factor certainly restricts the range of beetles which can live in arable. The bare ground is never completely covered even when the crop is fully grown, therefore causing less impediment to movement. This is probably one of the factors favouring the large Carabidae, which tend to be common in arable.

The Field Layer

Pasture stands apart from meadowland and arable with the lack of a complete component of the habitat. Pasture only possesses the ground zone according to the classification of Elton and Miller (1954: 485-6), vegetation less than 6 inches in height, it does not have a field layer. As the sweep-netting showed, beetles were making good use of the field layer in the daytime at Pixey Mead and Binsey Lane.

The presence of the field layer also has considerable influence on microclimate. The humidity is increased and the wind speed decreased towards the ground with tall herb cover (Waterhouse 1955). On short turf, there is considerable insolation (heating by the sun) at ground level; in tall grass on a sunny day, the warmest place is in the middle. With tall herb cover, the ground is better insulated at night. This means that in summer, there are much harsher conditions on the ground in pasture than in meadowland or a crop. Luff (1978: 60) suggested that the absence of larger diurnal species of Carabidae in open habitats was due to the difficulty in obtaining temporary shelter from insolation or desiccation. It is noticeable that the larger carabids were not nearly as abundant at Port Meadow as on the other sites, smaller species such as Loricera pilicornis and Bembidion guttula tended to predominate. Interestingly, the two most abundant carabids from the pitfall traps on Pixey Mead,

Pterostichus cupreus and P. versicolor, which together comprised 45% of the total, are large and diurnal (Thiele 1977: 266-7) whereas no large diurnal carabids were captured from either of the arable fields.

Crop Pests

Arable crops are now usually grown as monocultures but perhaps there were prehistoric equivalents of "drage", a mixture of oats and barley sown together. Monoculture tends to favour the outbreak of species in pest proportion. Cereals in Britain suffer from many serious pests, for example the wheat bulb fly, Leptohylemyia coarctata (Fall.) and aphids such as Sitobium spp. but there is no major coleopterous pest. In Eastern Europe, Zabrus tenebrioides (Goez.) is a serious pest which devours young cereal plants and climbs stalks to eat green grain (Thiele 1977: 120, 157). It is a very rare species in Britain (Lindroth 1974: 97). Helophorus nubilus F. was once recorded as a serious cereal pest in East Anglia, the larvae eating wheat shoots, (Petherbridge and Thomas 1936) but it is a grassland species which usually only gives trouble in freshly ploughed grassland (Gair et al. 1972: 44-5). Elateridae and chafers have already been mentioned, likewise Crepidodera ferruginea (Scop.) sometimes follows on from ploughed grassland (Gair et al. 1972: 45). Oulema melanopa (L.) is sometimes a serious cereal pest in the U.S.A. but in Britain is unimportant (Gair et al. 1972: 44, 141). Phyllotreta vittula Redt. is claimed to attack cereals (Gair et al. 1972: 106) but it is possible that it only spreads onto cereals when there is already a considerable infestation on cruciferous weeds in the crop. In general, the benefits Coleoptera derive from cereals are those of shelter while other invertebrate cereal pests provide a food source for some predatory beetles.

Some other crops fare worse from beetles, several species of Sitona attack Vicia faba (Fowler 1891: 216-17; Joy 1932: 177-8) and Sitona damage was noted on the leaves of the beans at Binsey Lane, but none was captured. A flea beetle, Longitarsus parvulus (Pk.) feeds on flax (Joy 1932: 415). Cultivated species of Brassica often suffer from damage by species of Phyllotreta (Ormerod 1890: 180-4; Stapley 1949: 173-8) while turnips seem particularly vulnerable to Ceutorhynchus pleurostigma (Marsh.) (Isaac 1923), Helophorus rufipes (Bosc.) and H. porculus Bed. (Petherbridge 1928: 677-8). However, all these species can complete their life cycles on wild plants.

Weeds and Phytophagous Species

Weeds within a crop are probably much more important to Coleoptera than the crop itself. Within a cereal field, areas of high weed cover, especially from horizontally spreading weeds, have more predatory ground beetles than areas with few weeds (Speight and Lawton 1976). The role of the weeds seems to be protecting the beetles from insolation during the day and desiccation throughout the day and night.

Weeds are also hosts for many phytophagous species. The sweep-netting showed considerable differences between the almost weed-free crop at North Hinksey, from which no phytophagous beetles were taken, and the weedy crop at Binsey Lane, where 43 individuals which fed on the foliage and flowers of the weeds were caught. However, the distribution of beetles amongst the weed species was very uneven. Ceutorhynchus florealis and the three species of Phyllotreta, in total 40 individuals, were feeding on Capsella bursa-pastoris while the dominant weed, Chenopodium album, did not seem to be suffering any beetle damage. The remaining three individuals, Meligethes aeneus, were probably feeding on the pollen of Sonchus asper. The purely phytophagous species from the pitfall trapping at Binsey Lane are likewise mostly species which tend to be associated with Cruciferae. The only solely phytophagous beetle from North Hinksey was the polyphagous weevil Barypithes aranaeiformis. Omniverous Carabidae will be mentioned later.

While a wide range of arable weed species are attacked by Coleoptera, the Cruciferae and Polygonaceae seem to stand out as families in which many of the members tend to carry large beetle populations. Along with Phyllotreta spp., many species of Ceutorhynchus favour cruciferous weeds (Freude et al. 1966: 208; Hoffman 1954: 880). Polygonum aviculare agg. seems to be the main host of Chaetocnema concinna (Marsh.) (Freud et al. 1966: 255), but it also attacks other Polygonaceae (and mangolds).

Grassland Herbs and Phytophagous Species

The phytophagous fauna of grassland is distinct from that of arable. Meadowland tends to carry higher populations of phytophagous Coleoptera which feed above ground than pasture. Some species seem particularly characteristic of meadowland. A diverse catch of phytophagous species was made by sweep-netting on Pixey Mead, and a similar range of species was caught in the pitfall traps. The host plants of the more fastidious species are: Trifolium pratense, Filipendula ulmaria, Plantago spp. and Rumex spp.

Apion trifolii was the most abundant beetle from the sweeping on Pixey Mead and many members of that genus are common in meadowland. The adults of A. trifolii hibernate under shelter and emerge in late April. They attack Trifolium pratense, feeding on the foliage (Jones 1950: 313-14). The eggs are laid in the clover flower heads before they open and the larvae feed on the developing seeds. The adults emerge soon after pupation and a second generation of eggs is laid in the flower heads. The adults of the second generation then migrate to hedges etc. to hibernate, very few hibernating in the grassland where they breed. Several other species of Apion attack clover, including A. dichroum Bed., A. apricans Hbst. and A. assimile Kirb. (Freeman 1967: 535). The life cycle of A. dichroum has been studied in great detail. It is univoltine and feeds on Trifolium repens L., but is otherwise similar to A. trifolii. A comparison was made between the populations of A. dichroum on grazed plots and ungrazed plots from which grazing animals had been excluded from the later spring onwards. On the ungrazed plots, the sward was much deeper and there was a greater density of clover flowers. Three times as many overwintered adults colonised the ungrazed plots per unit area as the grazed plots. The females were found to prefer flowers high up in the sward for oviposition. The risk of flowers being grazed while the larvae are developing means that the difference between adults leaving the grazed and ungrazed plots for hibernation would be considerably greater than a factor of three.

Many other species of Apion feed on grassland Papilionaceae. Morris (1967: 459-63) investigated the response of Apion loti Kirb., which feeds on Lotus corniculatus L. seeds, to the cessation of grazing on chalk grassland. The ungrazed plots were enclosed in March and sampling took place in July. More flowers of L. corniculatus were observed in the enclosures than on the grazed plots and there was an even more marked difference in the use of seed pods. A count of larvae, pupae and adults showed that there were about 90 times more individuals of A. loti in the ungrazed than on the grazed plots.

A few individuals of Sitona spp. were caught on Pixey Mead. These species also seem to be favoured by meadowland. Their larvae are subterranean, feeding on roots or root nodules of leguminous plants (Danthanarayana 1967) while the adults feed above ground on the leaves and flowers.

Only one phytophagous species was caught in the pitfall traps on Port Meadow, Notaris acridulus, and none by sweepnetting. It is a marshland weevil which was probably feeding on the Glyceria and Eleocharis (Hoffman 1958: 1435).

Ground-Dwelling Predators and Omnivores: Carabidae

The most abundant beetles from the pitfall traps were the Carabidae and Staphylinidae. Most carabid species are largely carnivorous, with a broad spectrum of prey, but many take some plant food, especially seeds, while adults of Amara spp. feed almost exclusively on seeds (Thiele 1977: 106-31). The diet of many of the Carabidae captured is given in Sunderland (1975) and Speight (1976: Table 1.1).

Some of the Carabidae from the four sites do not show any marked preference for either arable or grassland conditions, for example Pterostichus melanarius, one of the few large carabids caught on Port Meadow. It is mainly carnivorous, and feeds on a wide range of invertebrates including caterpillars, small carabids, slugs and spiders (Sunderland 1975: 508; Thiele 1977: 108-9). One of the carabids, Harpalus rufipes, however, stands out as very abundant in the two arable samples and rare from grassland samples. The adults and larvae feed on a variety of weed seeds, especially Chenopodium album, which it finds on the ground under the plants, but also including Stellaria media and Polygonum aviculare (Briggs 1965-66: 82, 89-90). H. rufipes larvae can be reared to adults and to lay eggs on a diet consisting exclusively of C. album seeds but at least the adults are usually omnivorous, being regarded by Thiele as 50% carnivorous. Aphids seem a particularly favoured prey along with the larvae of various invertebrates (Sunderland 1975: 509; Thiele 1977: 108-9). Briggs suggests H. rufipes to be adapted to conditions of arable cultivation in which areas of annual weed seeds occur (Briggs 1965-66: 90), very much the conditions at Binsey Lane, but, as the results from North Hinksey show, it is also adapted to modern weed-free cultivation, perhaps feeding on aphids fallen from the crop. Thiele (1977: 26-7) regards H. rufipes as characteristics of arable land and Greenslade (1965: 157) says that it is essentially an arable species.

Agonum dorsale was reasonably abundant from the pitfalls on the arable sites but was absent from the grassland samples. It was the second-most abundant carabid caught by Speight (1976: Fig. 3.1) in his study of beetles in winter wheat. Thiele (1977: 26-7) lists A. dorsale as particularly frequent in arable and not very common in grassland although Greenslade (1965: 166-7) regards it as both a grassland and arable species. A. dorsale is carnivorous, feeding mostly on aphids (Sunderland 1975: 511).

Some species of Amara usually live on open, rather sandy, bare or disturbed ground. In particular, A. apricaria (Pk.) A. bifrons (Gyll.) and A. tibialis (Pk.) all share this habitat, where they feed on weed seeds (Lindroth 1974: 87, 91-3). None was recorded from the pitfalls in this survey, which were all on alluvial clay loams, but on suitable soils they are present amongst weedy crops. Apart perhaps from occurring on eroded heathland (cf. Thiele 1977: 35), they are not grassland species.

Pterostichus cupreus and P. versicolor were very abundant from Pixey Mead but absent from the other sites, perhaps because the microclimate was unsuitable. They are probably omnivorous (Thiele 1977: 108-9). Another carabid, Amara communis, was also numerous on Pixey Mead. It has carnivorous larvae but the adults eat seeds (Thiele 1977: 125). In Britain, all three beetles are regarded as grassland species (Greensdale 1965: 156, 158), although P. cupreus is sometimes caught on arable land. The observation of Thiele (1977: 32) that P. cupreus is an arable species and only P. versicolor a meadowland species does not seem to be true for the British Isles. A. communis is regarded as very much a grassland species (Thiele 1977: 27), and only in meadowland will sufficient seed be set by the grassland plants to support a high population level.

The possible reasons for the reduced number of large Carabidae from Port Meadow have already been mentioned. The most abundant carabid from Port Meadow, Loricera pilicornis, occurs in a wide range of terrestrial habitats (Greenslade 1965: 156). Some of the species of Bembidion from Port Meadow such as B. guttula tend to live in wet grassland, marsh and moist woodland.

Ground Dwelling Predators and Omnivores: Staphylinidae

Staphylinidae were abundant in all the pitfall traps except those on Pixey Mead. Apart from specialised dung feeders, the Staphylinidae captured are either purely carnivorous or feed on both animal and fungal material. The diet of many of the Staphylinidae captured is given in Sunderland (1975) and Speight (1976: Table 1.1).

Philonthus cognatus was the most numerous staphylinid from Port Meadow. P. cognatus is a carnivore, (Eghtedar 1970) and although rare in the arable samples in this survey, it can be very abundant under arable conditions (Speight 1976: 62-3). Tachyporus signatus was very common in the pitfall traps on the arable sites but was also present in small numbers from the other two localities. It is a carnivore which is often abundant in arable and can withstand ploughing (Dunn 1960: 276; von Lipkow 1966; Speight 1976: 61-2). However, *T. signatus* occurs in many other terrestrial habitats, even dung. *Tachyporus* spp. can be very numerous in arable (Potts and Vickerman 1974: 157) but can also be abundant in grassland. Their feeding habits have been mentioned above (p.24).

Oxyteline Staphylinidae from the genus *Anotylus* present somewhat of a problem. *A. rugosus* (F.) and *A. sculpturatus* are certainly common in some arable habitats (Potts and Vickerman 1974: 157) but they are species which are normally regarded as denizens of dung and foul organic material (Horion 1963: 221, 230. Are the captures from arable (and grassland) merely individuals which formed part of the background rain of migratory Coleoptera? *A. rugosus* seems to be omnivorous (Eghtedar 1970); it has been recorded as a predator of the lettuce root aphid (Dunn 1960: 276) and of the immature stages of the cabbage root fly under arable conditions (Coaker and Williams 1963) as well as living in accumulations of decaying organic material/dung. The status of *A. sculpturatus* remains uncertain. Hinton (1944: 122) mentions that it feeds on dung but it is presumably also able to feed on decaying plant material.

The Community of Dung

Of the sites surveyed, dung of large herbivores was confined to Port Meadow. Each dropping will support an entire animal community (Mohr 1943) with members that are very mobile and good at finding their habitat quickly (Elton 1966: 331). Species of *Aphodius* can smell dung while they are in flight and will fly upwind to locate the source (Landin 1961: 182). Some of the beetles inhabiting dung are coprophagous, others are carnivorous on the eggs and larvae of Diptera, dung beetles and mites. The beetles show a marked succession throughout the life of an individual dropping, the species which require fresh dung moving on to feed and oviposit in new droppings after the dung has become more suitable for other species (Holter 1979: 317-18; Koskela 1972: 166-8; Koskela and Hanski 1977: 208-9).

A few dung beetles from the genera *Aphodius* and *Colobopterus* were caught in the pitfall traps on Port Meadow. Most species in these genera are confined for the purpose of breeding to the dung of large herbivores in the form of droppings in the field, although *C. erraticus* (L.) larvae have been found in human faeces, a few, for example *A. fimetarius* (L.), have larvae which can also feed on rotting vegetation and *A. villosus* Gyll. never occurs in dung (Landin 1961: 209-21). Whereas *Aphodius* spp. lay eggs in the dung, *Onthophagus* and *Geotrupes* spp. burrow into the soil below or near the edge of the dropping and store dung in galleries for their larvae (Elton 1966: 334-5). Some species of *Onthophagus* can breed in other material but *Geotrupes* is restricted to the dung of large herbivores. Species of the above-mentioned genera of scarabaeoid dung beetles are not restricted to the dung of any one large herbivore. *A. zenkeri* Germ. tends to be associated with deer dung in Britain (Britton 1956: 15), but this is because its larvae cannot tolerate the high temperature that results from the insolation of dung in open habitats, and in Sweden also occurs in cattle dung in woodland (Landin 1961: 136-8, 211). Some species do not like horse dung because it is too coarse, while cow droppings are particularly favoured because a crust tends to form on the surface which reduces desiccation (Landin 1961: 144, 192-3). Many species of *Aphodius* can co-exist in one field of droppings (Elton 1966: 333) and up to 100 adults can be found in one dung pat (Holter 1979: 317).

One species of *Sphaeridium* and three species of *Cercyon* were captured by the pitfall traps on Port Meadow. Their larvae are carnivorous while the adults are coprophagous (Koskela and Hanski 1977: 207). *Megasternum* and *Cryptopleurum* spp. have similar habitats. Many of the species from these four genera of Sphaeridiinae also occur in other sorts of decaying organic material and, unlike most species of *Aphodius*, *Colobopterus*, *Onthophagus* and *Geotrupes*, are able to breed in dung heaps as well as droppings in the field. Several species of Staphylinidae from Port Meadow occur in dung. Both adults and larvae of *Platystethus arenarius* will feed exclusively on dung (Hinton 1944: 126-8) while some species of *Philonthus* are carnivores in dung. *Anotylus sculpturatus* has already been mentioned along with *A. rugosus*, another species which often occurs in dung although it was not caught in this survey. Like the Sphaeridiinae, many of the Staphylinidae of dung also live in manure heaps and other sorts of decaying organic material.

Other Groups of Coleoptera

Only a few groups of Coleoptera from the pitfall trapping remain to be described. Three individuals of *Nicrophorus vespillo*, a carrion beetle, had perhaps been attracted by the smell of decaying insects. A few members of the Lathridiidae, Cryptophagidae and Phalacridae were present in some of the samples. They all tend to feed on moulds on plant material.

The sweep-netting on Pixey Mead captured several species which as adults tend to congregate on flowers: *Cantharis* spp. *Rhagonycha* spp. and *Selatosomus* spp., where they copulate, feed on pollen and perhaps eat other insects visiting the flowers. These beetles also congregate on flowers in hedgerows and sometimes on disturbed ground, Umbelliferae flowers being particularly favoured.

Some of the beetles which fell into the pitfall traps would have been strays from other habitats. The only obvious stray is *Brachypterus urticae*, a nettle feeder, from Binsey Lane. *Urtica dioica* L. was however, abundant on some of the nearby allotments.

Population Structure

All four localities had relatively diverse faunas of Coleoptera, even the wheatfield at North Hinksey, despite its lack of suitable plants for phytophagous beetles. In some other surveys, a much reduced insect species diversity has been found on grazed compared to ungrazed grassland (Morris 1969: 484-5; Morris 1971). The biomass and numbers of Coleoptera were higher from Pixey Mead than Port

Meadow, which is perhaps what would be expected because grazing was not taking such a large proportion of primary production (Hutchinson and King 1980: 380-3). The greatly reduced numbers of Coleoptera from the pitfall traps at Binsey Lane compared with North Hinksey, however, was unexpected. A possible explanation is that many wolf spiders (Lycosidae) were caught in the pitfall traps at Binsey Lane. Perhaps the spiders were in part replacing some of the Carabidae and Staphylinidae.

Where an ecosystem has been simplified, it becomes more susceptible to damaging outbreaks of a few species (Pimentel 1961). Such an outbreak of aphids was certainly developing on the wheat at North Hinksey until it was sprayed. Whether or not it can be regarded as an outbreak, *Tachinus signatus* made up 65% of the Coleoptera from the North Hinksey pitfalls, more than twice the percentage of any other species in any of the pitfall traps. Sweep-netting suggested a more diverse and balanced population of phytophagous Coleoptera on Pixey Mead than at Binsey Lane, where numbers of *Ceutorhynchus floralis* were high.

LIMITATIONS ON THE INTERPRETATION OF ANCIENT ARABLE/PASTORAL LANDSCAPES FROM COLEOPTERAN EVIDENCE.

The survey has shown that not only are there considerable differences between arable and grassland but that there can also be great differences between meadowland and pastureland. Depending upon the type of management, the differences between meadowland and pastureland can be less extreme (e.g. Morris 1973); it must also be remembered that meadows are often grazed after the hay has been cut. Animals can be folded on arable fields following harvesting and their droppings will attract beetles. Grassland in an outfield system may occasionally be ploughed for crops.

When it comes to using Coleoptera from archaeological deposits to obtain information about past agricultural landscapes, a number of other limitations arise as well as some blurring of the differences between meadowland, pasture and arable:

1. Some beetles live in other habitats as well as grassland or arable. A major problem here is that, since cereals do not have serious coleopterous pests and the beetles which are pests of other arable crops mostly occur on agricultural weeds, it is not possible to tell from beetle evidence alone whether it is a community from an arable habitat or certain types of weedy disturbed ground. Other forms of archaeological evidence must be taken into account.

2. Some species are restricted to particular soils, have a restricted range or have different ecologies in different parts of their range. For example some Carabidae show a habitat shift in the cool wet climate of Argyll, with what are woodland species in Berkshire living in grassland (Greenslade 1968). Various species of *Amara* which only occur on sandy soils have already been mentioned.

3. Different beetles have different dispersive powers and differing susceptibilities to being trapped in archaeological deposits. Southwood divided insect movement into two types, trivial and migratory, the trivial being confined to the habitat of the population. The level of migratory movement (by definition, away from the habitat) was found to be positively correlated with the degree of impermanence of the habitat, although both arable and grassland were regarded as temporary habitats because they are not climax vegetation (Southwood 1962: 174-5, 200-2). Propellor trapping of carabid beetles at Silwood, Berkshire, in a locality close to arable, grass heath and woodland showed flight to be most frequent in beetles from arable, followed by grassland, while no woodland species were caught (Greenslade and Southwood 1962). This was true both in terms of number of individuals and number of species captured.

4. The deposits examined are death assemblages containing Coleoptera from many different habitats. Information on the arable and pastoral aspects of the landscape will come from the "background fauna" of Kenward (1978: 6-7), an attempt must be made to eliminate from the consideration those individuals which lived in the deposit or in its immediate vicinity.

5. Some of the beetle species which feature prominently in the survey of modern habitats are not found in sufficient quantity in archaeological deposits to have much interpretative value. The pitfall trapping suggests that *Pterostichus cupreus*, *P. versicolor* and *Fleutiauxellus quadripustulatus* would all be good indicators of grassland if only they were more abundant in archaeological deposits.

6. Many beetles are not identifiable to species from fragments. Most species of *Ceutorhynchus*, including *C. floralis*, are very hard to identify from their individual sclerites and likewise, many species of *Amara*, including *A. communis*, must be identified from genitalia if fragmented. Thus several potentially useful groups of Coleoptera are lost due to problems of preservation.

7. The information from Coleoptera about an area is likely to be disjointed. Coleoptera from a given locality do not become as widely dispersed as the pollen from many plants in that place. For information about non-wetland landscapes, one is mostly forced to reply upon insects preserved in localised waterlogged deposits in the terrestrial environment: ancient river and stream channels, man-made features such as wells and ditches. Different features will tend to have different trapping characteristics. It is rare to find long sequences of the sort which make pollen analysis so valuable. Sometimes, in order to obtain sufficient Coleoptera for interpretation, the results from several different deposits on a site must be combined, and they are unlikely all to have been of exactly the same date.

CHARACTERISTIC GROUPINGS OF ARABLE AND GRASSLAND COLEOPTERA

Despite the above limitations, Coleoptera from archaeological deposits give at least some information on arable/pastoral ratios.

The relevant groupings of beetles used by Robinson (1981: 279-82) are considered and some new groupings proposed.

Pasture/Dung

Beetles from the genera Geotrupes, Colobopterus, Aphodius and Onthophagus are good indicators of grazed grassland. They are commonly found in archaeological deposits, most of them occur on a variety of soils, they fly readily and sufficient of them are restricted to the dung of large herbivores in the field (rather than manure heaps). Aphodius testudinarius (F.) has been recorded from manured fields (Horion 1958: 158) but this is not a suitale habitat for most of the species.

Dung/Foul Organic Material

Cercyon spp., Megasternum sp., Cryptopleurum spp., Anotylus rugosus, A. sculpturatus and Platystethus arenarius have been grouped as species of dung/foul organic material. They are common, widespread, species with good powers of dispersal but as well as occurring in dung in the field, they will also live in dung heaps and other accumulations of decaying organic material. Therefore they can only be used to reinforce the evidence for pasture from the group of dung beetles.

Especially Meadowland

The abundance of many weevils from the genera Apion and Sitona seems to be positively correlated with the presence of meadowland. They are widespread, can be quite common in archaeological deposits and have extensive pre- or post-hibernation flights (Southwood 1962: 193). Caution must be applied to the interpretation of this group. Sitona species which will attack Vicia faba have already been mentioned. The species of Sitona which feed on Ulex (Joy 1932: 176-7) are recognisable and can be eliminated from the group. Few of the meadowland species of Apion, which mostly feed on members of the family Papilionaceae, are readily identifiable from fragments, but many of the species which feed on other plants (Fowler 1891: 130-1) for example mallows and nettles, can be identified and so excluded from the total.

On Roots in Grassland

A new grouping with larvae which feeds on roots in both pasture and meadowland comprises various Elateridae and Scarabaeidae, in particular, Phyllopertha horticola (L.), Hoplia philanthus (Fuess.), Serica brunnea (L.), Omaloplia ruricola (F.), Agrypnus murinus (L.), Athous spp. and Agriotes spp. Various other beetles could be added to this list if they but occurred in archaeological deposits. Adults of some of these species will fly to trees and shrubs to feed (although adult feeding is not always essential, Raw 1951-52: 644-5) but oviposition is in grassland. Only a few individuals from a population of P. horticola take long migratory flights, but sufficient of the remaining population take shorter flights for the dispersal of this species in relation to archaeological deposits to count as good (Milne 1958-59: 690-1; Milne 1960-61: 374-6). Some of these scarabs and elaterids are widespread, living in most sorts of soils unless they are waterlogged for long periods.

Sandy Disturbed Ground/Arable

Amara apricaria, *A. bifrons* and *A. tibialis* formed a group used in Robinson (1981: 267) to suggest arable. They are indeed species of bare ground and arable, but they are restricted to sandy and gravelly soils. They do not occur on heavy soils. Fortunately, these three species are some of the few members of the genus *Amara* which can be identified from fragments. They are all capable of flight (Lindroth 1945: 105, 109, 172).

General Disturbed Ground/Arable

Harpalus rufipes and *Agonum dorsale* are carabids which are very much more abundant on arable/weedy disturbed ground than in other habitats. They are widespread species, inhabiting a great range of soil types and they are encountered reasonably frequently in archaeological deposits. This new grouping might seem an ideal indicator of arable, but unfortunately, the dispersive power of *H. rufipes* is variable. In France, *H. rufipes* flies readily on warm nights but at East Malling, in Kent, it was only occasionally caught in light traps between 1953 and 1963. On a few nights, however, averaging less than one per year, it was trapped in very high numbers. In all these instances, air temperature exceeded 18°C at 22hrs G.M.T., a rare occurrence in S.E. England. Flight of *H. rufipes* appears to be uncommon in S.E. England and the large numbers of this species which can be caught by pitfall trapping suggest crawling to be its principle means of dispersal (Briggs 1965-66: 82). *H. rufipes* from archaeological deposits must therefore be regarded as giving rather local information.

Arable Weeds

A final new grouping, species which tend to feed on arable weeds, does not seem to be of much value (see below). It is hard to find suitable species which are common in archaeological deposits, readily identifiable and suitably restricted to weeds of disturbed ground. *Phyllotreta* spp. and *Chaetocnema concinna* comprise this grouping but their food plants include some grassland Cruciferae and Polygonaceae. Various species of *Phyllotreta* attack *Rorippa* spp. (Cooter and Cribb 1975: 91), plants likely to have been growing in or at the edge of many of the type of waterlogged deposits in which insect remains are preserved. Another problem with insects which feed on annual weeds is that they do not establish a balance with their hosts. One year, a beetle may be very abundant and defoliate its host, in the following year the beetle may be rare.

THE USE OF SPECIES GROUPINGS OF COLEOPTERA TO ASSESS ARABLE/PASTORAL RATIOS AROUND ARCHAEOLOGICAL SITES

The proportion of individuals of Coleoptera from the pitfall traps falling into each of the species groupings given above is shown in Fig. 3:1 as a percentage of the total beetles from each site. Obviously only trends, rather than values, can be compared with results from archaeological deposits because the trapping method gives results dominated by Carabidae and Staphylinidae. However, two points can be made from Fig. 3:1. Even for a species grouping with abundant individuals (General Disturbed Ground/Arable), there was quite a large difference in the percentage belonging to the grouping in different examples of that habitat. Therefore, it is only substantial differences that are likely to be of interpretative significance when comparing archaeological assemblages. Secondly, a reasonably high percentage of the Dung/Foul Organic Material grouping occurred away from either of those two habitats. In most cases, it seems that a grouping will have some representation even if the habitat it is used to indicate is absent.

For archaeological samples, the choice of deposits which accumulated under water and the exclusion from calculations of any species which spends part of its adult life under water goes some way towards eliminating differences due to very local conditions. The number of individuals in each species grouping for a sample (or group of samples) can then be expressed as a percentage of the total terrestrial Coleoptera in that sample. Not all the species used are abundant in archaeological deposits, so the sample ought to contain at least a few hundred terrestrial individuals. It must be remembered that what values constitute a high or a low percentage will be particular to each species grouping.

Results from some Iron Age and Roman Sites

The results of these methods for several Iron Age and Roman rural archaeological sites are given in Fig. 3:2. Substantial insect assemblages have been identified from these sites, which all seem to have been set in relatively unwooded landscapes. The waterlogged deposits examined were in the bottoms of wells or ditches. For those sites on which pollen analyses were also done, the arable/pastoral ratio was calculated from the average pollen percentages after Turner (1964: 81-2). All the sites used the products of both arable and pastoral agriculture, as shown by the presence of carbonised grain, quern fragments, sheep bones and cattle bones.

The Sites

Fisherwick (SK187082)

Lowest gravel terrace of Tame, Staffordshire. Iron Age settlement (Smith 1979). Coleoptera by P.J. Osborne, pollen by J.R.A. Greig. Samples used: 1j, 301j, 302f. 2nd century bc.

Mingies Ditch (SP391059)

Windrush floodplain, Oxfordshire. Iron Age settlement (Robinson 1981: 256-9). Coleoptera by M.A. Robinson, pollen by G.W. Dimbleby. Samples used: Sample groups 1 and 2 (Robinson unpublished). c.3rd century bc.

Farmoor (SP443057)

Thames floodplain and first gravel terrace (Iron Age), first gravel terrace (Roman), Oxfordshire. Iron Age and Roman settlement (Lambrick and Robinson 1979; Robinson 1981: 259-65). Coleoptera by M.A. Robinson, pollen by G.W. Dimbleby. Samples used: see Robinson (1981: 280-1). Iron Age - c. 2nd century bc, Roman - 2nd-4th centuries AD.

Appleford (SU523936)

First terrace of Thames, Oxfordshire. Roman settlement (Hinchliffe and Thomas 1980; Robinson 1981: 265-6). Coleoptera by M.A. Robinson. Samples used: all. 2nd-4th centuries AD.

Barton Court Farm (SU510978)

Second terrace of Thames, Oxfordshire. Roman villa (Miles, forthcoming; Robinson 1981: 266-8). Coleoptera by M.A. Robinson, pollen by J.R.A. Greig. Samples used: all Roman except 832 spit 4 moss. 4th century AD.

Mount Farm (SU582967)

Third terrace of Thames, Oxforshire. Roman settlement (Lambrick 1979: 113-115). Coleoptera by M.A. Robinson, pollen by J.R.A. Greig. Samples used: 2nd-4th century AD.

Interpretation

The proportion of beetles belonging to the Arable Weeds grouping seems to vary from site to site without being related to the importance of arable, while the values for Dung/Foul Organic Material seem remarkably consistent. The other five groupings, however, do seem to be giving information on the arable/pastoral ratio.

The results from Fisherwick, Mingles Ditch and Farmoor Iron Age suggest that these sites were surrounded by a landscape predominantly of pasture. Percentages for the Pasture/Dung grouping were very high for Fisherwick and Farmoor Iron Age. The percentages of the On Roots in Grassland grouping were high for Mingies Ditch and Fisherwick, while there was evidence for seasonal flooding at Farmoor (Lambrick and Robinson 1979: 109-10) which may have made conditions less suitable for the underground larvae of these beetles. Beetles from the two Disturbed Ground/Arable groupings were either very rare or absent from these three sites. In contrast, the Mount Farm site seems to have been set amongst arable fields. The Disturbed Ground/Arable groupings were well represented but the percentage for Pasture/Dung was very low.

It is somewhat harder to place the remaining sites. Appleford gave intermediate values for all the useful groupings except Especially Meadowland, which was twice as high from Appleford as from any of the other six sites. Possibly grassland was important and there were hay meadows in the vicinity of the site. The percentage for Pasture/Dung from Roman Farmoor, at over 20% was high enough to

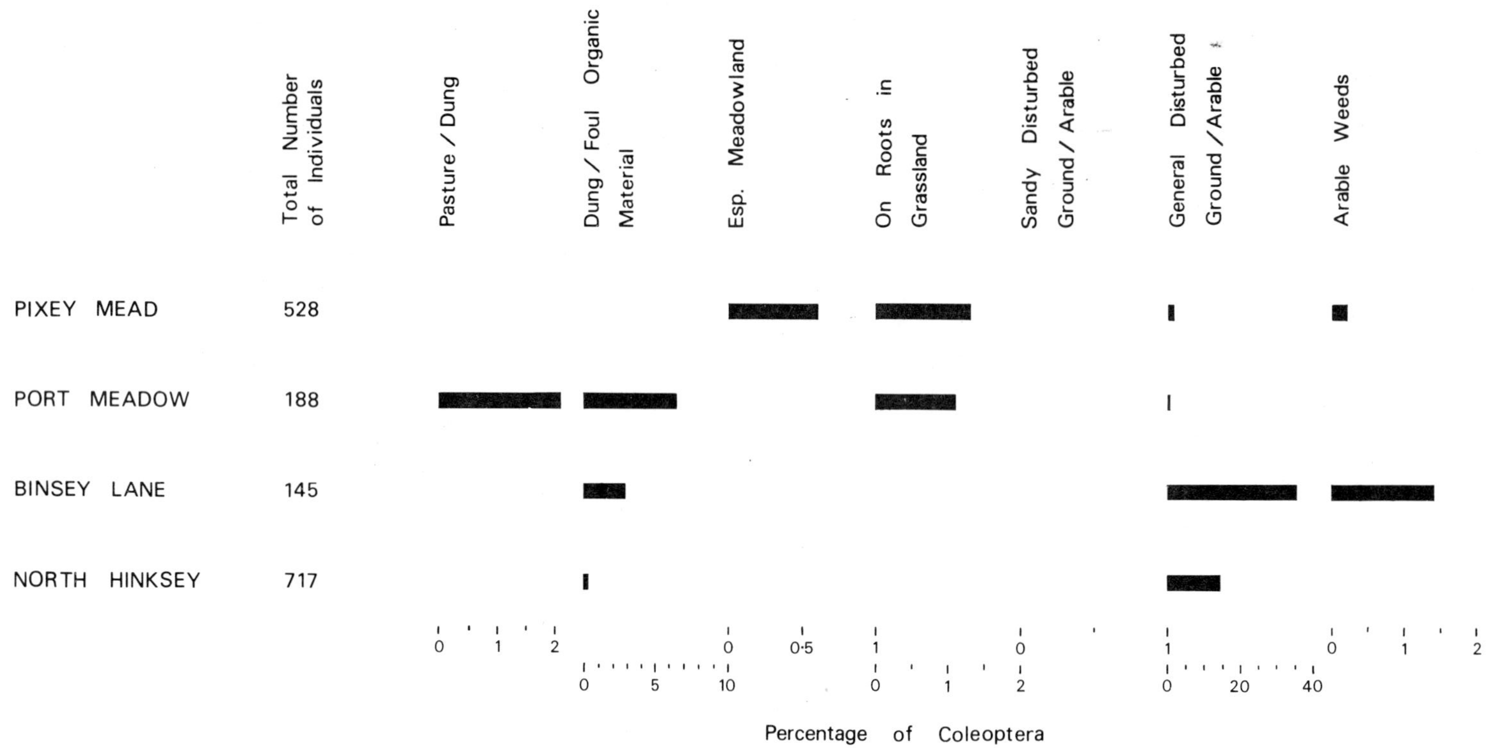

Fig. 3:1 Percentage of Coleoptera from the four modern sites sampled falling into the habitat groupings.

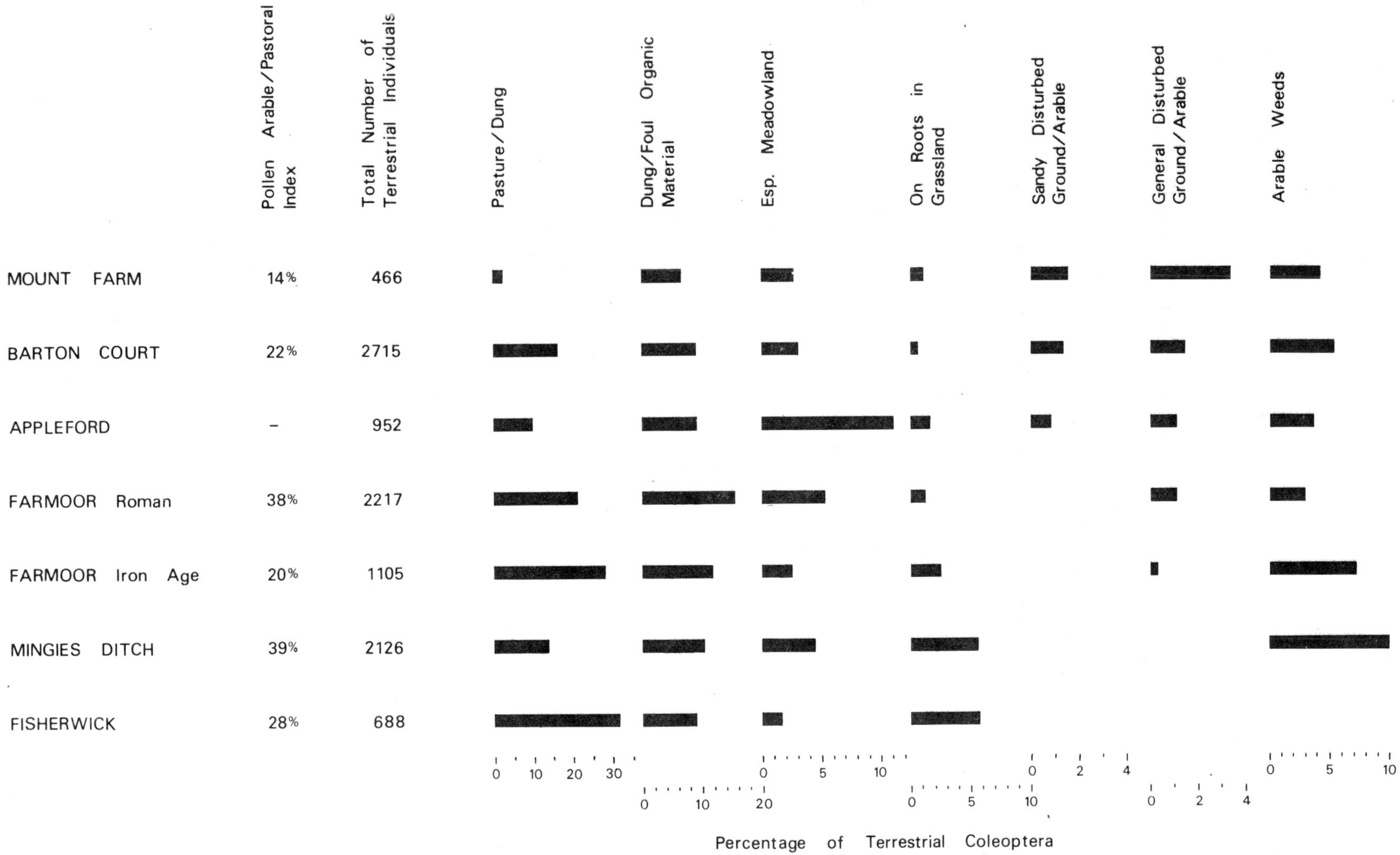

Fig. 3:2 Percentage of terrestial Coleoptera from various Iron Age and Roman sites falling into the habitat groupings compared with Turner's arable/pastoral index.

show the importance of pastureland. The value for General Disturbed Ground/Arable was no more than might be expected from bare ground around a settlement but there could also have been a significant presence of arable. The results from Barton Court suggest that both pasture and disturbed ground/arable were important.

It is possible to have some confidence in the interpretation suggested for Fig. 3:2 because the results from the various groupings agree with each other. Fisherwick had the highest values for both Pasture/Dung and the On Roots in Grassland whereas the two Disturbed Ground/Arable groupings were absent. Mount Farm had the lowest percentage for Pasture/Dung and the second lowest percentage for On Roots in Grassland but the highest values for the Disturbed Ground/Arable groupings.

These results should not only be considered in isolation. The soil conditions at Mount Farm and Barton Court, well drained sandy loam on high gravel terraces, might be expected to favour arable. The situation of the Iron Age sites at Farmoor and Mingies Ditch would be expected to favour grassland. The results can also be compared with the results of pollen analysis for most of the sites. Turner's arable/pastoral index tends to have a value below 15% in arable regions and above 50% in pastoral regions (Turner 1964: 81-2). Most of the percentages fall between the two values, but, apart from Iron Age Farmoor, show the same trends suggested by the Coleoptera, for example the low value for the pollen index at Mount Farm supporting the interpretation of arable and the higher value of the index for Mingies Ditch supporting the interpretation of pasture. At Farmoor Iron Age, the low value of the ratio was due to one pollen sample having a high percentage for Chenopodiaceae while another had a large percentage of Compositae-Liguliflorae. The Chenopodiaceae pollen was probably a local factor because seeds of *Atriplex* sp. were very abundant in that deposit. Liguliflorae can also be numerous in grassland; it is Compositae-Tubuliflorae which are better indicators of arable and their pollen was not abundant in the Farmoor Iron Age samples.

The use of the species groupings of Coleoptera must not be regarded as a substitute for the consideration of the whole insect death assemblage. Trends shown by the groupings can be reinforced or weakened when a community approach is taken to the assemblages. It must be established whether the full range of species which make up the community suggested by the indicator species is present, and whether the species are consistently represented in all the samples. For example, the presence of pastureland was postulated at Barton Court, although the percentage of the Pasture/Dung grouping was not as high as for some of the other sites. The community approach confirms the presence of pasture. The following dung beetles from the super-family Scarabaeoidea were present in all three samples from the site:

Geotrupes sp.

Aphodius contaminatus (Hbst.)

A. foetidus (F.)

A. paykulli Bed.

A. prodromus Brahm.

A. rufipes (L.)

A. cf. *sphacelatus* (Pz.)

Oxyomus sylvestris (Scop.)

This shows an appropriately diverse fauna of Aphodius species, it was not the case that the value for the Pasture/Dung grouping was due to a single species being very abundant in one sample. All the Barton Court samples also contained a full range of other inhabitants of the droppings of large herbivores including species such as Onthophilus striatus (Forst.), Gyrohypnus angustatus Step. and Philonthus spp. (all species which at present are not included in any grouping).

CONCLUSIONS

Coleoptera cannot yet be used to obtain an arable/pastoral ratio from archaeological deposits in the true sense but they can show the importance of pastureland very well. Coleoptera are also able to give limited information about meadowland and disturbed ground/arable. The only serious problems with assessing the importance of pasture from beetles are that it is unknown over what range the information is applicable and how much any tendency for domestic animals to congregate around a site would distort the results.

Unfortunately, it does not seem possible to separate evidence of weedy disturbed ground from evidence for arable. What is more, the evidence about disturbed ground/arable given by Coleoptera seems to be from a smaller radius around the deposit than that given about grassland.

The groupings of Coleoptera have mostly proved useful and it ought to be possible to create some more of them. The species composition of some of the groupings, however, may need to be altered for use in some parts of England.

For the archaeological sites used in the survey, the Coleoptera gave as much information on arable/pastoral ratios as the pollen. The waterlogged deposits present on them were not as suitable for providing such information from pollen as, for example, a long peat bog sequence would have been. Also, pollen analysis is less time consuming than insect investigations. Pollen and insect studies should, however, be used to complement each other. From dung beetles, Coleoptera are able to give details of pasture which pollen cannot provide, while pollen is capable of giving information of a more regional nature. It is ironic that most European cereals have neither wind-blown pollen nor distinctive coleopterous pests from which the palaeoecologist can deduce arable land.

ACKNOWLEDGEMENTS

I would like to thank Dr. M. R. Speight of the Forestry Department, Oxford University, for a useful discussion about coleopteran ecology in relation to agriculture and for the use of his thesis. I am grateful to Miss Alison McDonald of the Botany Department, Oxford University for identifying the grasses in my survey areas on Port Meadow and Pixey Mead.

Bibliography

Applebaum, S. (1972) Roman Britain. In: The Agrarian History of England and Wales vol 1: II (ed. Finberg, H.P.R.) Cambridge: University Press pp. 1-277

Atty, D.B. (1967) The beetle occupants of a tussock. Entomologist's Monthly Magazine 103 184-186

Baker, H. (1937) Alluvial meadows; a comparative study of grazed and mown meadows. Journal of Ecology 25 408-420

Briggs, J.B. (1965-66) Biology of some ground beetles (Col., Carabidae) injurious to strawberries. Bulletin of Entomological Research 56 79-93

Britton, E.B. (1956) Coleoptera:Scarabaeoidea. Royal Entomological Society of London. Handbook for the Identification of British Insects Vol. 5 pt. 11

Coaker, T.H. and Williams, D.A. (1963) The importance of some Carabidae and Staphylinidae as predators of the cabbage root fly Erioischia brassicae (Bouché). Entomologia Experimentalis et Applicata 8 156-164

Coope, G.R. and Osborne, P.J. (1968) Report on the coleopterous fauna of the Roman well at Barnsley Park, Gloucestershire. Transactions of the Bristol and Gloucestershire Archaeological Society 86 84-87

Cooter, J. and Cribb, P.W. ed (1975) A coleopterist's handbook. London: The Amateur Entomologists' Society.

Danthanarayana, W. (1967) Host specificity of Sitona beetles. Nature London 213 1153-1154

Dunn, J.A. (1960) The natural enemies of the lettuce root aphid, Pemphigus bursarius (L.). Bulletin of Entomological Research 51 271-278

Edwards, E.E. and Evans, J.R. (1950) Observations on the biology of Corymbites cupreus F. (Coleoptera, Elateridae). Annals of Applied Biology 37 249-259

Eghtedar, E. (1970) Zur Biologie und Ökologie der Staphyliniden Philonthus fuscipennis Mannh. und Oxytelus rugosus Grav. Pedobiologia 10 169-179

Elton, C.S. (1966) The pattern of animal communities. London: Methuen

Elton, C.S. and Miller, R.S. (1954) The ecological survey of animal communities: with a practical system of classifying habitats by structural characters. Journal of Ecology 42 460-496

Evans, A.C. (1944) Observations on the biology and physiology of wireworms of the genus Agriotes Esch. Annals of Applied Biology 31 235-250

Evans, J.G. (1972) Land snails in archaeology. London: Seminar Press.

Fowler, W.W. (1891) The Coleoptera of the British Islands. 5 London: Reeve and Co.

Freeman, B.E. (1967) The biology of the white clover seed weevil Apion dichroum Bedel (Col. Curculionidae). Journal of Applied Ecology 4 535-552

Freude, H. Harde, J.W. and Lohse, G.A. (1966) Die Käfer Mitteleuropas. 9 Krefeld: Goecke and Evers.

Gair, R. Jenkins, J.E.E. and Lester, E. (1972) Cereal pests and diseases. Ipswich: Farming Press

Girling, M.A. (1982) Fossil insect faunas from forest sites. In: Archaeological aspects of woodland ecology. Symposia of the Association for Environmental Archaeology No. 2 (ed. Bell, M. and Limbrey, S.) Oxford: British Archaeological Reports pp. 129-146

Greenslade, P.J.M. (1964) Pitfall trapping as a method for studying populations of Carabidae (Coleoptera). Journal of Animal Ecology 33 301-310

Greenslade, P.J.M. (1965) On the ecology of some British carabid beetles with special reference to life histories. Transactions of the Society for British Entomology 16 149-179

Greenslade, P.J.M. (1968) Habitat and altitude distribution of Carabidae (Coleoptera) in Argyll, Scotland. Transactions of the Royal Entomological Society of London 120 39-54

Greenslade, P.J.M. and Southwood, T.R.E. (1962) The relationship of flight and habitat in some Carabidae. The Entomologist 95 86-88

Hinchliffe, J. and Thomas, R. (1980) Archaeological investigations at Appleford. Oxoniensia 45 9-111

Hinton, H.E. (1944) Some general remarks on sub-social beetles, with notes on the biology of the staphylinid Platystethus arenarius (Fourcroy). Proceedings of the Royal Entomological Society of London (A) 19 115-128

Hoffman, A. (1954) Coleoptérès Curculionides. II Faune de France 59 Paris: Lechevalier

Hoffman, A. (1958) Coleoptérès Curculionides. III Faune de France 62 Paris: Lechevalier

Holter, P. (1979) Abundance and reproductive strategy of the dung beetle Aphodius rufipes (L.) (Scarabaeidae). Ecological Entomology 4 317-326

Horion, A.D. (1958) Faunistik der Mitteleuropäischen Käfer. 6 Überlingen-Bodensee: Aug. Feyel.

Horion, A.D. (1963) Faunistik der Mitteleuropäischen Käfer. 9 Überlingen-Bodensee: Aug. Feyel.

Hutchinson, K.J. and King, K.L. (1980) The effects of sheep stocking level on invertebrate abundance, biomass and energy utilization in a temperate, sown grassland, Journal of Applied Ecology 17 369-387

Isaac, P.V. (1923) The turnip gall weevil Ceuthorhynchus pleurostigma, Marsh. (Coleoptera, Curculionidae). Part 1, life history and bionomics. Annals of Applied Biology 10 151-170

Jones, J.M. (1950) Clover seed weevils. Annals of Applied Biology 37 313-320

Joy, N.H. (1932) A practical handbook of British beetles. 1 London: Witherby

Kenward, H.K. (1975) Pitfalls in the environmental interpretation of insect death assemblages. Journal of Archaeological Science 2 85-94

Kenward, H.K. (1976) Reconstructing ancient ecological conditions from insect remains; some problems and an experimental approach. Ecological Entomology 1 7-17

Kenward, H.K. (1978) The analysis of aracheological insect assemblages: a new approach. Archaeology of York 19/1 London: Council for British Archaeology

Kenward, H.K. (1979) Megathrips lativentris (Heeger) (Thysanoptera: Phlaeothripidae) from Bronze Age deposits at Thorne Moor, Yorkshire. Entomologist's Monthly Magazine 114 25-26

Koskela, H. (1972) Habitat selection of dung - inhabiting staphylinids (Coleoptera) in relation to age of the dung. Annales Zoologici Fennici 9 156-171

Koskela, H. and Hanski, I. (1977) Structure and succession in a beetle community inhabiting cow dung. Annales Zoologici Fennici 14 204-223

Lambrick, G.H. (1979) Berinsfield, Mount Farm. Council for British Archaeology Group 9 Newsletter 9 113-115

Landin, B.-O. (1961) Ecological studies on dung-beetles. Opuscula Entomologica Supplementum 19 1-227

Lewis, T. (1961) Records of Thysanoptera at Silwood Park, with notes on their biology. Proceedings of the Royal Entomological Society of London (A) 36 89-95

Lindroth, C.H. (1945) Die Fennoskandischen Carabidae. 1 Göteborg: Elanders.

Lindroth, C.H. (1974) Coleoptera, Carabidae. Royal Entomological Society of London Handbooks for the Identification of British Insects 4 pt. 2

Lipkow, E. von (1970) Biologisch-ökologische Untersuchungen über Tachyporus Arten und Tachinus rufipes (Col., Staphyl.). Pedobiologia 6 140-177

Luff, M.L. (1965) A list of Coleoptera occurring in grass tussocks. Entomologist's Monthly Magazine 101 240-245

Luff, M.L. (1968) Some effects of formalin on the numbers of Coleoptera caught in pitfall traps. Entomologist's Monthly Magazine 104, 115-116

Luff, M.L. (1978) Diel activity patterns of some field Carabidae. Ecological Entomology 3 53-62

Miles, D. (Forthcoming) Archaeology at Barton Court Farm, Oxon.

Milne, A. (1958-59) Biology and ecology of the garden chafer, Phyllopertha horticola (L.).IV The flight season: introduction and general aspects. Bulletin of Entomological Research 49 685-699

Milne, A. (1960-61) Biology and ecology of the garden chafer Phyllopertha horticola (L.).VII The flight season: male and female behaviour, and concluding discussion. Bulletin of Entomological Research 51 538-578

Mohr, C.O. (1943) Cattle droppings as ecological units. Ecological Monographs 13 275-298

Morris, M.G. (1967) Differences between the invertebrate faunas of grazed and ungrazed chalk grassland. I. Responses of some phytophagous insects to cessation of grazing. Journal of Applied Ecology 4 459-474

Morris, M.G. (1969) Differences between the invertebrate faunas of grazed and ungrazed chalk grassland. III. The heteropterous fauna. Journal of Applied Ecology 6 475-487

Morris, M.G. (1971) Differences between the invertebrate fauna of grazed and ungrazed chalk grassland. IV. Abundance and diversity of Homoptera - Auchenorhyncha. Journal of Applied Ecology 8 37-52

Morris, M.G. (1973) The effects of seasonal grazing on the Heteroptera and Auchenorhyncha (Hemiptera) of chalk grassland. Journal of Applied Ecology 10 761-780.

Omerod, E.A. (1890) Manual of injurious insects and methods of prevention. London: Simpkin, Marshall, Hamilton and Kent

Osborne, P.J. (1969) An insect fauna of late Bronze Age date from Wilsford, Wiltshire. Journal of Animal Ecology 38 555-566

Osborne, P.J. (1978) Insect evidence for the effect of man on the lowland landscape. In: The effect of man on the landscape: the lowland zone. Council for British Archaeology Research Report 21 (ed Limbrey, S. and Evans, J.G.) London pp. 32-34

Petherbridge, F.R. (1928) The turnip mud beetles (Helophorus rugosus Ol. and Helophorus porculus Bedel). Annals of Applied Biology 15 659-678

Petherbridge, F.R. and Thomas, I (1936) Damage to wheat by Helophorus nubilus F. Annals of Applied Biology 23 640-648

Pimentel, D. (1961) Species diversity and insect population outbreaks. Annals of the Entomological Society of America 54 76-86

Pitkin, B.R. (1976) The hosts and distribution of British thrips. Ecological Entomology 1 41-47

Pollard, E. (1968) Hedges III The effect of removal of the bottom flora of a hawthorn hedgerow on the Carabidae of the hedge bottom. Journal of Applied Ecology 5 125-139

Potts, G.R. and Vickerman, G.P. (1974) Studies on the cereal ecosystem. Advances in Ecological Research 8 107-197

Raw, F. (1951-52) The ecology of the garden chafer Phyllopertha horticola (L.) with preliminary observations on control measures. Bulletin of Entomological Research 42 605-646

Robinson, M.A. (1981) The Iron Age to early Saxon environment of the Upper Thames Terraces. In: The environment of man: the Iron Age to the Anglo-Saxon period (ed Jones, M. and Dimbleby, G.) Oxford: British Archaeological Reports British Series 87 pp. 251-286

Smith, C. ed. (1979) Fisherwick. Oxford: British Archaeological Reports British Series 61

Southwood, T.R.E. (1962) Migration of terrestrial arthropods in relation to habitat. Biological Reviews 37 171-214

Speight, M.R. (1976) Studies on the ecology of ground and rove beetles (Coleoptera: Carabidae and Staphylinidae) in winter wheat. (York: University, D.Phil. thesis.)

Speight, M.R. and Lawton, J.H. (1976) The influence of weedcover on the mortality imposed on artificial prey by predatory ground beetles in cereal fields. Oecologia (Berlin) 23 211-223

Stapley, J.H. (1949) Pests of farm crops. London: Spon.

Sunderland, K.D. (1975) The diet of some predatory arthropods in cereal crops. Journal of Applied Ecology 12 507-515

Thiele, H.-U. (1977) Carabid beetles in their environments. Berlin: Springer-Verlag.

Turner, J. (1964) The anthropogenic factor in vegetational history I. Tregaron and Whixall Mosses. New Phytologist 63 73-90

Vickerman, G.P. and Sunderland, K.D. (1975) Arthropods in cereal crops: nocturnal activity, vertical distribution and aphid predation. Journal of Applied Ecology 12 755-765

Waterhouse, F.L. (1955) Microclimatological profiles in grass cover in relation to biological problems. Quarterly Journal of the Royal Meterological Society 81 63-71

White, E. (1960) The natural history of some species of Aphodius (Col., Scarabaeidae) in the Northern Pennines. Entomologist's Monthly Magazine 96 25-30

Table 3:1: PIXEY MEAD FLORA

Ranunculus acris L.

R. repens L.

Cerastium holosteoides Fr.

Trifolium dubium Sibth.

T. repens L.

T. pratense L.

Lotus cf. *pedunculatus* Cav.

Vicia cracca L.

Lathyris pratensis L.

Filipendula ulmaria (L.) Maxim.

Potentilla reptans L.

Sanguisorba officinalis L.

Silaum silaus (L.) Schinz

Rumex acetosa L.

Primula veris L.

Rhinanthus minor L.

Prunella vulgaris L.

Plantago lanceolata L.

Chrysanthemum leucanthemum L.

Centaurea nigra L.

Leontodon autumnalis L.

L. hispidus L.

Taraxacum sp.

Carex spicata Huds.

Festuca rubra L.

Lolium perenne L.

Poa trivialis L.

Dactylis glomerata L.

Bromus cf. *thominii* Hard.

Hordeum secalinum Schreb.

Trisetum flavescens (L.) Beauv.

Arrhenatherum elatius (L.) Beauv.

Holcus lanatus L.

Phleum pratense L.

Table 3:2: PORT MEADOW FLORA

Ranunculus repens L.

R. flammula L.

Rorippa nasturtium-aquaticum (L.) Hayek

Stellaria palustris Retz.

Potentilla anserina L.

Oenanthe fistulosa L.

Myosotis scorpioides L.

Veronica catenata Pennell

V. scutellata L.

Mentha aquatica L.

Juncus articulatus L.

Eleocharis palustris (L.) Roem. & Schult.

Glyceria fluitans (L.) R.Br.

Agrostis stolonifera L.

Alopecurus geniculatus L.

Table 3:3 BINSEY LANE WEED FLORA

Ranunculus repens L.	*P. persicaria* L.
Papaver somniferum L.	*P. convolvulus* L.
Capsella bursa-pastoris (L.) Medic.	*Rumex obtusifolius* L.
Agrostemma githago L.	*Solanum nigrum* L.
Stellaria media (L.) Vill.	*Veronica persica* Poir.
Chenopodium album L.	*Senecio vulgaris* L.
Atriplex hastata L.	*Cirsium arvense* (L.) Scop.
Mercurialis annua L.	*Centaurea cyanus* L.
Euphorbia peplus L.	*Sonchus asper* (L.) Hill.
Polygonum amphibium L.	*Agropyron repens* (L.) Beauv.

Table 3:4: NORTH HINKSEY WEED FLORA

Myosotis arvensis (L.) Hill	*Anisantha sterilis* (L.) Nevski
Galium aparine L.	*Hordeum vulgare* L.

Table 3:5: COLEOPTERA FROM PITFALL TRAPS - as a percentage of the total catch for each site.

	PIXEY MEAD	PORT MEADOW	BINSEY LANE	NORTH HINKSEY
CARABIDAE				
Carabus granulatus L.	0.2	-	-	-
Nebria brevicollis (F.)	1.0	2.1	0.7	-
Notiophilus biguttatus (F.)	-	-	1.4	-
Loricera pilicornis (F.)	-	11.2	0.7	1.4
Dyschirius globosus (Hbst.)	0.4	-	-	-
Clivina fossor (L.)	1.5	-	0.7	0.3
Trechus micros (Hbst.)	0.2	-	-	-
Asaphidion flavipes (L.)	-	-	-	0.3
Bembidion properans Steph.	-	4.3	-	-
B. gilvipes Stm.	2.1	-	-	-
B. biguttatum (F.)	-	0.5	1.4	-
B. guttula (F.)	0.8	10.1	-	0.3
B. lunulatum (Fourc.)	-	4.3	-	0.4
Pterostichus anthracinus (Pz.)	0.2	-	-	-
P. cupreus (L.)	20.1	-	-	-
P. madidus (F.)	-	-	0.7	0.4
P. melanarius (Ill.)	5.7	1.1	7.6	1.8
P. nigrita (Pk.)	-	0.5	-	-
P. strenuus (Pz.)	-	-	0.7	0.3
P. vernalis (Pz.)	1.1	-	-	-
P. versicolor (Stm.)	25.4	-	-	-
Agonum dorsale (Pont.)	-	-	2.8	3.4
A. marginatum (L.)	-	1.6	-	-
Amara communis (Pz.)	14.6	-	-	-
A. plebeja (Gyll.)	-	-	0.7	-
A. similata (Gyll.)	-	-	7.6	0.1
Harpalus rufipes (Deg.)	1.3	0.5	32.4	10.9
H. cf. *rufibarbis* (F.)	-	-	-	0.1
Demetrias atricapillus (L.)	-	-	0.7	-

	PIXEY MEAD	PORT MEADOW	BINSEY LANE	NORTH HINKSEY
HYDROPHILIDAE				
Helophorus brevipalpis Bed.	-	4.8	-	-
H. rufipes (Bosc d'Antic)	-	-	0.7	-
Sphaeridium scarabaeoides (L.)	-	1.6	-	-
Cercyon atomarius (F.)	-	1.1	-	-
C. melanocephalus (L.)	-	1.1	-	-
C. pygmaeus (Ill.)	-	0.5	-	-
Megasternum obscurum (Marsh.)	-	-	0.7	-
LEIODIDAE				
Ptomaphagus subvillosus (Goez.)	-	-	-	0.1
P. varicornis (Rosen.)	-	-	-	0.1
Sciodrepoides watsoni (Spence)	-	-	-	0.3
Catops morio (F.)	-	-	-	0.1
SILPHIDAE				
Nicrophorus vespillo (L.)	0.4	-	-	0.1
STAPHYLINIDAE				
Omalium rivulare (Pk.)	-	-	-	0.1
Platystethus arenarius (Fourc.)	-	1.1	-	-
P. nodifrons (Man.)	-	2.1	-	-
Anotylus sculpturatus (Gr.)	-	1.1	2.1	0.3
Stenus canaliculatus Gyll.	-	2.1	0.7	-
Rugilus orbiculatus (Pk.)	-	-	-	0.1
Xantholinus longiventris Heer	-	0.5	-	-
Philonthus cognatus Steph.	0.6	12.8	-	0.8
P. fimetarius (Gr.)	-	-	-	0.1
P. laminatus (Cr.)	0.2	2.7	-	0.1
P. marginatus (Strom)	-	1.1	-	-
P. politus (L.)	-	-	0.7	-
Gabrius cf. *pennatus* Shp.	0.4	1.1	-	-
Staphylinus brunnipes F.	-	-	-	0.1
S. fuscatus Gr.	0.6	-	-	-
Tachyporus chrysomelinus (L.)	0.2	1.1	-	-

	PIXEY MEAD	PORT MEADOW	BINSEY LANE	NORTH HINKSEY
T. hypnorum (F.)	0.8	2.7	-	0.7
T. nitidulus (F.)	-	-	0.7	-
T. obtusus (L.)	-	-	-	0.3
T. solutus Er.	-	1.6	0.7	0.6
Tachinus laticollis Gr.	-	-	-	0.1
T. signatus Gr.	0.2	8.5	16.6	65.0
Aleochara curtula (Goez.)	-	-	-	0.1
Aleocharinae gen. et sp. indet.	0.2	7.5	3.5	7.8
PSELAPHIDAE				
Pselaphus heisei (Hbst.)	0.2	-	-	-
SCARABAEIDAE				
Coloboptorus fossor (L.)	-	1.6	-	-
Aphodius sphacelatus (Pz.)	-	0.5	-	-
BYRRHIDAE				
Simplocaria semistriata (F.)	-	-	-	0.1
ELATERIDAE				
Fleutiauxellus quadripustulatus (F.)	6.1	-	-	-
Selatosomus incanus (Gyll.)	0.4	-	-	-
Agriotes obscurus (L.)	1.3	1.1	-	-
Adrastus pallens (F.)	-	-	0.7	-
CANTHARIDAE				
Cantharis nigra (De G.)	0.2	-	-	-
C. rufa L.	0.2	-	-	-
NITIDULIDAE				
Brachypterus urticae (F.)	-	-	0.7	-
CRYPTOPHAGIDAE				
Atomaria spp.	-	2.6	-	1.5

	PIXEY MEAD	PORT MEADOW	BINSEY LANE	NORTH HINKSEY
PHALACRIDAE				
Phalacrus fimetarius (F.)	-	-	-	0.1
Stilbus testaceus (Pz.)	-	-	-	0.1
COCCINELLIDAE				
Adalia bipunctata (L.)	-	-	0.7	0.1
Propylea quattuordecim-punctata (L.)	-	-	1.4	-
LATHRIDIIDAE				
Stephostethus lardarius (DeG.)	-	-	0.7	-
Enicmus transversus (Ol.)	-	-	2.8	0.3
PYROCHROIDAE				
Pyrochroa serraticornis (Scop.)	0.2	-	-	-
CHRYSOMELIDAE				
Phyllotreta exclamationis (Thunb.)	0.2	-	-	-
P. undulata Kuts.	-	-	1.4	-
Longitarsus luridus (Scop.)	3.6	-	-	-
L. melanocephalus (De G.)	0.6	-	-	-
L. pratensis (Pz.)	0.2	-	-	-
L. suturellus (Duft.)	-	-	0.7	-
APIONIDAE				
Apion trifolii (L.)	0.2	-	-	-
CURCULIONIDAE				
Barypithes araneiformis (Schr.)	-	-	-	0.7
Sitona sulcifrons (Thunb.)	0.4	-	-	-
Notaris acridulus (L.)	-	6.4	-	-
Ceutorhynchus erysimi (F.)	-	-	0.7	-
C. floralis (Pk.)	-	-	6.2	-
C. marginatus (Pk.)	0.2	-	-	-
Mecinus pyraster (Hbst.)	0.2	-	-	-
TOTAL NUMBER OF INDIVIDUALS	528	188	145	717
TOTAL NUMBER OF SPECIES	38	34	32	38

Table 3:6: COLEOPTERA FROM SWEEP-NETTING - number caught.

	PIXEY MEAD	PORT MEADOW	BINSEY LANE	NORTH HINKSEY
HYROPHILIDAE				
Helophorus brevipalpis Bed.	-	-	1	-
STAPHYLINIDAE				
Tachyporus chrysomelinus (L.)	1	-	-	-
T. solutus Er.	1	-	-	-
Aleochara lanuginosa Gr.	-	1	-	-
ELATERIDAE				
Fleutiauxellus quadripustulatus (F.)	1	-	-	-
Selatosomus incanus (Gyll.)	4	-	-	-
Adrastus pallens (F.)	-	-	1	-
CANTHARIDAE				
Cantharis nigra (De G.)	7	-	-	-
C. rufa L.	3	-	-	-
Rhagonycha femoralis (Brul.)	-	-	1	-
NITIDULIDAE				
Meligethes aeneus (F.)	-	-	3	-
M. cf. *nigrescens* Steph.	1	-	-	-
COCCINELLIDAE				
Adalia bipunctata (L.)	-	-	-	1
Coccinella septempunctata L.	1	-	-	-
LATHRIDIIDAE				
Corticaria fuscula (Gyll.)	1	-	-	-
BRUCHIDAE				
Bruchus loti Pk.	1	-	-	-

	PIXEY MEAD	PORT MEADOW	BINSEY LANE	NORTH HINKSEY
CHRYSOMELIDAE				
Galerucella tenella (L.)	1	-	-	-
Phyllotreta consobrina (Curt.)	-	-	2	-
P. nigripes (F.)	-	-	7	-
P. undulata Kuts.	-	-	2	-
Apthona lutescens (Gyll.)	1	-	-	-
Longitarsus suturellus (Duft.)	1	-	-	-
APIONIDAE				
Apion curtirostre Germ.	1	-	-	-
A. ervi Kirb.	1	-	-	-
A. trifolii (L.)	11	-	-	-
CURCULIONIDAE				
Phyllobius viridiaeris (Laich.)	2	-	-	-
Barynotus obscurus (F.)	1	-	-	-
Sitona suturalis Steph.	1	-	-	-
Hypera nigrirostris (F.)	1	-	-	-
Ceutorhynchus floralis (Pk.)	-	-	29	-
Mecinus pyraster (Hbst.)	1	-	-	-
Gymnetron pascuorum (Gyll.)	1	-	-	-
TOTAL NUMBER OF INDIVIDUALS	44	1	46	1
TOTAL NUMBER OF SPECIES	22	1	8	1

4. EARLY LAND USE ON DARTMOOR - PALAEOBOTANICAL AND PEDOLOGICAL INVESTIGATIONS ON HOLNE MOOR

David Maguire, Nicholas Ralph and Andrew Fleming

INTRODUCTION

The palaeoenvironmental studies reported in this paper form part of the work of the Dartmoor Reave Project. The project was designed to investigate the socioeconomic and environmental context of Bronze Age land boundaries, ruined walls known as 'reaves', on Dartmoor, a granite massif in Devon, south-west England. Most reaves occur on moorland just beyond the edge of present-day enclosed land, in a zone lying between about 200-275 m and 400 m above sea level. Because they are so extensive and well-preserved, it has been possible to recognise that the reaves form a coherent territorial pattern, which is reported in detail elsewhere (Fleming 1978, 1983). The pattern is one of communities whose territories were located radially around the perimeter of Dartmoor - and, in one case, in the Upper Dart basin, a natural geographic unit surrounded by the hills of the North Moor, and South Moor, and the East Moor. Each community had one large block of enclosed land, usually at least 500 ha in extent, known as a 'parallel reave system' because its dominant boundaries are long co-axial reaves. Within the parallel systems there are circular houses with stone footings, walled enclosures (sometimes in close association with houses), narrow lanes and 'transverse reaves' which, with the parallel reaves, demarcate rectilinear parcels or 'fields'. These reave systems are not unlike the 'cohesive' Celtic field systems of the southern English chalklands (Bradley 1978: 268) but they are much more extensive; particularly noteworthy are two very large (over 3000 ha) parallel reave systems on the eastern side of Dartmoor - the Rippon Tor system and the Darteet system which meet each other at right angles. Part of the latter system crosses Holne Moor, on the south-eastern side of Dartmoor, and it is here that we have undertaken detailed archaeological and palaeoenvironmental studies.

In these systems the parallel reaves run up to a 'terminal reave' which is a Bronze Age 'head dyke' separating the enclosed land from the open pastures. Beyond the terminal reave, single long-distance reaves head for the high moors; they divide the lower moorland pastures among the different communities in such a way that it is easy to see which pasture zone goes with which block of enclosed land. The pasture zones contain settlement sites, ranging from small, scattered huts which are easily labelled 'shielings' to thickly-walled enclosed 'villages' like Grimspound or Riders Rings (see Spooner and Russell 1967: Figs 37, 40). The higher pastures seem to have been subject to intercommoning; at any rate they are set apart from the lower moorland by 'contour reaves'. Some of these appear to be unfinished, but one on the western side of Dartmoor, stretches for at least 10 km. We seem to be dealing with an ancient landscape that has been comprehensively subdivided according to simple principles. Radiocarbon determinations from two recently-excavated sites some 14 km apart (Shaugh Moor - see Smith et al. 1981; Holne Moor - see Burleigh 1981) and the interdependence and inter-relatedness of different components of the layout suggest that this land division stemmed from a single decision taken around 1300 bc. It is probable that this subdivided landscape with its pattern of dispersed settlement within the enclosed fields extended well below the zone of high preservation on today's moorland. Parallel reaves run across most of the heathy commons on the hill tops to the east of the main moorland massif; furthermore, in some places, both to the east and to the south of Dartmoor, the modern field patterns have, in part, fossilised, and so preserved, the main elements of parallel reave systems.

Recognition of the extensive character of the 'cultural landscape' during the second millennium bc is a recent event. As such it post-dates the most extensive palaeobotanical studies undertaken on Dartmoor (Simmons 1964), and it is now difficult to reconcile the environmental 'reconstruction' presented by Simmons (1969: Fig. 3), which showed cleared zones in 'relict woodland' in the vicinity of known prehistoric settlements and 'forest' elsewhere, with the fully settled and clearly open landscape inhabited by the reave builders. Although Simmons' pollen sampling sites lay well away from the more heavily-settled moorland fringe, it must, nevertheless, be worthwhile to reconsider some of these early pollen diagrams and to ask how far their detailed interpretation might be altered in the light of present knowledge of second millennium land division and settlement. The value of new investigations whose sampling strategy can take account of the increased archaeological knowledge is also evident; the work on Holne Moor, like the recent studies on Shaugh Moor (Beckett 1981; Balaam 1983), is a step in this direction.

The study area on Holne Moor is located around NGR SX 681713, on a plateau some 300-350 m above sea level (Fig. 4:1). To the north lies the gorge of the river Dart; to the south, Holne Moor is sheltered to some extent by the higher land of Holne Ridge (450 m above sea level) which forms the north-east corner of the South Moor. With minor exceptions on its north-eastern periphery, the whole of the study area is underlain by a coarse-grained and strongly acidic granite of late Carboniferous or early Permian age. However, everywhere this lies concealed beneath several metres of Pleistocene solifluction deposits of a type described by Waters (1964) as the 'main head'. In addition to granite, this loamy, stony soil parent material contains minor loessial additions (Harrod et al. 1974) similar to those noted in other parts of southern England (Perrin et al. 1974). The major physiographic features of much of Dartmoor have been attributed to a combination of early Pleistocene marine and Tertiary sub-aerial erosion (Brunsden and Gerrard 1970) and these

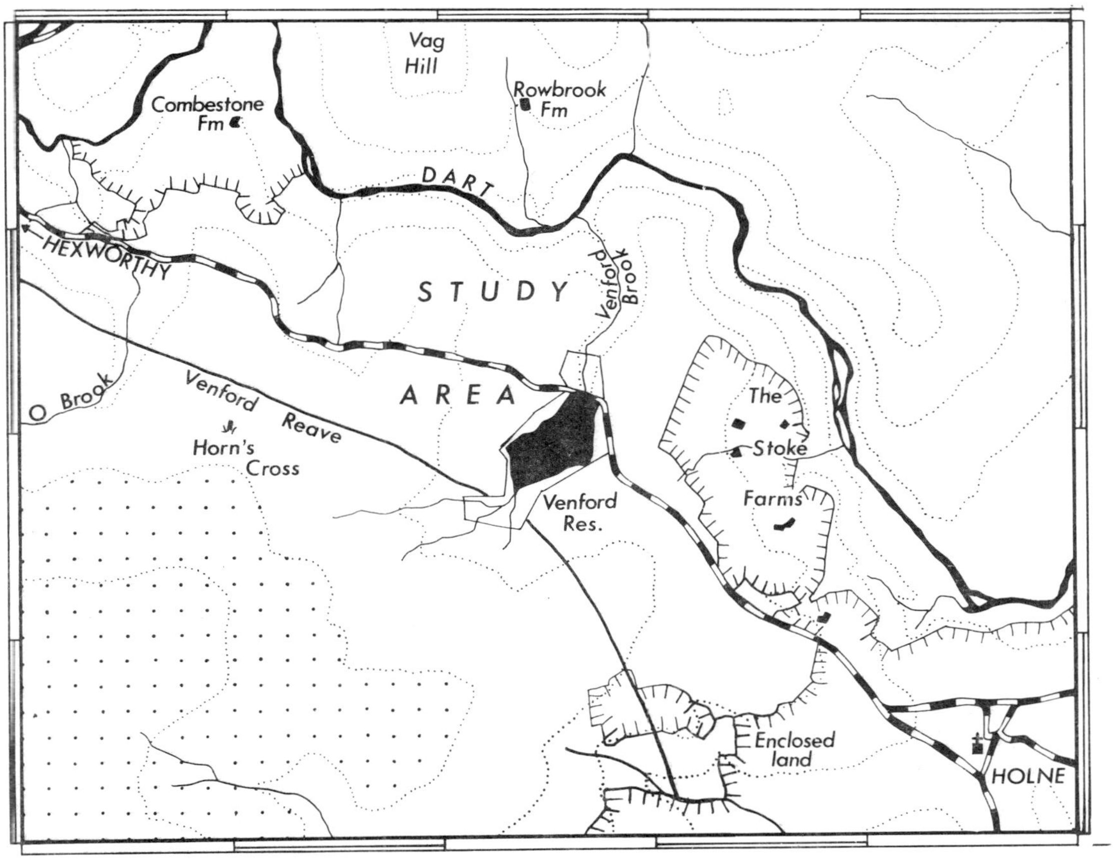

Fig. 4:1 General map of the Holne Moor study area and its location. Contours every 61 m (200 feet). Land over 396 m (1300 feet) is stippled. The border is marked in kilometres.

authors indicate Tertiary ages for erosional surfaces corresponding in height to those within the study area. Although the river Dart and, in its lower stretches, the Venford Brook have made deep incisions in the mature landscape represented by the plateau surfaces, the latter still retain the gentle forms of this earlier landscape. Slopes on the plateau surfaces rarely exceed 4° and with few exceptions have either convex or linear forms. The latter tend to dominate the upper slopes leading out of the flat plateau top while the former are most clearly seen at the margin of the study area where they lead into the steep slopes of the incised valleys.

The contemporary climate of Holne Moor, like that of all of Dartmoor, is a product of a blending of the oceanic pattern of the south-western peninsula with the more rigorous conditions of an upland plateau. Perkins (1970) and Brunsden and Gerrard (1970) provide concise descriptions of the resultant wet, windy but relatively mild weather. High humidity is characteristic and is associated with persistent low cloud and hill fog. Annual average precipitation on Holne Moor (1941-70) totalled 1773 mm, but there is considerable variation around this mean value; during the last thirty years annual precipitation on Holne Moor has varied from levels comparable with lowland stations in eastern England (e.g. 1953 - 1118 mm) to levels not out of place on the higher peaks of the Atlantic seaboard (e.g. 1974 - 2675 mm). The dominant winds are westerlies. In summer most weather arrives from the south-west; in winter colder, north-westerlies are far more frequent.

As noted earlier, Holne Moor lies mostly within the southern part of one of the largest parallel reave systems - the Dartmeet system; its terminal reave, the Venford Reave, runs roughly east-west across the moor, and parallel reaves leave it to head north to the edge of the Dart gorge, where they die out; reaves on much the same alignment recommence on the other side of the gorge. A large part of Holne Moor was also enclosed by medieval farmers; their field boundaries sometimes incorporate reaves, but for the most part this very different system of land enclosure has overridden and in places destroyed the earlier pattern of settlement. Both the prehistoric and the medieval land boundaries and settlements have been examined in detail and mapped at a scale of 1:1000 (Fig. 4:2), and since 1976 a programme of selective excavation and investigation of soil characteristics has been undertaken by two of the authors (A.F. and N.R.). Archaeological excavation (by A.F) has been concentrated in one area; two prehistoric houses and their surrounds have been examined (sites B and F), and detailed investigations of reaves have been carried out at six places (sites B, C, E, F, G and I). Two medieval boundaries have also been sectioned (site H and North Lobe).

Archaeological evidence of prehistoric land use on Dartmoor is not easy to obtain, and often poses interpretational problems. Animal bones have not survived in what is now a very acid soil (pH 4.0-4.5), although the discovery of animal footprints (Smith *et al.* 1981: Plate 14) has been shown to be possible in favourable conditions. Carbonised seeds have been recovered. Soil samples from inside the house on site F, which radiocarbon dates (Burleigh 1981) suggest was occupied from *circa.* 1300-1100 bc, have produced carbonised remains of wheat, barley and the horse-bean, *Vicia faba*,

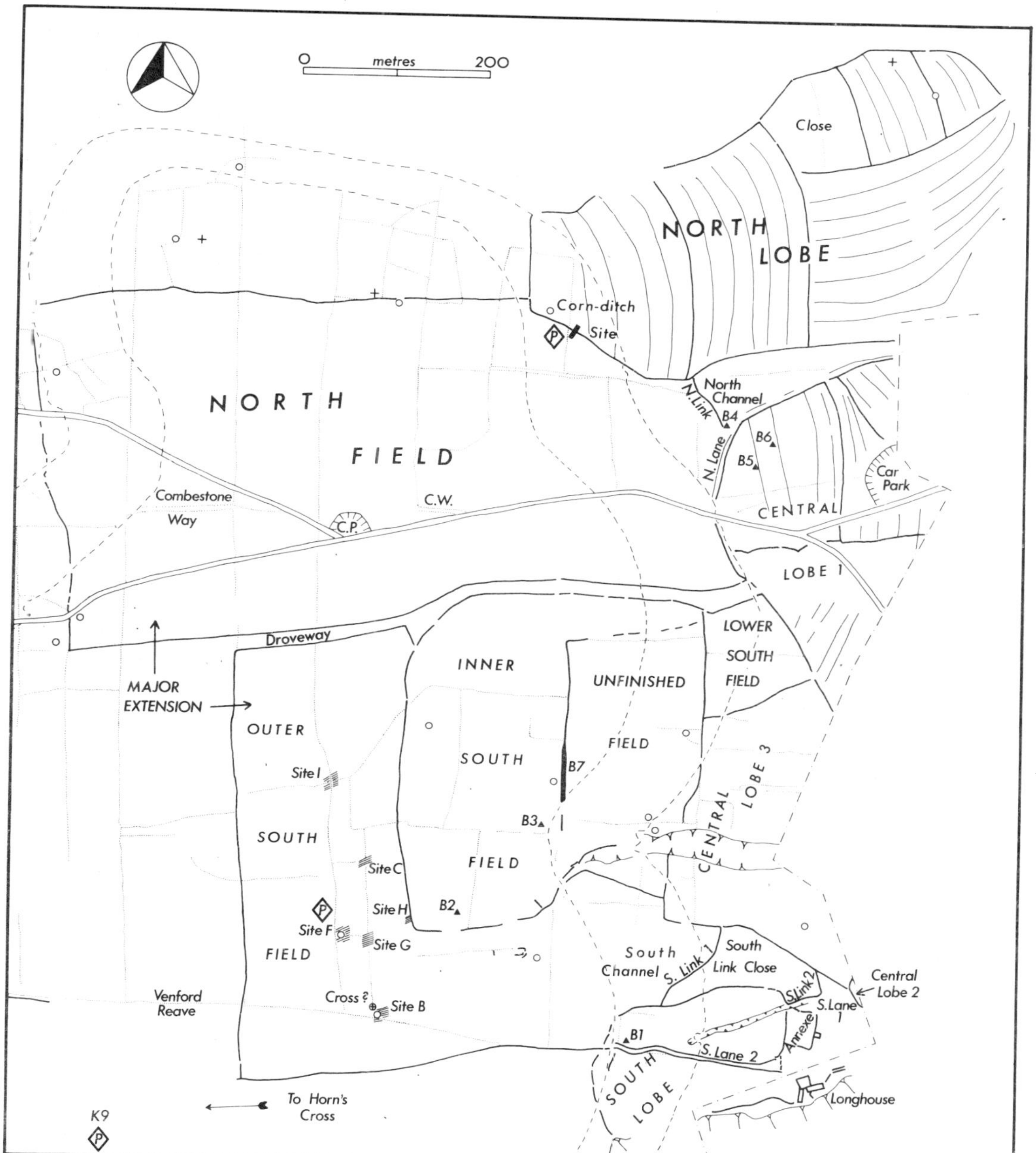

Fig. 4:2 Map of the study area showing prehistoric and medieval boundaries, archaeological excavation sites and pollen sampling locations. Solid lines - medieval boundaries; dotted lines - prehistoric reaves; open circles - prehistoric houses; crosses - prehistoric ceremonial monuments; hatching - medieval plough-marks on excavation sites; solid triangles - rabbit buries; open diamonds (with P) - pollen sampling locations.

together with some 'weed' seeds (Martin Jones: personal communication). However, it is of course possible that these plants were grown elsewhere than on Holne Moor, and it is one of the ironies of our work that the sub-fossil cereal pollen grains recovered from the earth fill of the house-wall on this site (see below) constitute, in a strict sense, better evidence for local cultivation than these carbonised remains. There are one or two lynchets on Holne Moor which are likely to be prehistoric in date, although the masking and smoothing effect of a thin peat cover, and the knowledge that medieval cultivations could have occurred, makes positive identification impossible in most cases. Only when two identical, unmodified prehistoric boundaries traverse the same slope, one showing a profile significantly different from that of the other, may one tentatively infer that the difference is caused by lynchetting - and even in this case, questions may be asked about the nature of any earlier boundaries underlying the reaves.

Better evidence of prehistoric land use has been obtained during the excavation of reaves; these have shown that in several places, and notably on sites G and B, stones have been piled up against reave faces in a manner suggesting prehistoric field clearance (cf. Fleming 1977). It may well be that the stones and boulders used in the construction of reaves also testify to the effects of cultivation; stone clearance problems would normally be expected to increase after the earliest period of tillage, and the excavations have also shown that fences and banks, the latter perhaps surmounted by hedges, were employed as boundaries before the stone reaves were built. A similar sequence is apparent in house construction; the earliest houses, of timber, were later replaced or refurbished by houses with stone wall-footings. The relatively frequent use of wood for the construction of houses, fences and other stake-built structures, perhaps mainly during the earlier phases of settlement, may indicate the existence of a coppice not too far away from these sites.

The layout of the Dartmeet parallel reave system is neither the first nor the last archaeologically-attested event on Holne Moor. A triple stone row, a cist, at least one burial cairn and two probable ring cairns were built in or near the study area, and most or all of these ceremonial monuments should pre-date the reave system. There is also a chain of 'shieling' settlements to the south of the terminal reave, and these too may reasonably be assumed to be earlier than the field system of the late second millennium bc. It is thus likely that, whatever the environment on Holne Moor immediately before the reaves were laid out, there was an earlier phase when the moor was relatively open land used as pasture. Finally, as noted earlier, much of the study area was enclosed by medieval farmers, and there is also evidence of post-medieval tinning and warrening. The development of medieval settlement and enclosure has been studied in detail and published by two of the authors (Fleming and Ralph 1982); only a brief summary of these events can be offered here.

It has been suggested that in the late Saxon period three farmsteads were established in the valley of the Venford Brook. These settlements had fields on both sides of the Brook; on the western side, in the study area, they had what we call 'Lobes', fields and field systems with curving boundaries thrusting out onto the moor (Fig. 4:2), interrupted by 'strolls' - large funnel-shaped gaps between the Lobes through which animals could pass to and from the moor. In the northern part of the study area, where there may have been a small hamlet, the Lobes were cultivated as sub-divided arable, in strips, although it has not been established that these subdivisions were primary features of the medieval colonisation. Lynchets are also present in the two more southerly Lobes, and although in some cases these lynchets may include a prehistoric element, there is no doubt that all the Lobe lands came under the plough at some point during the medieval period - a conclusion which is supported by the soil studies outlined below.

Towards the middle of the 13th century, the Forest Law which had been imposed after the Norman Conquest ceased to apply to the purlieus of Dartmoor, and on what had become the Commons of Devon it became possible for commoners to take in further land for outfield use. We have suggested that it was at this time that commoners from the two local manors of Holne and Stoke constructed what we have called the Major Extension - essentially two very large enclosures separated by a Droveway some 15 m broad. As can be seen in Fig. 4:2, the excavation sites discussed earlier all lie within the southern enclosure known as the Outer South Field, and all of these sites revealed plough furrows scored into the surface of the mineral soil horizon beneath the thin peat which blankets most of the study area. On site H the furrows were found to post-date the Inner South Field bank and ditch, which seem to have been constructed during the earliest stages of the Major Extension, and so must be a product of the medieval outfield tillage. The pattern of furrows suggests that this outfield may also have been ploughed in strips or lands, through these divisions are not visible on the present-day surface (Ralph and Fleming: forthcoming).

The Major Extension appears to have been rapidly abandoned, presumably by the mid-14th century, but the Lobes continued in use; their boundaries were probably refurbished in late medieval times. However by the 16th century, and perhaps a good deal earlier, the Venford farms and fields must have been largely abandoned. The fields were slighted by tin workings, and rabbit buries were built in and near them; the material for the buries was sometimes obtained by robbing disused medieval hedgebanks. It is probable that at least one late- or post-medieval tin-mill was built in the valley, but this too is likely to have gone out of use before the Civil War. Today the heathery pastures of the study area only provide sustenance for Scottish Blackface sheep, a small herd of Galloway cattle and a dozen moorland ponies.

Since 1976 the soils and, to a lesser extent, the vegetation of Holne Moor have been subjected to intensive investigations (by N.R) designed to reveal the influence of prehistoric and medieval land use upon the development of soils within the study area. This work included high density field survey of the contemporary soils, extensive quantitative analysis of phosphorus and organic matter in the soils of the abandoned fields and droveways, detailed study of two buried palaeosols of prehistoric and medieval age, and the initiation of the palaeobotanical studies which form the principle subject of this paper. Pollen has been extracted from the two buried palaeosols, their overburdens, three modern soil profiles and from the sediments in a medieval ditch. In order to understand our assessment of the chronology of these soil pollen samples, it is necessary to provide here a brief description of the sampled soils, an outline of pedogenesis within the study area and of the way in which it has been affected by medieval land use.

PEDOGENESIS AND LAND USE ON HOLNE MOOR

The dominant soils of all parts of the study area lying outside the Lobes and smaller medieval enclosures are either Iron-pan stagnopodzols of the Hexworthy Series (Clayden 1964) or Ferri-humic stagnopodzols of the Rough Tor Series (Harrod et al. 1976); within the Lobes most soils are of podzol or podzolic type and include soils similar to the Humic brown podzolic soils of the Moorgate Series and the Typical brown podzolic soils of the Moretonhampstead Series (Clayden 1964; 1971; Harrod et al. 1976); in addition there are soils whose characteristics mark them as intergrades between the podzolic soils and the stagnopodzols. An up-to-date description of the general featues of these soil types has recently been presented by Hogan (1982), who also discussed their classification and modern land use potential. The genesis of Dartmoor's soils has been discussed by Clayden and Manley (1964). They suggested that a change to heath vegetation and cooler, wetter conditions at the end of the Bronze Age initiated changes in a progenitor 'brown earth' soil. In some areas a surface layer of 'mor' humus formed, which subsequently led to iron-pan formation and thus the Hexworthy Series visible today. In other areas these soil developments had not occurred, and they postulated that

> '... the brown earths of today (i.e. the soils of the Moretonhampstead Series) may occupy more recently cleared woodland or may have been in continual agricultural use since forest clearance; they also occur generally in lower rainfall areas, where surface waterlogging is less likely to occur....' (1964: 134-135).

Twenty years ago no palaeosol or soil pollen studies had been undertaken on Dartmoor, and Clayden and Manley's ideas about soil history were, therefore, highly speculative. However, their views are in general well-supported by the pedological and palaeobotanical studies on Holne Moor, though these allow a more refined estimate of both the influence of land use and the time scale of soil transformations.

Credible evidence for the nature of the early soil development on Dartmoor is still not abundant; a soil buried around 1300 bc beneath the house wall on site F (Fig. 4:3) is at the moment the only palaeosol of this age which has been subjected to thorough quantitative physical and chemical analysis. Although many of the morphological features of this buried soil (e.g. soil colour, structure, horizons) are not unlike those of the Humic brown podzolic soils within the study area (e.g. the soils of the North Lobe) and so conform more or less to the predictions of Clayden and Manley's model, there are several indications that the palaeosol has been

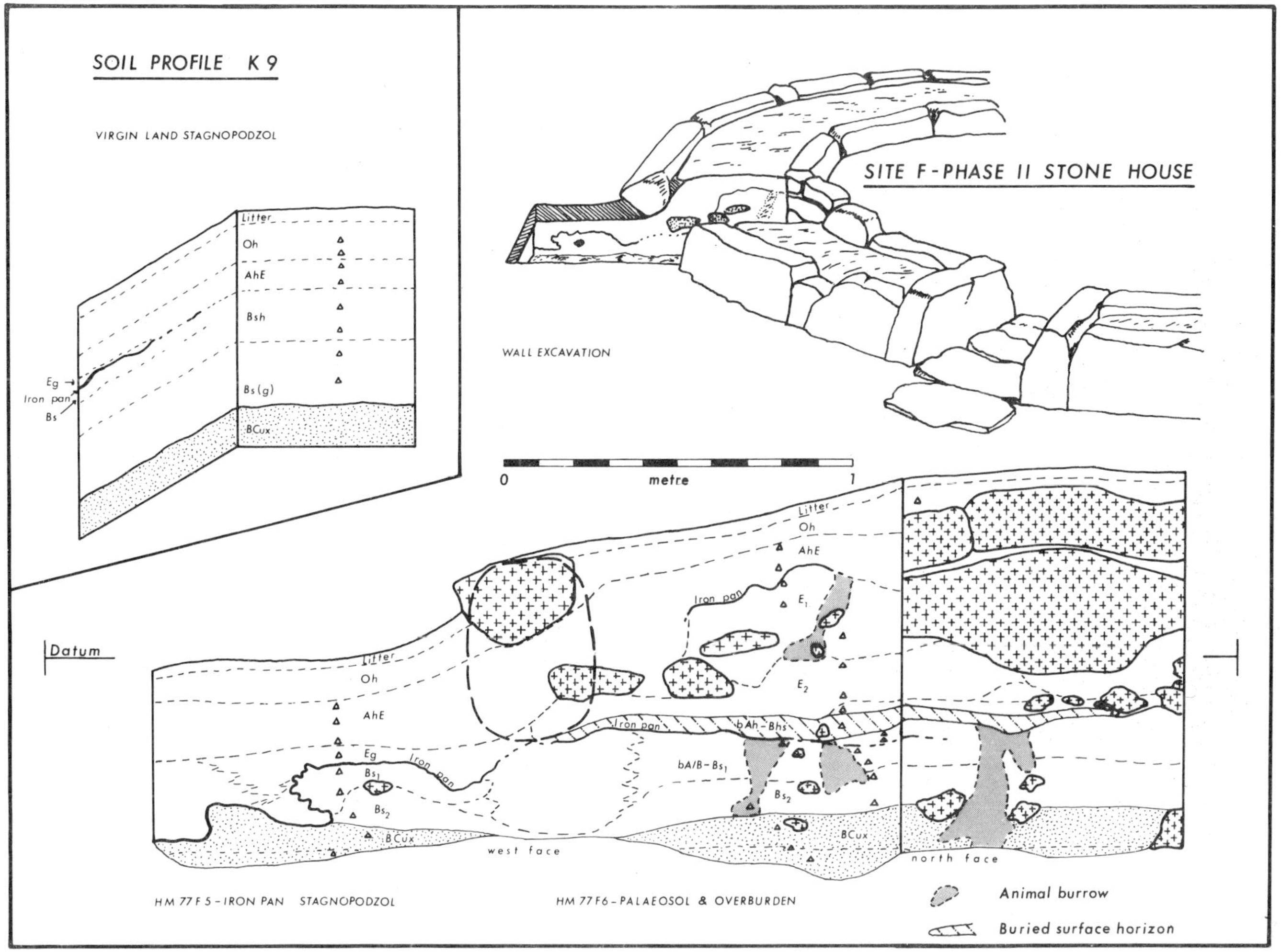

Fig. 4:3 Diagram showing pollen soil sampling at site F - prehistoric palaeosol, overburden and adjacent control profile - and at location K9 - virgin land stagnopodzol. For conventions see also Fig. 4:4.

affected by soil processes after its burial and it is not easy to distinguish relict features from those which should be attributed to pedometamorphic processes. The latter are certainly responsible for the fragmentary 'incipient' iron-pans in both the overburden and the palaeosol (similar post-burial iron deposition was noted by Staines (1972) during his examination of what seems to have been a comparable palaeosol beneath a house wall at Chagford). It can also be shown that the creation of well-developed eluvial horizons in the overburden involved losses by translocation, and that this may have altered, inter alia, the iron, organic matter and phosphorus content of the underlying palaeosol horizons.

Despite the problems of interpretation posed by post-burial alteration (which will be discussed in greater detail elsewhere), it can be argued that the soil buried by the reave builders not only lacked the peat horizon and iron-pans seen in the modern stagnopodzols but also possessed some structural and phosphorus characteristics similar to those of the present-day brown podzolic soils. This interpretation is fully in accord with the data recovered from a second, much better-preserved palaeosol buried some 2500 years later beneath the 'corn-ditch' boundary of the North Lobe (Fig. 4:4), and with several other medieval palaeosols beneath boundaries and tinning waste. Collectively these buried soils indicate that the depth of peat which has accumulated on the stagnopodzol soils of Holne Moor has at least doubled since early medieval times. The corn-ditch builders of the 11th century buried a soil with about 8-9 cm of peat (a figure that makes allowance for post-burial changes due to compression and loss of organic matter); today virgin soils on Holne Moor are typically covered by 15-20 cm of peat. Backward projection of the peat accumulation rate during historic time suggests that peat may have started to form around 700 BC, a very similar date to that estimated by Maltby and Crabtree (1976) for the same event on nearby Exmoor. Note, however, that although highly plausible the Holne Moor date has been obtained through calculations which incorporate several unproven assumptions. The development of iron-pans within the study area seems to have occurred much later. No pan was present in the soil beneath the corn-ditch boundary despite an ubiquitous presence in surrounding stagnopodzols, but pans are present in some of the Lobe soils, where in many cases they must have formed since cultivation ceased, probably in the 15th century. The very thin, soft and fragmentary character of many of the pans in the study area suggests that pan formation may be an active process today.

Medieval cultivations in both the Lobes and the outfields of the Major Extension have had a substantial affect on the properties of their surface soils; a full account of this is in preparation (Ralph and Fleming: forthcoming). The legacy of tillage takes a variety of forms; some of this variety appears to reflect differences in land preparation techniques and/or the frequency and longevity of tillage, and some seems to be due to differences in the extent to which post-agricultural pedogenesis has re-imposed stagnopodzol features on the medieval arable soils. The clearest and simplest example of the way in which soils have been altered by anthropogenic intervention is found in the subdivided fields of the North Lobe. Here the 'proto-stagnopodzol' soil buried by the Lobe boundary was transformed within

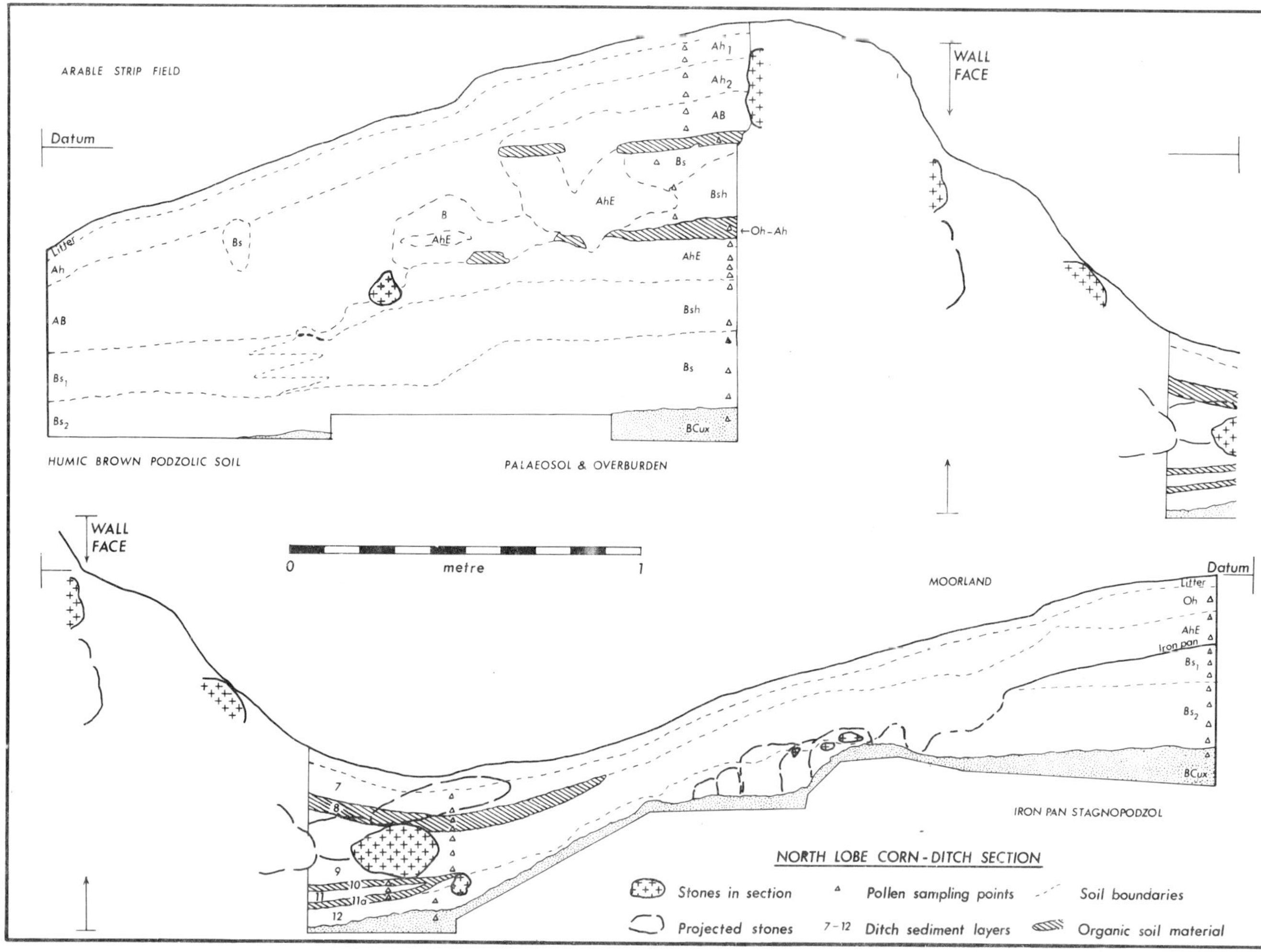

Fig. 4:4 Diagram showing soil pollen at the North Lobe corn-ditch excavation - early medieval palaeosol, overburden, nearby control profile and ditch sediments.

the field into a brown podzolic soil (Fig. 4:4); the thin peat cover of early medieval times was completely destroyed and has not yet re-formed. Perhaps it never will. In other Lobes (e.g. Central Lobe 1 - Fig. 4:2), probably abandoned somewhat earlier than the North Lobe, a new, often very thin peat now blankets the plough strips. In some of these areas the old plough soil takes the form of an exceptionally thick and light-coloured Ah/E horizon beneath a mere 5-9 cm of peat. In the outfields two patterns are evident. In some areas, the peat has a relatively high content of mineral soil; loss-on-ignition values (LOI) range from 25-50%, compared to a typical range of 55-80% in the virgin soils beyond the bounds of the outfields. It is thought that in these 'Low LOI' zones brief and probably relatively late outfield episodes (16th century?) with minimal land preparation simply mixed the peat horizon with the underlying mineral soil; the process would also have led to organic matter losses, and the present peat is usually significantly thinner than the peat in virgin land. The Iron-pan stagnopodzols adjacent to the North Lobe corn-ditch are of this type, and the profile sampled for pollen as a 'control' at this site (Fig. 4:4) may have been affected by these processes.

In other parts of the outfields a thinning of the peat is the principal sign of human intervention; LOI values span a slightly wider range than those in virgin land, and include some lower values. However, in general the peat in these zones has much the same amount of mineral material as the peat outside the enclosures. In this case it is thought that the peat visible today has re-formed after a single episode of tillage at the time when the outfield boundaries were constructed (13th century); total destruction of the original peat cover seems likely, and this could reflect the employment of more intensive methods of land preparation (e.g. paring and burning) and/or more prolonged cultivations. Some 12-15 cm of peat now covers these zones, which include the area surrounding site F and thus the 'control' profile adjacent to the prehistoric palaeosol (Fig. 4:3). It is worth mentioning that, due to the depradations of peat cutters and a fairly high level of natural variability in peat depth and quality, the recovery of these sort of patterns requires very dense sampling over substantial areas, with particular attention being paid to the way in which soil properties change at land boundaries.

The soil evidence recovered using these techniques has backed up the information about medieval land use obtained by more traditional archaeological field surveying (see Fleming and Ralph 1982), and has contributed important new evidence for the character of outfield cultivation and the sequence of land abandonment within the study area; equally important in this context, it enables us to understand the likely age of the pollen in both surface peats and, in some cases, the upper mineral soil beneath.

PALAEOBOTANICAL SAMPLING

The establishment of the reave systems on Dartmoor presents the palaeobotanist with an exceptional opportunity. The location and extent of cleared land at a specific period is now known for this region. Subject to the availability of suitable deposits, it would be possible to take samples for pollen analysis according to a carefully designed strategy which would take into account the proximity and character of systems of enclosure, the likely direction of prevailing winds, topographic factors, etc. This should allow us to obtain a very much clearer idea of the ability of pollen analysis to transmit a picture of past land use as opposed to broad regional trends in vegetational communities. We do not claim that the research undertaken thus far on Holne Moor represents an adequate response to this opportunity. Deep blanket bog deposits are not present in or close to the study area, and preliminary investigations suggest that the small valley and flush bog deposits that are present may be too young or too disturbed to provide useful pollen samples from the prehistoric landscape. They may, however, prove useful for investigation of medieval land use, and it is hoped that further work using such deposits will be undertaken. For the moment it has been necessary to rely on soil pollen analysis with all that this must imply in terms of the inexact chronology of samples and the problems posed by differential destruction of pollen within the soil.

To maximise our chances of understanding the chronology of samples, attention has been concentrated on the pollen content of the two palaeosols and their overburdens, since in these cases we do have approximate dates for the pollen in the buried surface horizons and for the inception of new pollen profiles in the over burdens (prehistoric surface soil: circa 1300 bc - overburden: circa 1100 bc; medieval surface soil and overburden: circa AD 1000). Interpretation of the pollen content of overburden materials often poses problems because it is difficult to assess the extent to which their present pollen content still reflects an initial, inherited pollen load. However, as will be shown below, this problem can be eased by concurrent physical and chemical analysis of overburden samples, since such analyses can provide independent evidence for the likely origin of these materials. We have also studied three control profiles. As noted above, two of these lay immediately adjacent to the palaeosol sites; a third, the Moorland profile (K9), was located in the extreme south-west of the study area some 150 m beyond the furthest bounds of the medieval outfield and the parallel reave system (see Fig. 4:2). It was hoped that this site, lying in virgin land to windward of the prehistoric and medieval enclosures, would yield a pattern of pollen deposition that was significantly different from the patterns in the palaeosols, their overburdens and adjacent control profiles, and that the differences would in some measure reflect the impact of land use practices on pollen deposition.

There are some slight differences in the properties of the soils at each of the sampling locations. The Moorland profile, K9, shown diagramatically in Fig. 4:3, is a stagnopodzol whose characteristics are, with one exception (a slightly thinner than usual Oh horizon), entirely typical of the soils in virgin areas. The surface peat is underlain by strongly humose Ah/E and Bsh horizons, and by a very stony light-coloured Bs horizon showing signs of weak gleying; the latter probably reflects the impermeability of an indurated BCux horizon which in this profile was encountered at a depth of 40 cm below the mineral soil surface. An iron-pan underlain by a thin (6 cm) sesquioxide-enriched layer was visible in one of the four faces of the soil pit, and in this face gleyed patches were present in the Ah/E immediately above the iron-pan; samples were not taken from this face. If we are right to assume that pollen input to the mineral soil ceases when a surface peat forms, then all the pollen in the mineral soil of this profile probably predates circa 700 BC. Conversely, the pollen in the Oh horizon should span the period 700 BC to the present-day.

On site F, where the prehistoric palaeosol was sampled, the adjacent control profile (Fig. 4:3) is an Iron-pan stagnopodzol with an unusually well-developed pan which has led to the creation of a clearly defined gleyed zone beneath the Ah/E horizon. Below the pan, Bs_1 and Bs_2 horizons were sampled, and in this profile the BCux horizon was encountered at the atypically shallow depth of 30-35 cm below the mineral soil furface. Medieval cultivations severely disturbed and may have entirely destroyed the original peat in this area; the pollen in the present Oh horizon could at best provide a (disturbed?) record of deposition since the 14th century, and, on this account, has been ignored. Pollen may have entered the mineral soil during and immediately after the outfield episode, and in this case we can only safely infer that the mineral soil pollen is likely to be of 14th century or earlier date. The palaeosol, like the control profile, has a very shallow depth of mineral soil, and there are in fact only minor differences in the physical and chemical properties of BCux and Bs_2 samples drawn from these two profiles. However the B horizons of the palaeosol have been penetrated by animal burrows originating in the buried Ah horizon; the latter, to judge from the similarity of phosphorus and organic matter values, appears to have been the source for the present infill of the burrows. A radiocarbon determination on the abundant charcoal (oak and probably pine - identifications by J. Hillam) in the burrow infill and the buried surface soil returned a date of 4810±240 bc (BM - 1604); this, of course, only provides a terminus post quem for the infilling of the burrows and tells us nothing about the age of the pollen in the burrow infill. However the stratigraphic evidence clearly indicates that the burrows and their infilling pre-date the construction of the house wall around 1300 bc.

Comparison of the upper horizons of the control profile and the buried soil reveals large differences. The nature, origin and significance of these differences cannot be considered properly here. However it should be noted that quantitative physical and chemical analyses indicate that it is unlikely that the palaeosol was significantly truncated prior to burial, and this accords with the archaeological evidence; the buried zone lay close to the wall of the earliest, timber house on the site and may well have consisted of bare, trodden earth which required no preparation (e.g. turf removal) before construction of the second house began. As noted earlier, the wall fill covering the palaeosol has been strongly leached; iron, organic matter and phosphorus have all be lost. This substantial alteration of the overburden materials since their emplacement makes it difficult to assess their original character, but consideration of their residual iron content (as indicated by ignited colours) and their stone content suggests that a reversed stratigraphy is present. Layer 4 seems to have been obtained from the surrounding surface soil (Ah), while layers 2 and 3 include material from greater depths (A/B and possibly upper B horizon); however, none of the overburden samples has a stone content as high as samples from the lower B horizon. The surface and near-surface origin of these materials implies a substantial initial load of inherited pollen, which could still dominate the pollen spectra; pollen input to the overburden is unlikely to have been substantial while the house remained roofed (*circa*. 1300-1100 bc), and may well have been cut-off by peat growth early in the first millennium BC.

The medieval palaeosol and its control profile lie some 700 m north-north-east of site F at the western boundary of the North Lobe (Fig 4:2). A partial excavation of this corn-ditch boundary was undertaken (by N.R) in 1978 (Fig 4:4); samples for physical, chemical and pollen analysis were taken from the palaeosol, the overburden, the ditch sediments and the undisturbed profile at the southern end of the ditch excavation. Here too the control profile is an Iron-pan stagnopodzol, and its features are typical of the soils in the surrounding part of the outfield, the North Field, though they contrast strongly with the soils in the North Lobe, to the east of the boundary. The principle differences between this control profile and that on site F are the absence of a gleyed zone above the iron-pan, which may reflect the slightly steeper slope of the land near the corn-ditch (*ca* 5-6° compared with 2-3° on site F), the slightly greater depth of mineral soil above the BCux horizon, and the difference noted earlier in the way in which the Oh horizon has been affected by cultivation. Although disturbed by outfield tillage, the peat in this area does not seem to have been destroyed. The mineral soil horizons could, therefore, provide a record of pollen deposition little or not affected by pollen input after 700 BC. A single sample from the disturbed peat has also been analysed.

The palaeosol beneath the bank is a well-preserved example of an earlier stage in the development of the present moorland stagnopodzols with which it shares many features (colour, structure, differentiation of horizons). Here too, quantitative phosphorus analyses suggest that the profile was intact at the time of burial. The main differences between this 'proto-stagnopodzol' and the modern soils are the absence of an iron-pan (and correspondingly more diffuse boundary between the Ah/E horizon and the B horizon) and the much thinner peat horizon. The pollen record from the mineral horizons should cover a period of time similar to that of the mineral horizons in the virgin profile K9, but pollen in the peat covers a more restricted span - *circa* 700 BC to *circa* AD 1000. The palaeosol is buried by just over 0.5 m of bank material, an overburden of similar depth to that covering the prehistoric palaeosol. In the lower half of the bank, the component soil materials from which it was built could be identified in the field using colour, texture and organic content as criteria; some of the individual clods of earth thrown up during construction were still identifiable. Subsequent phosphorus, organic matter and particle size analyses have confirmed field designations of the stratigraphic units shown in Fig 4:4 by demonstrating that the turves and B horizon materials in the lower half of the overburden still have much the same properties as their counterparts in the soil below the bank. The pollen spectra of these materials exhibits a similar pattern (see below), and it is clear that most or all of their present pollen was emplaced at the time of bank construction.

By contrast, in the upper half of the bank a new soil profile of Humic brown podzolic character, similar to the soils in the North Lobe, is present. New chemical characteristics have developed, and a new pollen profile has been imposed during the past thousand years; only particle size analysis (particularly stone content) reveals that the material in this part of the overburden was obtained from excavation of the very stony lower B and BCux horizons and thus that the amount of inherited pollen was probably very small. Although by no means representing a precise reverse stratigraphy, the bank evidently comprises the material excavated from the adjacent ditch. Some of this material has fallen back into the ditch, but the remaining volume of the bank corresponds closely to the volume of the present ditch, implying that little material has eroded down the ramp into the arable field.

The sediments which accumulated in the ditch (Fig 4:4) fall into three stratigraphic groups. Layers 10-12 pre-date a wall collapse, from which much of the material in layer 9 may be derived; the upper layers, 7 and 8, represent resumed silting and organic accumulation on a new, more southerly ditch axis. The wall collapse could correspond in time to a period when wall maintenance had ceased and the fields in the North Lobe had been abandoned; if so, layer 9 may seal the only sediments that are likely to preserve a pollen record dating to the period of intensive use of the fields.

To summarise, we have eight pollen profiles. Three control profiles provide us with 'norms' for landscape locations; first, virgin land to windward of all enclosures; secondly, a site within the reave system where the surface peat has been very seriously disturbed by medieval outfield cultivation; thirdly, a less disturbed site, also within the reave system but in this case adjacent to an intensively-used, long-lived medieval arable field. The two buried soils should at least provide us with snapshots of pollen depostion at or shortly after the beginning of each of the major agricultural ventures on Holne Moor. The overburden profiles are more complex; it seems likely that most of the pollen within the house wall fill on site F dates to the earliest stage of the prehistoric episode. The corn-ditch bank profile must be split in two; the pollen in the lower half is perhaps mainly of methodological interest, since it can supply information about the mobility and longevity of pollen in displaced soil materials, but the pollen in the upper half, where no peat surface has formed, should provide a continuous record of pollen deposition since early medieval times. The earliest part of this record should cover most or all of the period when the subdivided arable fields of the North Lobe were in use, and part of this period may also be represented by the pollen recovered from the basal layers of our final pollen profile - the ditch sediments.

POLLEN ANALYSIS

Methods

Eighty-two samples for pollen analysis were collected and prepared for counting at the University of Sheffield. Treatment of a known weight of oven-dry sample (Dimbleby 1961; Jorgensen 1967) included the KOH, HF, acetolysis, safranin staining and mounting in glycerol jelly stages of Faegri and Iversen (1975). Counting and identification (by D.M) was undertaken at a magnification of X 400, with critical identification at X 1000 using oil immersion and phase contrast units, on a Vickers M15C microscope. Pollen identification and terminology in the main follows Moore and Webb (1978), except for Gramineae which follows Faegri and Iversen (1975)[1]. Plant names follow Clapham et al (1962).

Although pollen concentration data is available, all values in this paper are expressed as a percentage of the pollen sum, which, unless otherwise indicated, is equal to the total number of dry land pollen and spores. Since glycerol jelly was used as a mounting medium, a correction has been applied to the size of Gramineae pollen grains, using the mean size of Corylus avellana grains ($\bar{x}$ = 28.9, n = 1261, S_{n-1} = 2.45 and K = 1.16; see Maguire: this volume).

Pollen diagrams

The results of pollen analysis on all the profiles studied are illustrated in Figs 4:5-4:10. Each of these diagrams has been divided into Local Assemblage Zones (LAZ) on the basis of changes in the frequencies of pollen and spores (sensu West 1970). The names of the LAZ and the sequences at each profile are listed in Tables 4:1 and 4:2. Our estimate of the chronological relationships between the various LAZ at each site is presented in Table 4:3. For the most part this chronological assessment flows directly from the arguments considered earlier, but our treatment of the two profiles which were disturbed by cultivation in medieval times require further consideration.

Table 4:1 Local Assemblage Zones in profiles K9, HM77 F5 and 6

Moorland Profile K9

HM M1 (0-17.5 cm) Oh

Ericaceae-type - Gramineae - Quercus - Corylus/Myrica

HM M2 (17.5-30 cm) Ah/E, upper Bsh

Gramineae - Polypodiaceae - Quercus - Corylus/Myrica

HM M3 (+ 30 cm) lower Bsh

Quercus - Polypodiaceae - Gramineae - Corylus/Myrica

Field Profile HM 77 F5

HM F1 (15-33) Ah/E, Eg

Gramineae - Corylus/Myrica - Polypodiaceae

Hut Profile HM 77 F6

HM H1 (0–10 cm) Oh
Gramineae - Ericaceae-type - Cyperaceae

HM H2 (10–62 cm) Wall fill, buried Ah
Gramineae - Polypodiaceae - Corylus/Myrica - Quercus

HM H3 (+ 62 cm) Old animal burrow beneath buried Ah
Quercus - Polypodiaceae - Corylus/Myrica - Gramineae

Table 4:2 Local Assemblage Zones in the Moorland by Corn-ditch profile, the Corn-ditch profile and the Ditch sediments

Moorland by Corn-ditch Profile

HM MC1 (0–14 cm) Oh, upper Ah/E
Ericaceae-type - Gramineae

HM MC2 (14.5–30 cm) lower Ah/E, Bs_1
Gramineae - Polypodiaceae

HM MC3 (+ 30 cm) Bs_2
Polypodiaceae - Gramineae

Corn-ditch Profile

HM CD1 (0–32 cm) Ah_1, Ah_2, A/B and O/A (old turf) in overburden
Ericaceae-type - Gramineae - Cyperaceae

HM CD2 (32–47 cm) Bs, Bsh material in overburden
Gramineae - Polypodiaceae

HM CD3 (47–60 cm) Buried Oh, upper Ah/E in palaeosol
Gramineae - Ericaceae-type

HM CD4 (60–83 cm) middle and lower Ah/E, Bsh in palaeosol
Gramineae - Polypodiaceae

HM CD5 (+ 83 cm) Bs, BCux in palaeosol
Polypodiaceae - Gramineae

Ditch Profile

HM D1 (0–7 cm) upper layer 7
Ericaceae-type - Gramineae - Cyperaceae - Picea - Pinus

HM D2 (7–34 cm) lower layer 7, layers 8–11
Ericaceae-type - Gramineae - Cereal-type

HM D3 (+ 34 cm) layer 12, BCux at base of ditch
Gramineae - Ericaceae-type

Table 4:3 Chronological relationships between soil pollen profiles on Holne Moor

PROFILE:	Moorland K9	Field HM77 F5	Hut HM77 F6	Moorland by Corn-ditch	Corn-ditch Palaeosol	Corn-ditch Overburden	Ditch Sediments
LAZ:	HM M -	HM F -	HM H -	HM MC -	HM CD -	HM CD -	HM D -
							- 1 -
ERICACEAE-type dominant	- 1 -			- 1 -		- 1 -	
			- 1 -				- 2 -
circa 1000 AD					**********		- 3 -
					- 3 -		
circa 700 BC		ONSET OF PEAT FORMATION					
ERICACEAE-type increase							
GRAMINEAE	- 2 -	- 1 -	- 2 -	- 2 -	- 4 -	- 2 -	
circa 1300 BC dominant		**********					
QUERCUS - POLYPODIACEAE - GRAMINEAE	- 3 -		- 3 -	- 3 -	- 5 -		

********** = Palaeosol Surface horizon

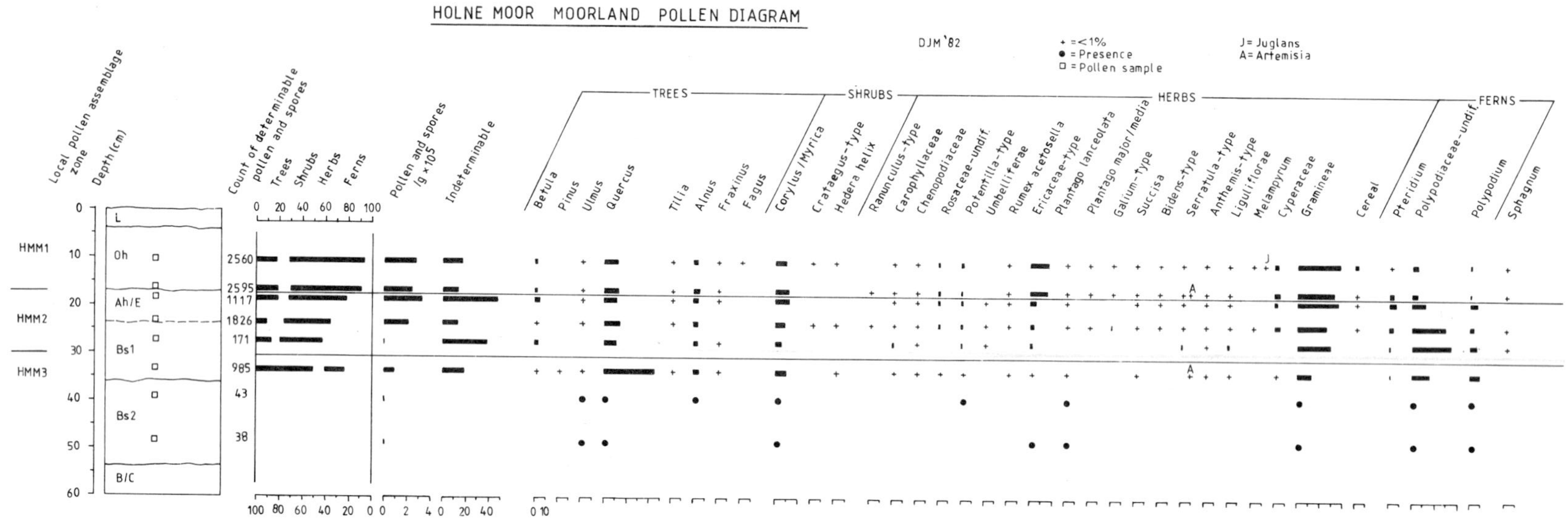

Fig. 4:5 Pollen diagram for the Moorland profile (K9), a virgin land stagnopodzol.

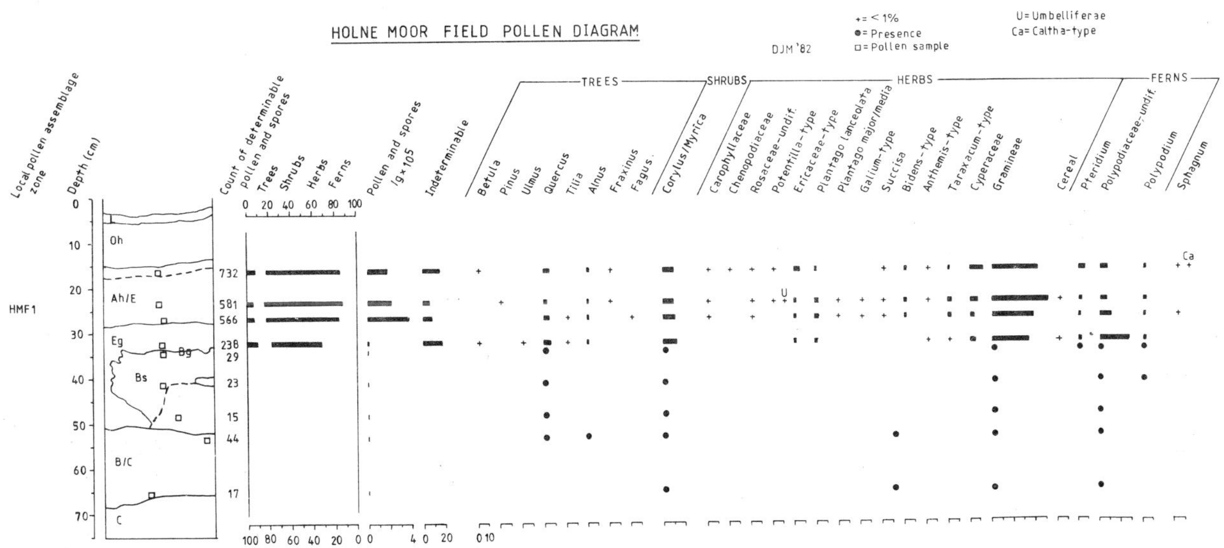

Fig. 4:6 Pollen diagrams for the Field control profile, an Iron-pan stagnopodzol on Site F (HM77 F 5).

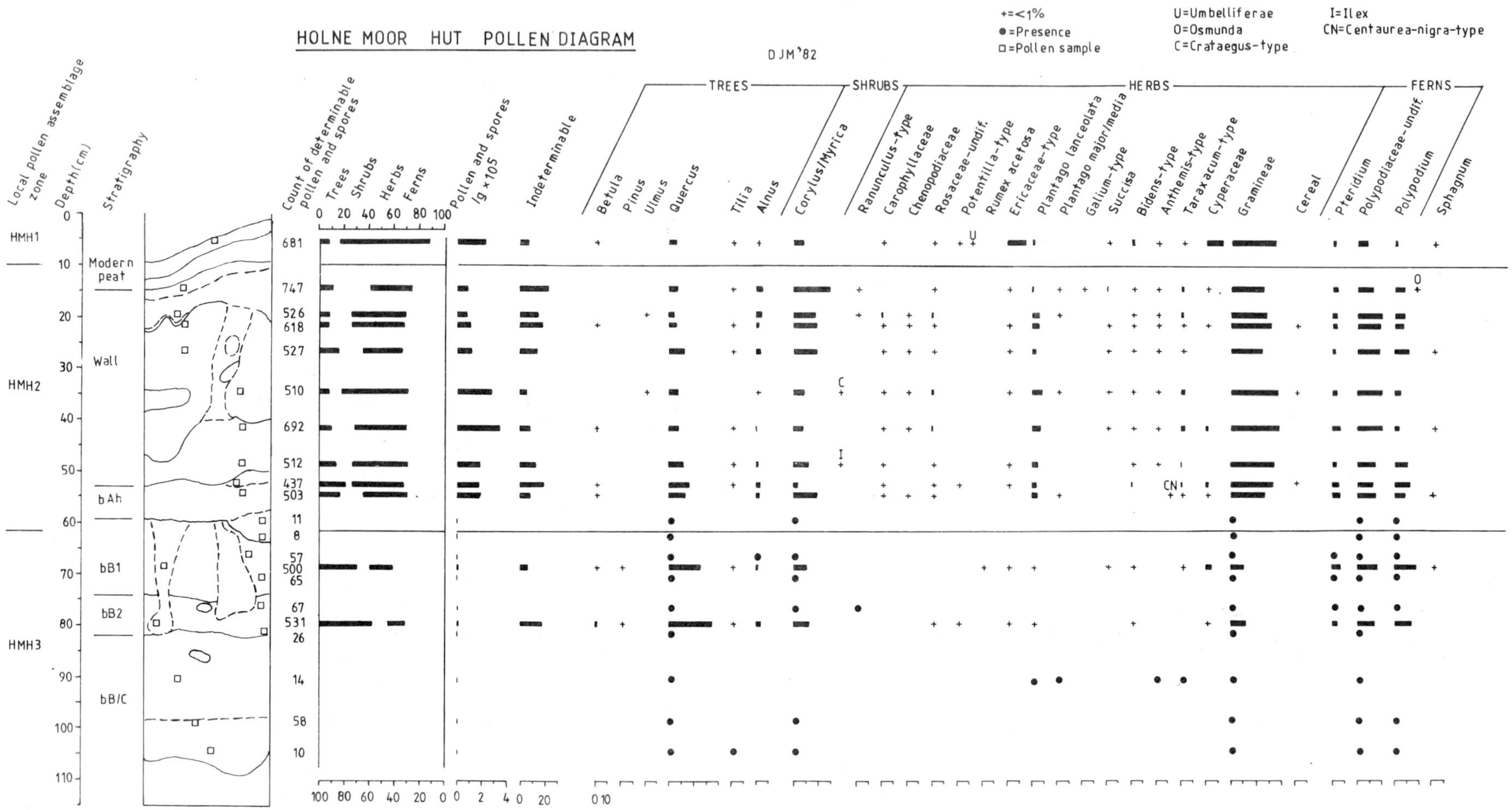

Fig. 4:7 Pollen diagram for the Hut profile, a buried prehistoric palaeosol and its overburden on Site F (HM77 F 6).

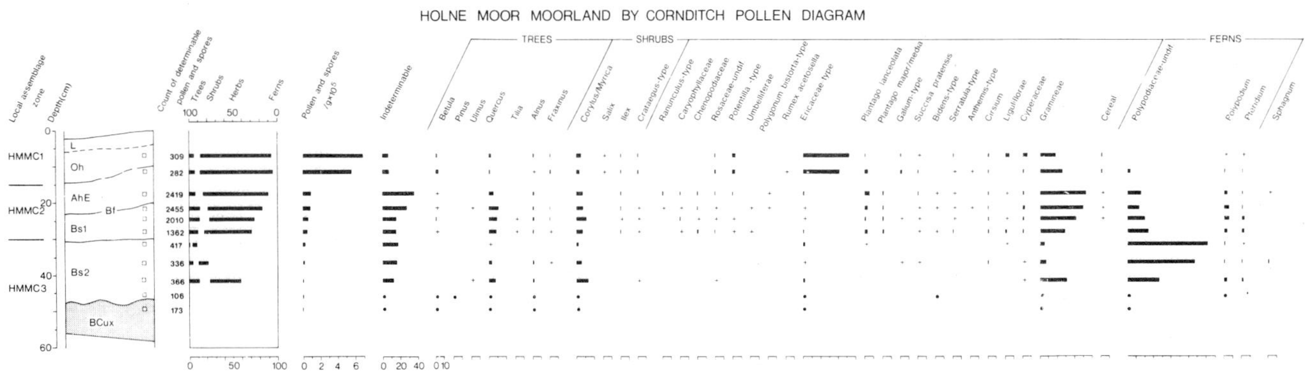

Fig. 4:8 Pollen diagram for the Moorland by Corn-ditch control profile, an Iron-pan stagnopodzol in the North Field.

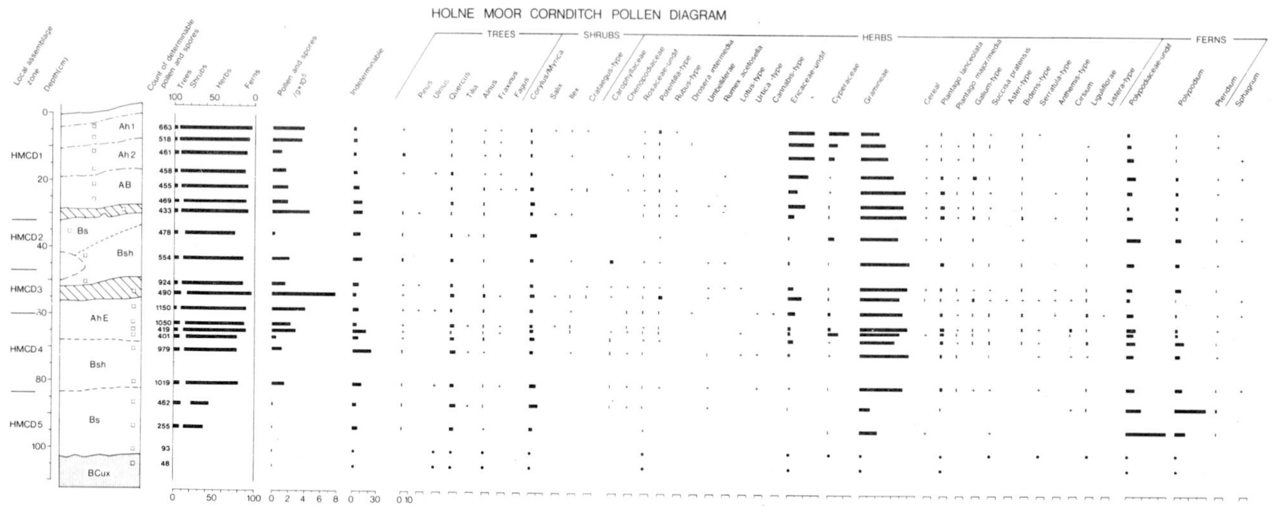

Fig. 4:9 (above) Pollen diagram for the Corn-ditch profile, a buried medieval palaeosol and its overburden at the North Lobe boundary. Fig. 4:10 (below) Pollen diagram for the Ditch sediments adjacent to the North Lobe boundary.

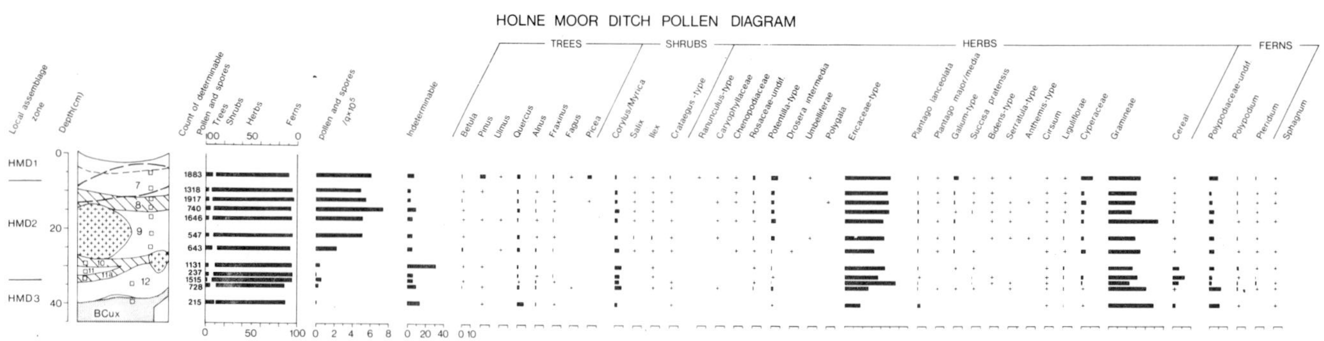

In both cases there were grounds for believing that pollen could have penetrated the mineral soil profile after the initial formation of a cover of peat; it was felt that this was more likely to have affected the Field profile (Fig. 4:6) than the Moorland by Corn-ditch profile (Fig 4:8). However, comparison of each of these pollen diagrams with that of the virgin Moorland profile (Fig 4:5) suggests that in fact the reverse may be the case. The pattern of pollen within the Ah/E and Eg horizons of the Field profile (LAZ - HM F1) is broadly similar to the pattern in the Ah/E and upper Bsh horizons of the Moorland profile (LAZ - HM M2); in each of these LAZ, Gramineae are dominant, Polypodiaceae decline, Ericaceae-type begin to expand in the uppermost sample, and small quantities of Quercus and Corylus/Myrica pollen are continuously present. If these LAZ are broadly time equivalent, and if, as we believe, the Moorland profile has not been disturbed, then we may attribute the pollen in the Field profile to a period not later than *circa* 700 BC. The few grains of cereal type pollen in HM F1 have therefore been labelled 'probably prehistoric', and this uncertainty must remain. A similar comparison of the LAZ in the Moorland by Corn-ditch profile (Fig 4:8) suggests that HM MC2 is broadly equivalent to HM M2 in the Moorland profile; however, in the latter, the M1-M2 boundeary coincides with the Oh-Ah/E interface, whereas in the Moorland by Corn-ditch profile, the MC1-MC2 boundary lies within the Ah/E horizon. It seems that in this case the pollen in the uppermost sample from the mineral soil post-dates a very substantial increase in Ericaceae-type pollen. Although all other relevant diagrams from Holne Moor suggest that a modest increase in the proportion of ericaceous vegetation preceded the establishment of a peat cover, the values of over 40% Ericaceae-type pollen in HM MC1 must, on the evidence of the Corn-ditch profile (Fig 4:9) and the Ditch profile (Fig 4:10), be attributed to a period later than AD 1000. Accordingly, we have assigned the pollen in the upper Ah/E sample in the Moorland by Corn-ditch profile to the medieval period. Lower samples from this profile seem to contain much older pollen assemblages, but formally we must label the cereal-type grains from the lower Ah/E horizon in this profile 'probably prehistoric'.

Although the general pattern and chronology of changes in vegetation since *circa* 1300 bc can properly be inferred from our diagrams, the nature and timing of earlier changes are far less clear. At present there is no direct way of determining the age of the earliest zones in these diagrams, and we cannot even be sure that they all cover the same period of time. In addition to this chronological uncertainty, there are problems posed by the relatively small quantities of pollen present in some B horizon samples. In some cases values are so low that percentage statistics would not provide a reliable picture, and in these instances we have merely give an indication of the taxa present in the sample. The sites in the southern part of the study area are worst affected; there are no reliable data from the lower B horizon samples in the virgin Moorland profile, and both upper and lower B horizons in the shallow soils on site F failed to produce enough pollen for statistical analysis. Exceptionally, samples from the fill of the old animal burrow within the palaeosol produced reasonable counts, but it is particularly difficult to estimate the age of the pollen in such a context. In the northern part of the study area, we have obtained adequate counts from all but the lowest samples (BCux and the basal samples from the lower B horizon).

The low counts in B horizon samples may in part be due to poor preservation, and relatively poor conditions for pollen survival may also be a factor affecting the numbers of indeterminable grains and the frequencies of Polypodiaceae. The latter taxon presents further interpretational problems, since we do not know whether to attribute these spores to ephytic growth in woodland or to more open, wet moorland communities (see Staines 1979:32). The evidence in our diagrams is equivocal; in the Moorland profile (K9), an increase in Polypodiaceae is accompanied by a decline in Quercus and an increase in Gramineae (HM M3 - HM M2). However, later in the diagram, Polypodiaceae decline as both Gramineae and Ericaceae-type pollen increase. On the whole we are inclined to believe that most of the Polypodiaceae grains are derived from woodland, but that they may often be over-represented due to differential survival (the northern sites illustrate this well: LAZ - HM CD5 & HM MC3). The still significant representation of this taxon in the upper ditch sediments (LAZ - HM D1) and upper layers of the Corn-ditch overburden (LAZ - HM CD1) also suggest that the source of many of these spores may lie at some distance, perhaps in the Dart valley and the lower stretches of the Venford Brook valley. Vegetation in hedgerows represents another possible source.

We have not encountered any evidence that the iron-pans have affected the distribution of pollen and spores. In some cases this could simply reflect the thin and perhaps permeable character of the pans (e.g. the pan within the overburden on site F), but in other profiles, and probably in all the profiles, it must indicate that there has been minimal movement of pollen since pan formation. This accords with the soil evidence which indicated that the pans on Holne Moor may mainly have formed within the last few hundred years and certainly long after the sealing of the soil surface by peat formation. It is also consistent with the pollen distribution evident in both of the Dartmoor Iron-pan stagnopodzols studied by Staines (1979:Figs 7 & 8).

Table 3 indicates clearly our estimate of the pollen chronology within the Corn-ditch profile; the pollen assemblage from the lower part of the overburden (HM CD2) largely copies LAZ HM CD4 in the palaeosol, and we regard these zones as time equivalent. On the other hand, pollen in the upper part (HM CD1) largely copies the pattern in the Ditch profile, though the latter may not contain as complete a record of pollen deposition over the last thousand years as the former. It is not easy to interpret the Hut profile on site F. Although pollen frequencies fluctuate from sample to sample, all of the samples from the overburden are similar to the sample from the buried Ah horizon, and have been placed in the same LAZ (HM H2). If significant quantities of pollen did enter the overburden before peat sealed the surface, the pollen rain during such a period must have been very similar to that trapped within the buried surface soil. If, on the other hand, all of the pollen in the overburden was emplaced at the time of wall construction, then we must infer that a pollen rain similar to that in the buried soil had been accumulating for some considerable time prior to the construction of the stone house on site F; only the lower part of the overburden is likely to have been obtained from surface turves (see above). It is of course possible that there was little change in the pollen rain during a period lasting from sometime before the house was constructed until well after its abandonment. Whichever hypothesis is correct, one is entitled to regard the pollen in the overburden as prehistoric, and this is consistent with the insignificant quantity of Ericaceae-type pollen in all samples. The absence of this taxon from the buried Ah horizon also provides useful confirmation of the brown podzolic status of the prehistoric soil, since it is quite clear from these studies as well as those of Staines (1972; 1979) that Ericaceae-type pollen began to accumulate in the mineral soils of the Dartmoor stagnopodzols prior to the formation of a surface peat.

Having outlined the principles upon which our analysis is based we can now describe trends in vegetational history within the study area, and examine the incidence of taxa that are particularly relevant to assessment of past land use. We shall not attempt to provide an exhaustive discussion of each LAZ in all the profiles studied; further analysis of the pollen diagrams will be presented in a later paper (Maguire: forthcoming).

VEGETATIONAL HISTORY AND LAND USE INDICATORS

Vegetational history

Today much of the plateau of Holne Moor is covered by plant communities of a type best described as *Calluna-Molinia* moorland, though if the association-analysis hierarchy of Ward *et al*. (1972: Fig. 2) is employed, the rarity of deer-grass (*Trichophorum caespitosum*) would place the Holne communities into the 'drier' heath categories. Although a mixed community of *Calluna vulgaris* and *Molinia caerulea* with *Agrostis setacea* and *Erica tetralix* as the principle secondary components dominates most of the plateau, communities on the lower slopes, which were mainly enclosed by the Lobes of the medieval farmers, are more varied. In these areas, a wider variety of grasses and herbs is encountered and *Pteridium aquilinum* is far more common. In the North Lobe and parts of Central Lobe 1, where brown podzolic soils are found, the heath plants are very rarely observed. *Pteridium aquilinum* blankets these areas (with a density of 30-40 fronds/m^2 adjacent to the Corn-ditch sampling site); the principle ground flora are *Agrostis tenuis*, *Anthoxanthum odoratum*, *Galium saxatile*, *Potentilla erecta* and *Holcus lanatus* (unpublished quadrat surveys by Dr. S. Rodgers of Seale-Hayne Agricultural College - for further details see Ralph 1982). Much of the pattern of vegetation on Holne Moor can be explained as a response to moor-burning and/or the variation in soil properties arising from medieval farming.

Trees occur sporadically on the plateau and with very few exceptions are limited to the more sheltered spots on the southern, eastern and north-eastern slopes; for the most part these isolated trees are hawthorn (with occasional rowan) and typically they are found growing on old boundaries and, in a few cases, old tinning waste piles. It is tempting to see some of them as relicts or even direct survivors of ancient hedges on the boundaries. In addition to these few semi-natural trees, there is a plantation around the Venford reservoir containing spruce, larch and pine. The Dart gorge is clothed by oak woodland which in places reaches 300 m above sea level on the north facing slopes which provide the margin of our study area; here it succumbs to the combined pressures of grazing, moor-burning and exposure.

With one major exception, the most modern sample recovered (the upper sample in LAZ - HM D1 - Fig 4:10) reflects this landscape satisfactorily; Ericaceae-type and Gramineae pollen are dominant, but the oak woodland of the gorge and the plantation trees are also represented. The rare cereal grains are most likely to have blown in either from Rowbrook farm (1 km north-north-west across the gorge) or the Stoke farms (1 km east across the Venford Brook valley). The notable difference is the low frequency of Pteridium. It may be that bracken has recently become much more abundant; certainly other elements in the vegetation of the North Lobe, which lies only a few metres from the northern sampling sites, are better represented in these sites than in the southern sites (K9 and site F). For example, both Potentilla-type and Galium-type achieve a regular percentage of the pollen in HM MC1, HM D1 and D2, and HM CD1, but are absent or only more sporadically present in the equivalent LAZ in the southern sites. Potentilla erecta is three to four times more common in the North Lobe than elsewhere on Holne Moor, and Galium Saxatile is found only in the Lobe and in small grassy patches with bracken often associated with the disturbed soils by boundaries, rabbit buries and tinning piles. There is then some evidence that our diagrams may be capable of monitoring vegetation patterns within Holne Moor, but clearly it is likely to be very difficult confidently to identify this factor as one moves backwards in time.

Vegetational patterns over the past thousand years are best represented in the Corn-ditch profile (HM CD1) and the Ditch profile (HM D1, D2 and D3); throughout this span Ericaceae-type and Gramineae pollen have dominated the pollen rain. The proportion of tree pollen appears to have remained much the same throughout this period, and it seems that today's landscape may not be very different from that of early medieval times. Some decline in Corylus/Myrica and Gramineae pollen and an increase in the proportion of Ericaceae-type pollen may indicate a reduction in hazel scrub and a higher density of the heathland plants; regular moor-burning could account for the latter. The only really substantial change in the frequencies of any taxon occurs in the Ditch profile (Fig 4:10) where, in samples from beneath the collapse layer, cereal-type pollen frequencies reach values unequalled elsewhere in these diagrams. The incidence and frequency of cereal-type pollen and other agricultural 'indicator' species will be considered in detail below, but it is worth noting here that although sporadic cereal-type grains are found in nearly all samples from the upper half of the Corn-ditch overburden (HM CD1), they are never common. Moreover, there is, on the face of it, little evidence in the pollen spectra from these deposits to reflect the major agricultural episode that engulfed the eastern side of Holne Moor for some three or four hundred years.

The Corn-ditch palaeosol continues our sequence backwards; HM CD3 and the upper samples from HM CD4 record the beginning of the rise in ericaceous plants that led eventually to the moorlands of medieval and later times. Gramineae pollen is dominant throughout these two zones; tree and shrub pollen (Quercus, Alnus and Corylus/

Myrica) are slightly greater than in medieval and modern times, but there can be no doubt that open grassland supplied the bulk of the pollen in these samples. The sample from the buried Oh-Ah horizon must date to the period AD 1000 to *circa*. 700 BC. By this time ericaceous plants were supplying over 15% of the total dry land pollen and spores; in the earlier zone (HM CD4), which should pre-date 700 BC, the Ericaceae-type pollen only exceeds 5% in a single sample. The basal LAZ in the Corn-ditch profile (HM CD 5) is harder to interpret; the density of pollen drops substantially, Polypodiaceae values are much higher, Gramineae and herb values are substantially lower but tree taxa are much the same. If we assume that the Polypodiaceae are over-represented (see p. 82) and on this account exclude them from our assessment, we find that the proportion of tree and shrub pollen (Trees *circa* 20%, Shrubs *circa* 22%) is much greater than in later zones, but that herbs and grasses still supply more than half the pollen (*circa* 58%) in these two samples.

The Moorland by Corn-ditch pollen diagram (Fig 4:8) shows a similar pattern to the Corn-ditch diagram with the exception that the medieval and possibly post-medieval pollen record is confied to the uppermost two samples. Note however that in this profile the expansion of the Polypodiaceae in LAZ - HM MC3 occurs at the expense of all other taxa; a circumstance indicative of extensive pollen destruction in the upper half of the lower B horizon. The lowest sample from which statistically reliable quantities of pollen were recovered in this profile seems to have been less severely degraded; the proportion of tree, shrub, herb and grass pollen in this sample is practically identical to that in HM CD 5. It seems that none of the samples from these northern sites provides evidence of woodland in the immediate vicinity, but there is some indication that shrub and woodland may have been more extensively developed, perhaps in the Dart gorge and the Venford Brook valley, during the earliest period recorded in these profiles.

A slightly different picture emerges from study of the pollen diagrams from the southern sites (Figs 4:5, 4:6 and 4:7). The primary site here is the virgin moorland profile K9. It lies at the edge of the Venford Brook valley's upper stretch, about 1 km south-south-west of the Corn-ditch sites and some 325 m south-west of site F (see Fig 4:2). The overall sequence in this profile is similar to that in the northern sites; Gramineae and Ericaceae-type pollen dominate in samples from the Oh horizon, and Gramineae are also dominant in the upper samples from the mineral soil (LAZ - HM M2), where again the rise in Ericaceae-type pollen is visible. However, there is substantially more tree and shrub pollen in the Oh samples from this site (*circa* 20% and 10% respectively) than was found in equivalent samples from the northern sites. In both the Corn-ditch and the Moorland by Corn-ditch profiles tree pollen never exceeds 10% and is generally closer to 5% throughout the medieval period (HM CD1 and HM MC1). Today, areas to the south and west of the reservoir that lie within its perimeter fence and in consequence remain ungrazed carry a well-developed scrub with occasional trees; it seems possible that in medieval times similar vegetation extended further west into the relatively sheltered valley immediately south of this sampling site.

There is little difference in the proportion of tree pollen in the various LAZ that record the Gramineae-dominant phase at any of the sites (HM M2, HM F1, HM CD4, HM MC2 and HM H2); values in the southern sites mainly lie between 6% and 15%, and in the northern sites between 8% and 12%. These values are in general slightly higher than medieval and modern values implying that during this period, when the reave systems must have been constructed, occupied and perhaps abandoned, the valley woodlands may have encroached on the plateau to a slightly greater extent than they do today.

The major difference apparent in the southern sites occurs in LAZ - HM M3 and HM H3; each of these zones provides evidence for a period when woodland was far more extensively developed than it is today. Unfortunately the very low density of pollen in many of the B horizon samples from these southern sites (HM M3 is represented by a single sample, and HM H3 is based on the two samples from the burrow in the palaeosol) makes interpretation of this phenomenon hazardous. If either or both of these LAZ are, as one might reasonably expect, roughly time equivalent to HM CD5 and HM MC3 in the northern sites, then this could be taken as evidence of a difference in the extent of woodland development in the northern and southern parts of Holne Moor. More extensive woodland in the sheltered southern valley beyond the Venford Reave and the medieval enclosures (the Moorland profile could have lain within such a woodland), and a more open vegetation on the more windswept northern part of the plateau seems entirely reasonable. However it is also possible (if a little unlikely in view of the better preservation of pollen in the northern sites) that the southern sites have preserved a record from an earlier period than the northern sites, a period in which woodland cover may have been greater in all areas. Grass and herb pollen provide less than 20% of the pollen sum in the samples from HM M3 and HM H3; Quercus, Polypodiaceae and Corylus/Myrica are the main taxa, and if the Polypodiaceae are assumed to derive from ephytic growth (see p. 82), this must indicate the existence of a damp and probably stunted oak woodland at both these sampling sites. We cannot be sure that these two LAZ represent the same period, nor does the pollen evidence allow us reliably to estimate the date, the rate or the cause of the subsequent woodland decline. The only information on this aspect is provided by the burrow samples (HM H3) and the sample from the buried Ah horizon. Together these samples indicate a steady decline in tree pollen (circa 40%, 30%, 15%), rather than an abrupt change, but the validity of treating these samples as a simple sequence can be doubted.

The poor resolution of a general vegetational history of this kind coupled with the chronological problems associated with soil pollen make it very difficult to identify the impact of agricultural land use. We cannot demonstrate that this factor was responsible for any of the major changes - woodland decline, the formation of grasslands, the transition to today's moorland - that are evident in the pollen diagrams. Even the expansion and decline of farming in the medieval period, which involved the enclosure of large areas of what was to become arable land, has left little trace in the pollen record of the major taxa. However, the impact of this event and of prehistoric land use can be discerned in the records of minor taxa including cereal-type pollen and the so-called 'agricultural indicator' species.

Cereals and agricultural indicators

A total of 400 Gramineae pollen grains were designated as cereal-type using size statistics; 134 of these grains were further classified using Faegri and Iversen's (1975) scheme. The location and classification of all these grains is shown in Figs 4:11, 4:12 and 4:13. Table 4:4 summarises this data, and indicates the chronological assignments that we have made. The overwhelming bulk of the cereal pollen can be linked with the medieval farming of Holne Moor. Indeed, over half the grains designated as cereals were obtained from the three very thin lenses of sediment at the base of the Ditch profile, which seems to have acted as a very efficient trap for pollen produced during a period when the North Lobe lay under the plough. However, other medieval contexts also produced substantial counts. Interesting, and not a little disturbing is the high count for the virgin Moorland profile where the two Oh horizon samples contained over five times as many cereal-type pollen grains as the two equivalent samples from the Moorland by Corn-ditch profile; most of these grains came from the upper sample and probably reflect outfield tillage in the enclosures of the Major Extension some 150 m north-east of the site. The buried surface soil beneath the Corn-ditch also produced cereal pollen, indicating that farming activities were under way prior to the building of this boundary; elsewhere (Fleming and Ralph 1982), we have suggested that the present 'corn-ditch' form of the boundary may derive from an early post-Conquest revamp of an earlier Anglo-Saxon 'block-wall' boundary. Small amounts of cereal pollen occur in the upper sediments of the Ditch profile, above the collapse layer. The highest values are reached in the top two samples and, at least in the upper sample, this pollen must be derived from the nearby farms mentioned above. Cereal pollen is curiously rare in the upper part of the Corn-ditch profile (HM CD1) and the equivalent LAZ in the Moorland by Corn-ditch profile (HM MC1); this may partly reflect the disturbed character of the Oh horizon in the latter, and the non-peaty soil profile of the former - both circumstances could have reduced the chances of pollen survival. However, it could also indicate that tillage of the North Lobe, and perhaps some of the other Lobes, occurred relatively infrequently and that lengthy grass fallows were employed.

Roughly two-thirds of the cereal pollen from medieval contexts was classified as *Secale*-type (Table 4:4); most of the other grains whose classification was attempted were classed as Indeterminable (*sensu* Cushing 1967), and there were a few grains of *Triticum*-type. No *Hordeum*-type or *Avena*-type were identified. If our sampling sites lay at some distance from the arable fields, the predominance of the wind pollinated taxon, *Secale* could be wholly explained as a result of the better dispersal of *Secale* compared to the other self-pollinated cereals, and we would have to admit that this data was incapable of providing a reliable picture of the type of crops sown on Holne Moor. However, in not dissimilar circumstances, Austin *et al*. (1981), working at Okehampton, on the northern fringe of Dartmoor, recorded a majority of *Hordeum*, lesser quantities of *Secale*, and a few grains of *Triticum* and *Avena*. It is certainly

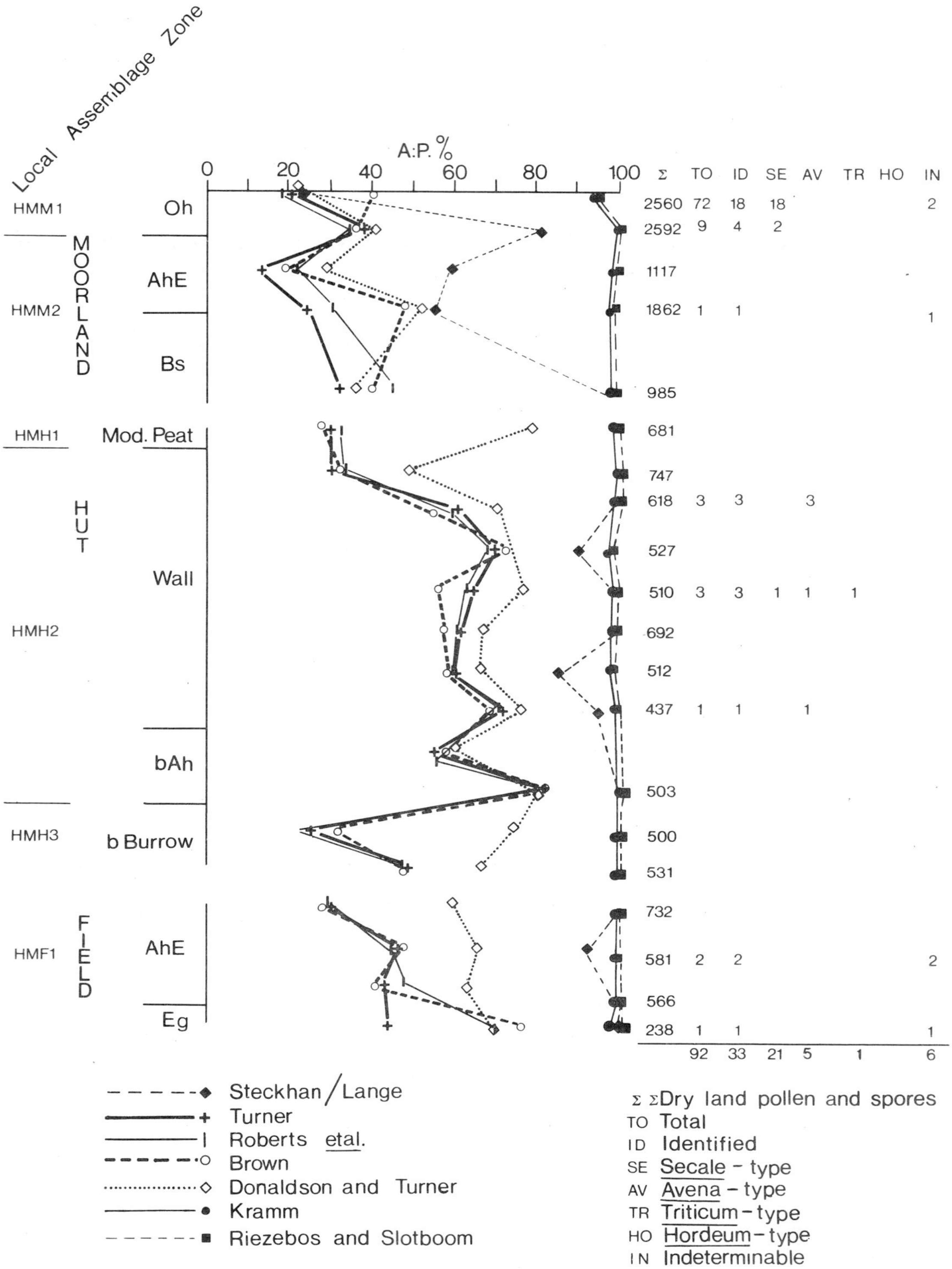

Fig. 4:11 Arable : Pastoral indicies and amounts of cereal-type pollen in the Moorland profile (K9), the Field control profile (HM77 F 5) and the Hut profile (HM77 F 6 - the prehistoric palaeosol and its overburden).

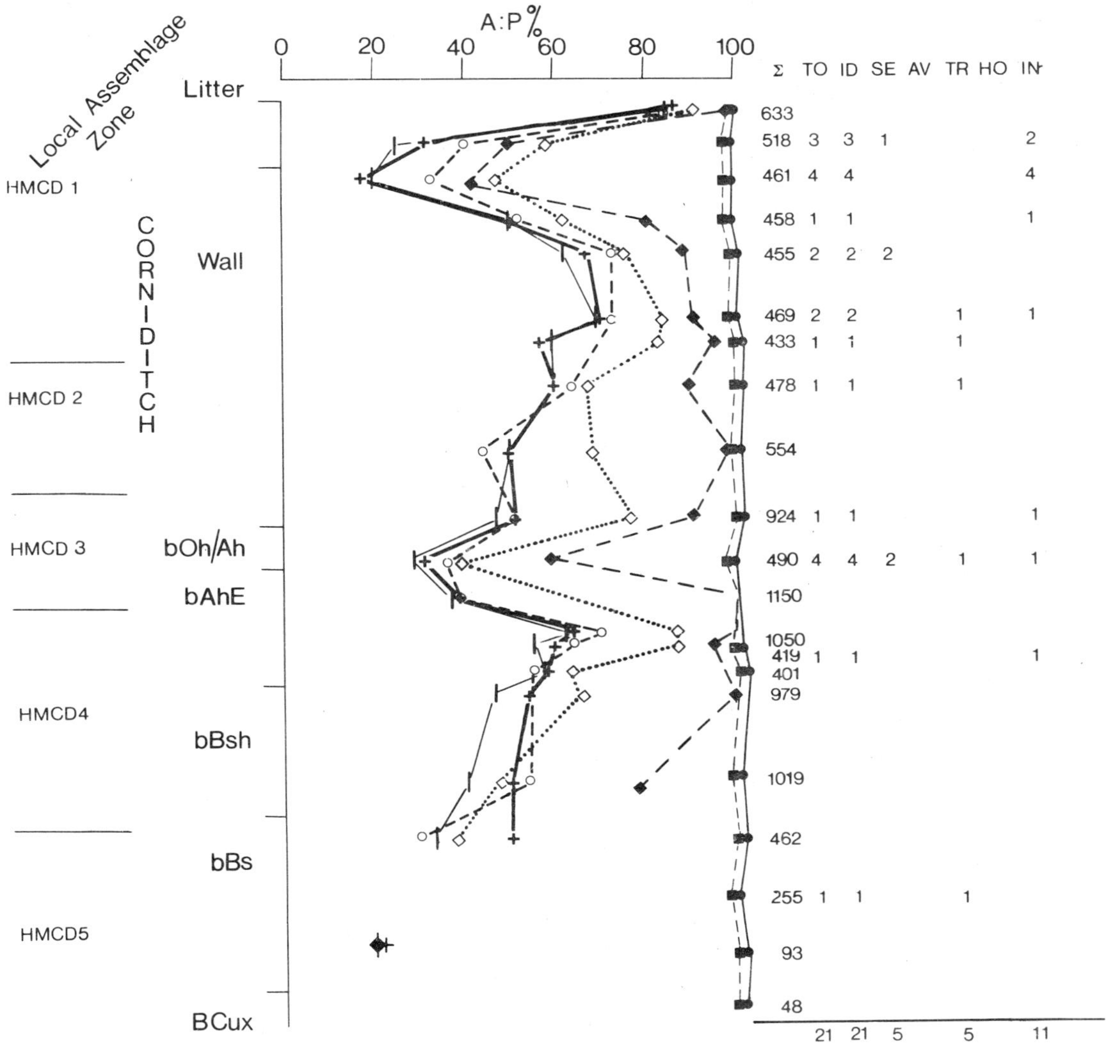

Fig. 4:12 Arable : Pastoral indices and the amount of cereal-type pollen in the Corn-ditch profile (the medieval palaeosol and its overburden).

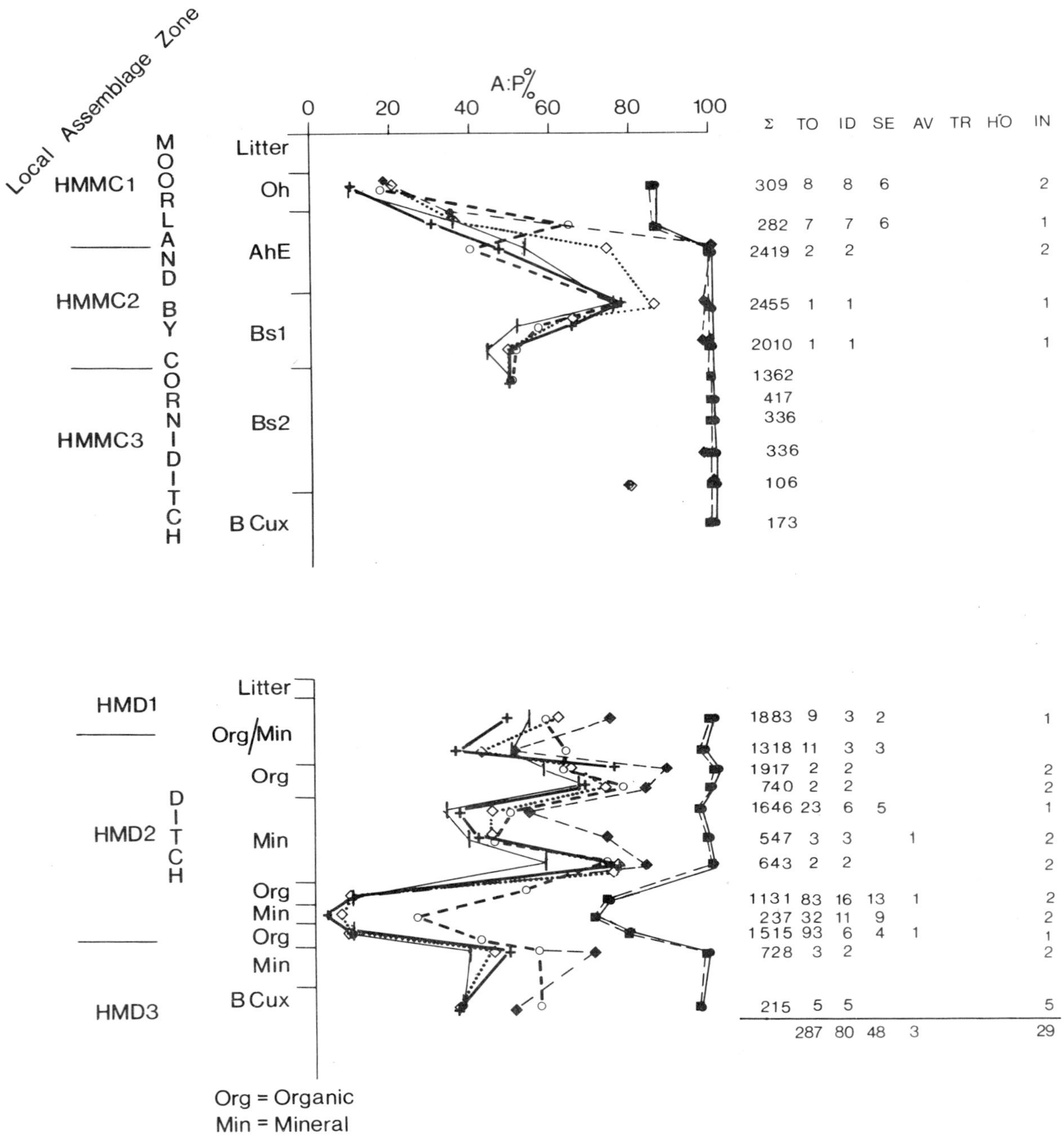

Fig. 4:13 Arable : Pastoral indices and amounts of cereal-type pollen in the Moorland by Corn-ditch control profile and the Ditch sediments.

Table 4:4 Summary of cereal-type pollen in Holne Moor soils

	TOTAL	IDENTIFIED	Secale	Avena	Triticum	Hordeum	Indeterminable
MEDIEVAL AND POST-MEDIEVAL							
Ditch							
Upper sediments (7-8)	24	10	5	-	-	-	5
Collapse layer (9)	28	11	5	-	1	-	5
Lower sediments (10-12) & BCux	216	40	26	-	2	-	12
Sub-total	268	61	36	-	3	-	22
Corn-ditch Overburden and Palaeosol	17	17	5	-	3	-	9
Moorland by Corn-ditch	15	15	12	-	-	-	3
Moorland K9	81	22	20	-	-	-	2
Total	381	115	73	-	6	-	36
PREHISTORIC AND PROBABLE PREHISTORIC							
Wall fill, site F	7	7	1	5	1	-	-
All other profiles	5	5	-	-	1	-	4
Probable prehistoric (see text)	7	7	-	-	1	-	6
Corn-ditch (lower overburden)							
Moorland by Corn-ditch (lower Ah/E)							
Field (Ah/E and Eg)							
Total	19	19	1	5	3	-	10
GRAND TOTAL	400	134	74	5	9	-	46

The locational sub-categories indicate profiles or contexts in which cereal-type pollen grains of a particular period were found; not all the cereal-type pollen in a given profile necessarily was allocated to the period indicated.

possible that, although the good dispersal properties of Secale pollen grains may have inflated the values for this taxon, the absence of Hordeum and Avena, and the rarity of Triticum in the Holne Moor samples may be due to infrequent or nil cropping of these cereals. The selection of rye (or oats) for cropping on very acid soils would make sense; barley and wheat are far less tolerant of soil acidity (Russell 1973: 661).

Very few cereal pollen grains were recovered from prehistoric contexts (Table 4:4), the single most important context being the wall fill on site F. All the grains from this location could be classified; five of them were of Avena-type, and Secale and Triticum were also noted. Once again, Hordeum-type could not be identified in any sample, but over half the grains attributed to this period were indeterminable. Although one cannot draw conclusions about cropping practices or the relative significance of arable activities from the tiny sample that has been extracted from the poor survival medium provided by mineral soils, these few grains do suggest that some arable cropping took place on Holne Moor prior to circa 700 BC; the wall fill samples could indicate cropping of oats in the earliest phase of the reave occupation. Only one cereal-type grain (Triticum) was recovered from any of the earliest LAZ in our diagrams - HM CD5 (Fig 4:12).

The pollen diagrams from Holne Moor generally have large numbers of agricultural indicators compared to diagrams from other parts of Dartmoor (cf. Caseldine and Maguire 1981; 1982). Plantago lanceolata commonly makes up 5-10% of the pollen sum in LAZ from the Ericaceae-dominant phase and the earlier Gramineae-dominant phase; the only significant exception occurs in the virgin Moorland profile where all samples contained less than 1% of pollen from this taxon. Regular percentage values for Taraxacum-type pollen in the profiles on Site F (HM H2 and HM F1) presumably reflect an area of rough waste ground surrounding the prehistoric house, and the common if somewhat erratic record of Bidens-type pollen both here and at the northern sites may be linked to the ditches that accompanied some boundaries in both the prehistoric and medieval periods. Pteridium values are mainly very low druing the Ericaeae-dominant phase, but reach significant percentage values in several of the samples from the Gramineae-dominant phase, including most samples from the wall fill on site F. If the erratic record of this spore producer is taken at its face value, then one might suppose that during this period at least some of the land within the pollen catchment of those sites remained rough grazing land in which bracken stands could flourish.

Several types of arable : pastoral ratios (Table 4:5) have been calculated for most of the Holne Moor samples, and the results of this analysis are shown in Figs 4:11, 4:12 and 4:13. A discussion of the derivation and interpretation of such ratios appears in a separate paper by one of the authors (Maguire: this volume). Low values are normally thought to indicate a dominance of arable agriculture and high values are thought to be associated with pastoral farming. Although ratios have been calculated for nearly all our samples, the ratios for some of the basal LAZ are statistically unreliable due to the very small number of grains involved in the calculations. It is immediately apparent from an examination of these diagrams that the indications provided by the different ratios do not always agree though the trends tend to be similar. There is also a notable contrast in the ratios from the northern and southern sites during the Gramineae-dominant phase. Thus most ratios in HM MC2, HM CD2 and HM CD4 tend to be greater than 50% (as do the ratios in samples from the wall fill on site F), suggesting a predominance of pastoral activities, but in the equivalent zones in the Moorland profile and the Field profile (HM M2 and HM F1) the ratios are mainly below 50% and there are some below 30%. While the relatively low values in HM F1 could reflect occasional arable use of the small prehistoric field in which this sampling site lies, it is somewhat harder to explain similar, and even lower, values in the virgin land profile; they could be taken to indicate that during the prehistoric period the (probably minor) arable activities on Holne Moor were concentrated in the southern, more sheltered zone, near the Venford Reave.

Table 4:5 Arable: Pastoral Indices

Steckhan (1961), Lange (1975)	*Plantago lanceolata* : *Plantago lanceolata* + cereal
Kramm (1978)	Non-cultivated Gramineae : non-cultivated Gramineae + cereal
Turner (1964)	*Plantago* : *Plantago* + Compositae + cereal + Cruciferae + *Artemisia* + Chenopodiaceae
Roberts, Turner & Ward (1973)	*Plantago lanceolata* + *Artemisia* + *Rumex* + Ranunculaceae : *Plantago lanceolata* + *Artemisia* + *Rumex* + Ranunculaceae + cereal + Chenopodiaceae + Cruciferae + *Vicia* + *Polygonum* + *Centaurea cyanus* + *Knautia* + *Trifolium* + *Centaurium*
Donaldson & Turner (1977)	*Plantago lanceolata* + *Rumex acetosa*/*acetosella*-type + *Rannunculus* spp. + Compositae-*Bellis*-type : *Plantago lanceolata* + *Rumex acetosa*/*acetosella*-type + *Ranunculus* spp. + Compositae-*Bellis*-type + Gramineae-cereal-type + Compositae-excluding *Bellis*-type + Cruciferae + *Polygonum* spp. + *Trifolium* spp.
Brown (1977)	*Plantago lanceolata* + *Plantago major*/*media* + *Remex acetosella* + *Rumex acetosa* : *Plantago lanceolata* + *Plantago major*/*media* + *Rumex acetosella* + *Rumex acetosa*

	+ Liguliflorae + Tubuliflorae + Cruciferae + Caryophyllaceae + Rumex/Oxyria-type.
Riezebos & Slotboom (1978)	Gramineae + Papilionaceae + Plantago lanceolata: Gramineae + Papilionaceae + Plantago lanceolata + cereal + Fagopyrum + Linum + Rumex + Artemisia + Centaurea

Although the utility of the ratios calculated for prehistoric samples seems questionable, the ratios for the medieval samples do clearly identify the major intake of arable land in early medieval times. The sediments beneath the collapse layer in the Ditch profile, which contained large numbers of cereal-type grains, exhibit extremely low ratios, and the same pattern is copied by the ratios from the upper part of the Corn-ditch overburden, which contained far fewer cereal grains. Low ratios also occur in the upper LAZ of the Moorland profile (HM M1) and the Moorland by Corn-ditch profile (HM MC1), and in both cases cereal pollen is present; these samples record episodes of outfield tillage. Even in this period there are some puzzling features. Although lowered ratios in the buried surface soil beneath the Corn-ditch (HM CD3), again accompanied by cereal pollen, are acceptable as indicators of the very earliest phase of medieval tillage, similar ratios in the upper sample from the buried Ah/E horizon, which contained no cereal pollen, should date to a much earlier period - pre circa 700 BC. It is perhaps more economical to assume either that the construction of the boundary involved some disturbance of the buried peat surface, which promoted pollen movement into the mineral soil, or that the ratio itself is in this instance, as must be the case with the upper burrow sample on site F (HM H3), giving a false picture.

The post-medieval period is represented by the uppermost sample from the Corn-ditch profile and the upper sediments in the Ditch; the ratios from these samples satisfactorily illustrate the pastoral use of Holne Moor that has continued down to the present day.

CONCLUDING DISCUSSION

The work described above has increased our understanding of the nature of former land use and its environmental context mainly, but not exclusively, for the medieval period (especially between about AD 1100 and AD 1350) and for the prehistoric period (around 1300 bc); it also has methodological implications which will be considered during our discussion of the two periods.

The historic period

In broad terms, the general environmental record for medieval and post-medieval times is unsurprising, although it is encouraging for the design of future studies that our samples appear to have monitored, to some extent, differences between areas of Holne Moor some 750 m apart. Two important features are the predominance of rye pollen among the medieval and post-medieval cereals, and the way in which the ditch of the North Lobe's corn-ditch has acted as a very effective pollen trap. Although the proportion of rye pollen to other cereal types in some of our samples may exaggerate the true position, there can be little doubt that rye was the principal crop grown during the episodes of cultivation recorded in the basal ditch sediments. This provides an interesting supplement to the historical records. Rye, wheat and oats seem to have been the staple crops on the estates of Tavistock Abbey and, perhaps, in much of lowland Devon during medieval times (Finberg 1951: 86-115). However, there are indications in these records and elsewhere that rye was more extensively cultivated during earlier medieval times; the pollen in the ditch sediments is almost certainly older than most or all of the (mainly 14th and 15th century) documents cited by Finberg. Alternatively, the absence of oats and the rarity of wheat in the Holne Moor samples could reflect differences in husbandry in lowland and upland areas. In the lowlands a rotation of wheat (dressed with sea-sand), rye and oats may have been practiced, while farmers on the acid, free-draining soils of the granite uplands, where it would have been difficult to obtain an adequate supply of sea-sand, may have chosen a more restricted rotation or simply have broken grass with occasional crops of rye.

Today, rye is rarely if ever grown on the moorland farms, but its abandonment by the latter may be quite recent; rye pollen is present in the uppermost sample from the ditch which includes pollen types that can only have been deposited in this century. This may also be an indication of differences between lowland and upland Devon. Finberg (1951:98) states that over much of Devon rye went out of cultivation in the 18th century, and rye was recorded for only three of the 257 Devon parishes who replied to the Milles enquiry of 1747-56 (Stanes 1969:54). There is no mention of rye in the 1801 Crop Returns for the parish of Holne (Michinton and Overton 1973); at least by then, barley (29%) appears to have replaced rye in a rotation with wheat (29%) and oats (26%). However most of the parish lies off the Moor and, in consequence, has both better soils and climate than our study area. The replacement of rye by barley in much of Devon during the 18th century may reflect improvements in access and/or the availability of burnt lime. Fussell (1955) indicates that the use of lime kilns was spreading in England during the late 16th and early 17th centuries, and Gawne (1970:55) notes the use of sea-sand in Widecombe in the 17th century. In any case, it would hardly be surprising if rye were chosen for outfield cropping in almost any period, and certainly in the 16th century, the youngest period when this type of land use is likely to have occurred on Holne Moor.

More of the kind of work done here and at Okehampton Park (Austin _et al_. 1981) will need to be completed before a proper synthesis of palaeobotanical and documentary records can be attempted. However, since crop returns tend to be concerned with relatively large manors, parishes, etc., containing varied ecological conditions, it is likely that for the moorland fringe palaeobotany will be more widely available, more precise and more generally useful than documentary records. It simply remains to exploit the opportunities which exist. These are legion, given the profusion of ditches, banks, tinners' heaps, rabbit buries and peat deposits around Dartmoor's long perimeter. Our work suggests that effective sampling can take place beneath corn-ditches and hedge-banks and in the ditches which accompany them, so that evidence for the high medieval period (circa AD 1100-1350) will be easiest to collect (see Fleming and Ralph (1982) for further discussion of the nature and chronology of these boundaries). Corn-ditches constructed during the period of Forest Law are potentially of considerable interest to the palaeobotanist, because by definition they were the external boundaries ('head dykes') of enclosed land, and had no hedges to filter lateral pollen movement or to provide their own pollen rain component. Earlier medieval boundaries are not yet well-known on Dartmoor; they apparently include block walls, which may be much less easy to sample and utilise than corn-ditches. Perhaps the pre-corn-ditch period will be best understood by analysis of a suitably-located peat column.

The prehistoric period

Most of the available environmental information covers the period immediately preceding the layout of the Dartmeet parallel reave system, around 1300 bc, and the period of its use. The most important questions concern the reliability of the general picture provided by the pollen evidence, and the extent to which this picture is confirmed, supplemented or contradicted by the archaeological evidence. Further comparisons can be made with the evidence from Shaugh Moor, some 14 km to the south-west (Smith et al. 1981; Balaam et al. 1982) where the environmental investigations also concerned the area near a terminal reave, in this case that of the Shaugh Moor parallel reave system. The latter has produced radiocarbon dates which suggest that it was laid out at the same time as the Dartmeet system (other arguments for seeing most of the Dartmoor reaves as part of a pattern laid out around 1300 bc, perhaps as a consequence of one political decision, are to be found in Fleming 1983).

At first sight it looks as if the Gramineae-dominant phase, with its relatively high arable: pastoral ratios and a lower cereal count than that obtained for medieval contexts, was one in which pastoralism was much more important than cereal cultivation. The wall fill of the site F house, with its contained cereal pollen, is likely, as explained above, to have come from the nearby ground surface or slightly deeper, and the cereal pollen probably relates to the first few decades of the Dartmeet system's life, the period when a wattle-and-daub house stood at site F, and the boundary to the west, which changes course to pass to the rear of this earlier house was an earthen bank, perhaps for a hedge.

In broad terms, the general importance of pastoralism in this part of the Dartmeet system seems hardly open to question. The system does after all have a well-defined terminal reave, or head dyke, beyond which lie the uplands of southern Dartmoor. Almost by definition, then, a substantial part of the land close to the southern sampling sites should have been devoted to grazing on a permanent or near-permanent basis. The Pteridium values of the house F wall-fill could imply relatively low grazing pressures in some areas, with cattle absent or of lesser importance than sheep, as Brown (1977:307) argued for a Bronze Age context on Bodmin Moor. The presence of a stake fence, presumably composed mainly of hurdles, as the first boundary along the line of the north-south reave immediately east of site F, may also hint at the regular presence of sheep rather than cattle. Lengthy discussion of the nature and function of the Holne Moor boundaries would not be appropriate here. However, gateways occur in both reaves and the boundaries which preceded them, so it is probable that they were definite barriers to animal movement, even if they were no taller than the one-metre walls of the Cotswolds, where 'they are fortunate in not having the black-faced ewe against them, and their field walls are seldom more than 39 inches high '(Rainsford-Hannay 1972:79). On Holne Moor there are a few short narrow lanes, sometimes leading to houses, and in one place an apparent informal hollow-way associated with a gateway; these suggest some not very highly organised efforts to control animal movement; speculatively it may be suggested that such movements involved sheep rather than cattle, and perhaps fairly small numbers at any one time.

On the other hand, the extent and importance of prehistoric cereal cultivation on Holne Moor may have been under-estimated by the evidence which we have been able to present. The striking contrast between prehistoric and medieval cereal pollen counts is somewhat exaggerated by the richness of the ditch sediments. With hindsight it is regrettable that we did not sample comparable ditch sediments which were available for the prehistoric period (there was a semi-circular ditch on the northern side of the site F house, and a ditch ran beside the earthern bank which preceded the reave just to the west). Our sampling strategy, in fact, was designed for first-approximation purposes, and some of its now apparent limitations simply reflect this fact.

It is formally correct, but possibly misleading, to consider the carbonised seed evidence from the interior of the site F house separately, on the grounds that the plants represented may not have been grown on Holne Moor. They may have been, and if so, Hordeum and Vicia would be added to the list of plants (Avena, Triticum and Secale) for which there is some evidence for prehistoric cultivation (Table 4:4). However the absence of Hordeum-type pollen in prehistoric contexts, as in medieval contexts, can also prompt the speculation that the carbonised grains of this cereal attest to trade with communities tilling soils more suitable for this crop. At the moment the extent of soil podzolisation on Holne Moor during the reave occupation is unclear; we are pursuing this issue both through further chemical analyses of the site F palaeosol and by investigating the nature and dating of sub-soil burrows which we suspect were created by moles. Whichever view one takes, it must be premature to talk confidently of crop rotation, and to make large claims for the role of cereals in the Holne Moor economy on the basis of the few grains available. However such possibilities must still, in the state of our evidence, be borne in mind.

The case for treating our present evidence with caution is strengthened by considering the implications of accepting Reynolds' (1981:109) high yields of Triticum in the Butser Hill experiments, and extrapolating them to a given prehistoric period. Mercer (1981:231-236) has used these figures to estimate the numbers of people who might have been surpported from a given ancient field system, but it would be equally possible to reverse the procedure, and estimate how much land would have been cropped at any one time if a given population, estimated by other means, was producing little or no surplus. In principle the effect of this approach would be to reduce the amount of land cropped at any given time, and perhaps increase the length of fallow periods, relative to expectations generated by assuming lower cereal yields. Maximum contrast might be achieved by comparing a subsistence economy cosseting small fields of high-yield cereals (the prehistoric strategy?) with a medieval economy in which large acreages of low-yield cereals were sown to produce a surplus extorted or encouraged by market or other incentives. Arable: pastoral ratios would presumably reflect the difference in arable acreage, but it would not be correct to conclude, in the first case, that cereals made only a minor contribution to total agricultural production or played a minor role in human diet, on the basis of the pollen evidence. It must also be remembered that the lower productivity per unit area of livestock production compared to arable production ensures that the ratio of pastoral to arable land will always be greater than the ratio of pastoral to arable products. Caution is clearly needed in extrapolating from pollen ratios to landscape, from landscape to agricultural production, and from agricultural production to human diet, on Holne Moor and elsewhere.

The people inhabiting the site F area, having initially made earthern banks and wooden fences, replaced those with reaves, and sometimes piled stones against the reave-faces. As far as can be seen, most of the prehistoric boundaries on Holne-Moor, whatever their beginnings, were reaves when they went out of use. A decision to construct reaves, which in some places took place part-way into the life of the boundary-system, would be explicable in terms of a need for field clearance, a need which, to judge from the evidence of the piled stones, continued after the reaves had been completed. In all these circumstances, it would be prudent to reserve judgement about the extent and importance of cereal cultivation on Holne Moor. Something of this ambiguity attaches also to the interpretation of the Shaugh Moor evidence. Beckett comments that 'in the region of Wotter Common the demise of trees and spread of weeds of pasture indicates a grazing pressure which has only been exceeded in very recent times' (sic). But he goes on to say: 'Bronze Age arable farming appears to have been on only a minor scale, but again the pollen evidence suggests that as much cultivation may have taken place at this time as at any time since, including the medieval period' (Beckett 1981:262). Like us, Beckett was working at the moorward end of a prehistoric field system.

From an archaeologist's point of view, the most striking phenomenon of Dartmoor's prehistory is the elaborate boundary system represented by the parallel reave systems and the long distance reaves beyond them which apparently divided up permanent grazing land. Although simple field systems, settlements, enclosures and even one or two reaves can be dated, with varying degrees of confidence, to the pre-reave period, 'prototypical' reave systems are rare or absent. It looks as if the reave pattern reflects a rather sudden and largely synchronous phase of boundary-making. It is interesting, though not perhaps surprising, to note the muted impact of this episode on the pollen record. On Holne Moor this is not convincingly documented, owing to the nature of our samples. But the Shaugh Moor area, before the reave system was laid out, seems to have included patches of heathland, bog, acid grassland and scrubby woodland in mosaic (Beckett 1981; Balaam 1982). This would be consistent with the existence of a rather ill-managed or casually used upland common under not very heavy grazing pressure. Beckett, following Simmons (1964:170) suggests that the Alnus decline contemporary with the reaves might have been associated with tin-streaming (1981:262), and Brown documented a comparable Alnus decline, apparently datable to this period, at Hawks Tor on Bodmin (1977:308). It is very tempting to associate the extensive, coherent Dartmoor land boundary pattern with a strong regional sociopolitical organisation. It is also noticeable that the use period of the reave systems (c. 1300-1100 bc?) coincides with that of the Wessex II (Camerton-Snowshill) and Arreton Down bronzes, with their high tin content (Britton 1963; Britton and Richards 1961). The apparently tighter control of land use on Dartmoor manifest at this time may have reflected a newly acquired interest, by the ruling elite, in local tin resources, even though the economy is unlikely to have been based on tin production. The appearance of larger houses, with thick-walled enclosures in areas like the Upper Plym valley would be consistent with this scenario. Nevertheless, it has to be said that the observed Alnus decline might simply reflect increased demands for fuel and timber, in a situation where casual intercommoning was being partially supplanted by a more all-the-year-round pattern of occupation, associated with a somewhat higher population.

As suggested above, the general vegetation reconstruction presented by Simmons (1969: Fig 3) has to be modified in the light of present knowledge of the location and size of reave systems. Nevertheless, the discussion may still revolve about the 'tonsure' model - that is to say, about the scale and timing of clearance which was expanding outwards and downwards. The evidence for the scale and whereabouts of Iron Age clearances on Dartmoor may be open to discussion (Caseldine and Maguire 1981:11) but it is notable that both on Bodmin Moor (Brown 1977) and Exmoor (Merryfield and Moore 1974) increased Iron Age clearance evidence is attributed to regional (i.e., off-moor) trends. A similar pattern has emerged from mid-Wales (Moore and Chater 1969) and from the East Moor, Derbyshire (Hicks 1972). Those who believe that in prehistoric times Dartmoor and related areas were 'marginal land' in the sense that they must have been occupied after the lowlands had been intensively settled, may have to develop a more complicated model.

Some of the detailed methodological implications of our work have been discussed above. In seeking to understand the implications of the Holne Moor results, we have been unable to avoid the complex interpretational problems presented by palaeobotanical evidence. It is impossible to arrange for all the archaeologist's questions to be answered; palaeoenvironmental interpretation is as much of an 'art' as the interpretation of excavational evidence. In one senes, our work demonstrates what may be achieved by a multilateral approach to problems of ancient land-use and landscape history. However, the picture presented is very much a first approximation. It will be clear to the reader that we have only just begun to exploit the possibilities presented by well-preserved field systems and associated settlements in this environment. It may seen banal and even ungrateful to demand more work and more data; but it does seem clear that taking this kind of study further, on Holne Moor or elsewhere, would be well worth while.

Notes

1. The taxon 'Polypodiaceae-undif.' has been abbreviated to 'Polypodiaceae' throughout the text of this article.

Acknowledgements

Some of the work reported in this paper was carried out during the tenure by one of us (N.R.) of a DES Major State Studentship and subsequently a Sheffield University Research Fellowship. Grants for the work of the Dartmoor Reave Project came from the British Academy, the Society of Antiquaries of London, the Dartmoor National Park Authority and the Ivan Margary Fund administered through the Council for British Archaeology. We thank the Dartmoor National Park Authority, the Holne Commoners and the South West Water Authority for permission to survey, sample and excavate on Holne Moor. We are also indebted to Christine Williams of Sheffield University, who collaborated with us during the initial stages of the palynological work and who carried out some of the laboratory preparation of pollen samples.

<u>Bibliography</u>

Austin, D., Daggart, R.H. and Walker, M.J.C. (1981) Farms and fields in Okehampton Park, Devon: the problems of studying medieval landscape. <u>Landscape History</u> 2 39-57

Balaam, N.D. (1982) Soil pollen analyses. In Balaam, N.D., Smith, K. and Wainwright, G.J., The Shaugh Moor Project: Fourth report - environment , context and conclusion. <u>Proceedings of the Prehistoric Society</u> 48 203-278

Balaam, N.D., Smith, K. and Wainwright, G.J. (1982) The Shaugh Moor Project: Fourth report - environment, context and conclusion. <u>Proceedings of the Prehistoric Society</u> 48 203-278

Beckett, S. (1981) Pollen analysis of the peat deposits. In Smith, K., <u>et</u> <u>al</u>., The Shaugh Moor Project: Third report - settlement and environmental investigations. <u>Proceedings of the Prehistoric Society</u> 47 205-273

Bradley, R. (1978) Prehistoric field systems in Britain and north-west Europe - a review of some recent work. <u>World Archaeology</u> 9 265-277

Britton, D. (1963) Traditions of metal-working in the Neolithic and Early Bronze Age of Britain: Part 1. <u>Proceedings of the Prehistoric Society</u> 29 258-325

Britton, D. and Richards, E.E. (1961) A study of the composition of Wessex Culture bronzes. <u>Archaeometry</u> 4 39-52

Brown, A.P. (1977) Late Devensian and Flandrian vegetation history of Bodmin Moor, Cornwall. <u>Philosophical Transactions of the Royal Society</u> B 276 251-320

Brunsden, D. and Gerrard, J. (1970) The physical environment of Dartmoor. In Gill, C. (ed) <u>Dartmoor: a new study</u>. Newton Abbot: David and Charles

Burleigh, R. Matthews, D. and Ambers, J. (1981) British Museum natural radiocarbon measurements XII. <u>Radiocarbon</u> 23 14-23

Caseldine, C.J. and Maguire, D.J. (1981) A review of the prehistoric and historic environment of Dartmoor. <u>Proceedings of the Devon Archaeological Society</u> 39 1-16

Caseldine, C.J. and Maguire, D.J. (1982) Dartmoor. In Bell, M.G. (ed) <u>The environmental archaeology of south-west England</u>. Bristol: University Geography Department

Clapham, A.R., Tutin, T.G. and Warburg, E.F. (1962) <u>Flora of the British Isles</u>. 2nd Edition. Cambridge: Cambridge University Press

Clayden, B. (1964) Soils of the middle Teign valley district of Devon. Bulletin 1. Rothamsted: Soil Survey of Great Britain, England and Wales

Clayden, B. (1971) Soils of the Exeter district (Sheets 325, 339). Memoir. Rothamsted: Soil Survey of Great Britain, England and Wales

Clayden, B. and Manley, D.J.R. (1964) The soils of the Dartmoor granite. In Simmons, I.G. (ed) Dartmoor essays. Devonshire Association for the Advancement of Science

Cushing, E.J. (1967) Evidence for differential pollen preservation in Later Quarternary sediments in Minnesota Revue Palaeobotany Palynology 4 87-101

Fleming, A. (1977) The Dartmoor reaves. Current Archaeology 55 250-252

Fleming, A. (1978) The prehistoric landscape of Dartmoor. Part 1: South Dartmoor. Proceedings of the Prehistoric Society 44 97-123

Fleming A. (1983) The prehistoric landscape of Dartmoor. Part 2: North and East Dartmoor. Proceedings of the Prehistoric Society 49 (forthcoming)

Fleming, A., Greeves, T. and Robinson, R. (forthcoming) Long cairns and chamber tombs on Dartmoor: some new discoveries. Proceedings of the Devon Archaeological Society

Fleming, A. and Ralph, N. (1982) Medieval settlement and land use on Holne Moor, Dartmoor: the landscape evidence. Medieval Archaeology 26 101-137

Fussell, G.E. (1955) Crop nutrition in Tudor and early Stuart England. Agricultural History Review 3 95-106

Gawne, E. (1970) Field patterns in Widecombe parish and the Forest of Dartmoor. Transactions of the Devonshire Association 102 49-69

Harrod, T.R., Catt, J.A. and Weir, A.H. (1974) Loess in Devon. Proceedings of the Ussher Society 2 554-564

Harrod, T.R. Hogan, D.V. and Staines, S.J. (1976) Soils in Devon II Sheet SX 65 (Ivybridge). Soil Survey Record 39. Rothamsted: Soil Survey of England and Wales

Hicks, S.P. (1972) The impact of man on the East Moor of Derbyshire from Mesolithic times. Archaeological Journal 129 1-18

Hogan, D.V. (1982) Upland soil survey in south-west England. Welsh Soils Discussion Group Report (1979) 20 135-161

Jørgensen, S. (1967) A method of absolute pollen counting. New Phytologist 66 489-493

Kramm, E. (1978) Pollen Analytische hockmooruntersuchungen zur floren-und siedlungs geschichte zwischen ems und hase. Abhandl. Landesmuseum fur Naturkunde Munster 40/4 44S

Lange, E. (1975) The development of agriculture during the first millennium A.D. Geologiska Foreningens i Stockholm Forhandlinga 97 115-124

Maltby, E. and Crabtree, K. (1976) Soil organic matter and peat accumulation on Exmoor: a contemporary and palaeoenvironmental interpretation. Transactions of the Institute of British Geographers NS 1 259-278

Mercer, R. (1981) Appendix. In Mercer, R. (ed) Farming practice in British prehistory. Edinburgh: University Press

Merryfield, D.L. and Moore, P.D. (1974) Prehistoric human activity and blanket peat initiation on Exmoor. Nature, London 250 439-441

Minchinton, W.E. and Overton, M. (1973) The 1801 crop returns for Devon. Devon and Cornwall Notes and Queries 32 197-203

Moore, P.D. and Chater, E.H. (1969) The changing vegetation of west-central Wales in the light of human history. Journal of Ecology 57 361-378

Moore, P.D. and Webb, J.A. (1978) An illustrated guide to pollen analysis. London: Hodder and Stoughton

Perkins, D.F. (1970) A note on the climate of Dartmoor. In Perkins, D.F. and Ward, S.D. (eds) Report on the Dartmoor Ecological Survey. Natural Environmental Research Council and the Nature Conservancy

Perrin, R.M.S., Davies, H. and Fysh, M.D. (1974) Distribution of late Pleistocene aeolian deposits in eastern and southern England. Nature, London 248 320-324

Rainsford-Hannay, F. (1972) Dry stone walling. Gatehouse of Fleet: Stewartry of Kirkcudbright Drystone Dyking Committee

Ralph, N. (1982) Assessment of ancient land use in abandoned settlements and fields - a study of prehistoric and medieval land use and its influence upon soil properties on Holne Moor, Dartmoor, England. Unpublished Ph.D. Thesis, University of Sheffield

Ralph, N. and Fleming, A. (forthcoming) Medieval settlement and land use on Holne Moor, Dartmoor: the pedological and archaeological evidence

Reynolds, P. (1981) Deadstock and livestock. In Mercer, R. (ed) Farming practice in British prehistory. Edinburgh: University Press

Riezebos, P.A. and Slotboom, R.T. (1978) Pollen analysis of the Husterbaach peat (Luxemborg): its significance for the study of sub-recent geomorphological events. Boreas 7 75-82

Roberts, B., Turner, J. and Ward, P. (1973) Recent forest history and land use in Weardale, N. England. In Birks, H.J.B. and West, R.G. (eds) Quaternary Plant Ecology. London: Blackwell

Russell, E.W. (1973) Soil conditions and plant growth. London: Longman

Simmons, I.G. (1964) Pollen diagrams from Dartmoor. New Phytologist 63 165-180

Simmons, I.G. (1969) Environment and early man on Dartmoor, Devon, England. Proceedings of the Prehistoric Society 35 203-219

Smith, K., et al. (1981) The Shaugh Moor Project: Third report - settlement and environmental investigations. Proceedings of the Prehistoric Society 47 205-273

Spooner, G.M. and Russell, F.S. (eds) (1967) Worth's Dartmoor. Newton Abbot: David and Charles

Staines, S.J. (1972) Soils and vegetation on Dartmoor: distribution and history. Unpublished M.Sc. Thesis, University of Bristol

Staines, S.J. (1979) Environmental change on Dartmoor. Proceedings of the Devon Archaeological Society 37 21-47

Stanes, R. (1969) Devon agriculture in the mid-eighteenth century: the evidence of the Milles enquiry. In Havinden, M.A. and King, C.M. (eds) The South-West and the Land. (Exeter papers in Economic History No. 2), Exeter University

Steckhan, H.U. (1961) Pollenanalytisch-vegetationgeschichtliche undersuchungen zur fruhen siedlungsgeschichte im Vogelsberg, Knull and Solling. Flora 150 514-551

Turner, J. (1964) The anthropogenic factor in vegetational history. I. Tregaron and Whixall mosses. New Phytologist 63 73-90.

Wainwright, G.J. and Smith K. (1980) The Shaugh Moor Project: Second report - the enclosure. Proceedings of the Prehistoric Society 46 65-122

Ward, S.D., Jones, A.D. and Manton, M. (1972) The vegetation of Dartmoor. Field Studies 3 505-534

Waters, R.S. (1964) The Pleistocene legacy to the geomorphology of Dartmoor. In Simmons, I.G. (ed) Dartmoor essays. Devonshire Association for the Advancement of Science.

West, R.G. (1970) Pollen zones in the Pleistocene of Great Britain and their correlation. New Phytologist 69 1179-1183

5. NEW APPLICATIONS OF PALAEOECOLOGICAL TECHNIQUES: INTEGRATING EVIDENCE OF ARABLE ACTIVITY IN POLLEN PEAT AND SOIL STRATIGRAPHIES, CEFN GRAEANOG, NORTH WALES

Frank Chambers

INTRODUCTION

In 1977-8, The Gwynedd Archaeological Trust Ltd. undertook the total excavation of both an enclosed hut-group (the site of Cefn Graeanog II, excavated by White, forthcoming) and a medieval farmstead (excavated by Kelly 1982) on Cefn Graeanog, Clynnog (NGR SH 454490) in advance of gravel extraction. During the course of the excavations, pollen samples were taken from soils beneath the two archaeological sites, from an accumulation of hillwash (colluvium) to the rear of the medieval farmstead and from a valley mire 200 m to the west of the excavations. Reports on the palynological investigations associated with the two sites will appear separately (Chambers 1982; forthcoming). This paper concentrates on pollen, peat and soil stratigraphic evidence for arable activity at Cefn Graeanog in the period since the Bronze Age. Emphasis is placed here on the integration of different lines of evidence to assist in reconstructing the agricultural history of the area.

The Study Area

Cefn Graeanog is the name of a farm at NGR SH 456489 on a ridge of fluvio-glacial gravel, 10 km south of Caernarfon, N. Wales (Fig. 5:1).

Glacial and fluvio-glacial activity in the Devensian cold stage resulted in the accumulation of deep deposits of fluvio-glacial gravel in the locality; these so disrupted the drainage pattern that the large expanse of post-glacial mire to the west and south of the excavations now drains both north, via the Afon Desach, and south via the Afon Dwyfach. The locality is more fully described by Kelly (1982).

Settlement History (after Kelly 1982; White 1978; White, Forthcoming)

There are indications of Bronze Age (or earlier) human activity on the gravel ridge in the form of a standing stone and two cairns; whilst a crouched inhumation found to the rear of the medieval farmstead apparently dates from the second millennium BC.

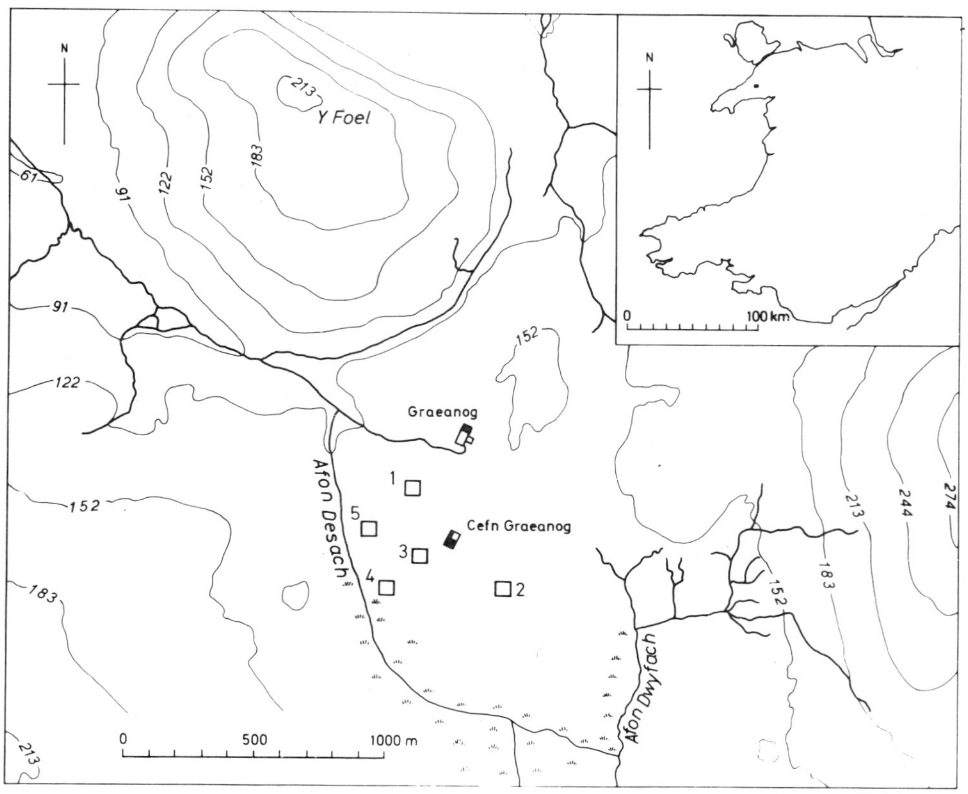

Fig. 5:1 Map showing location of Cefn Graeanog and archaeological sites (after Kelly 1982). 1 - Standing stone and two cairns; 2 - Cefn Graeanog I enclosed hut-group (Hogg 1969); 3 - Cefn Graeanog II enclosed hut-group (White, forthcoming); 4 - Round-cornered rectangular structures - possibly early medieval; 5 - medieval farmstead (Kelly 1982). Contours in metres (contour interval is 100 feet).

In addition to the recent major excavation of Cefn Graeanog II, an earlier excavation of another enclosed hut-group (Cefn Graeanog I) was undertaken in this area by Hogg (1969) (Fig. 5:1). The two enclosed hut-groups apparently date from Iron Age, Roman, and in the case of Cefn Graeanog II, post-Roman times also.

At Cefn Graeanog II, four periods are recognised (White, forthcoming). Period I was a cultivated field, apparently of the first millennium BC, upon which a stone-built round hut (J) was constructed. This constitutes Period II, which is estimated to correspond with the 4th to 2nd centuries BC. Hut J became a grass-grown ruin, before Period III - a curving enclosure wall with at least one hut - was built over it in the first century AD. Period IV commenced in the third or fourth century AD. This necessitated clearance of much of Period III and involved the construction of a roughly rectangular stone-walled enclosure some 30 m by 30 m with three circular (domestic) buildings, two rectangular (storage) buildings and an entrance barn/byre. It is believed that at least one domestic building remained in use until the 6th century AD (The foregoing is based on the text of a lecture delivered to the Society of Antiquaries, London by R.B. White. The full report of the excavation of Cefn Graeanog II is shortly to appear as Volume 2 of the Cambrian Archaeological Society's Monographs and Collections).

There is documentary evidence to support the view that after the ending of occupation of Cefn Graeanog II in the 6th century AD (White 1978) the land became escheat (to the Crown) and subsequently was the gift of Cadwalader, who reigned towards the later part of the 7th century AD. The fate of the land between then and the construction of the medieval farmstead is a matter for conjecture (Kelly 1982: 887).

The medieval farmstead had two phases of building. The earlier, timber-framed, structure was replaced by four cruck-built stone-walled buildings which continued in use until the first half of the 13th century AD. The hearthstone (aelwyd) in the farmstead yielded a remanent magnetic date of last firing of AD 1240±30 (Kelly 1982).

The recent settlements of Graeanog and Cefn Graeanog farms (Fig. 5:1) were established in post-medieval and modern times.

Aims

This paper relates palaeoecological evidence for local human activity (and particularly for arable cultivation) to the settlement history or sequent occupance of Cefn Graeanog in the period since the Bronze Age. Two new applications of palaeoecological techniques are outlined: (a) pollen analysis of colluvium and correlation with soils buried beneath settlements; (b) determination of humification of valley mire peat and its relation to evidence for woodland clearance, agricultural activity and woodland regeneration.

METHODS

The methods detailed below refer to laboratory techniques. Field sampling of buried soils and the valley mire was undertaken by members of G.A.T. Ltd. The samples and mire monolith were then submitted to the author for analysis in the Palaeoecology Laboratory, Plant Science Department, University College, Cardiff.

Pollen Analysis

Samples were prepared for pollen analysis according to Barber (1976), with silicone fluid as mounting medium. A pollen sum of 500 total land pollen grains (T.L.P.) was counted (except where indicated). Summary curves for arboreal pollen (AP), comprising trees and shrubs, and for non-arboreal pollen (NAP) are presented in Figs. 5:2 and 5:3. The shrub component is represented here by *Salix* (willow, sallow), *Sambucus* (elder) and *Corylus* or Coryloid (mainly hazel) pollen.

The term 'Agricultural types' (Fig. 5:3) refers to taxa under the heading 'Arable' in the mire pollen diagram (Chambers, forthcoming). It includes both cultivated (Cerealia, *Linum* cf. *usitatissimum*, *Secale* and Cannabiaceae) and non-cultivated (*Urtica*, *Centaurea nigra*-type, Chenopodiaceae, *Rumex* spp., *Plantago* spp., *P. lanceolata*) taxa. Ribwort plantain (*P. lanceolata*), often regarded as an indicator of pastoralism (cf. Turner 1964), is included because its association with the disturbed habitats caused by human activity make it a useful clearance indicator. The ruderal knapweeds (*Centaura nigra*-type) are also included because their distribution in scrub and woodland margins has been encouraged by human activities; they also might be regarded as weeds of pasture rather than arable land. Other groups with possible arable affinity (e.g. Labiatae, Caryophyllaceae, Cruciferae) were excluded due to uncertainties as to the species involved.

Radiocarbon Dating

All radiocarbon dates from the valley mire were made on the fine particulate (F) fraction (<250 microns) of the peat to reduce the possibility of contamination from the potentially more mobile humic acid and from modern rootlets. The samples were pretreated by the author and passed to Dr. P. Q. Dresser for dating by gas proportional counting in the Radiocarbon Laboratory, Cardiff. The quoted precisions (uncertainties) of the dates are one standard deviation and derive mainly from counting statistics. Radiocarbon dates were also obtained from charcoal samples taken during the medieval farmstead excavation and submitted to Harwell by G.A.T. Ltd.

Humus Determinations

Contiguous (0.5 cm) samples of the upper (non-wood) valley mire peat were subjected to humus determinations, employing the colorimetric method developed by Bahnson (Aaby & Tauber 1975). The results are expressed as percentage humification in Fig. 5:3. A curve produced from moving averages of three consecutive samples is also presented to aid visual interpretation.

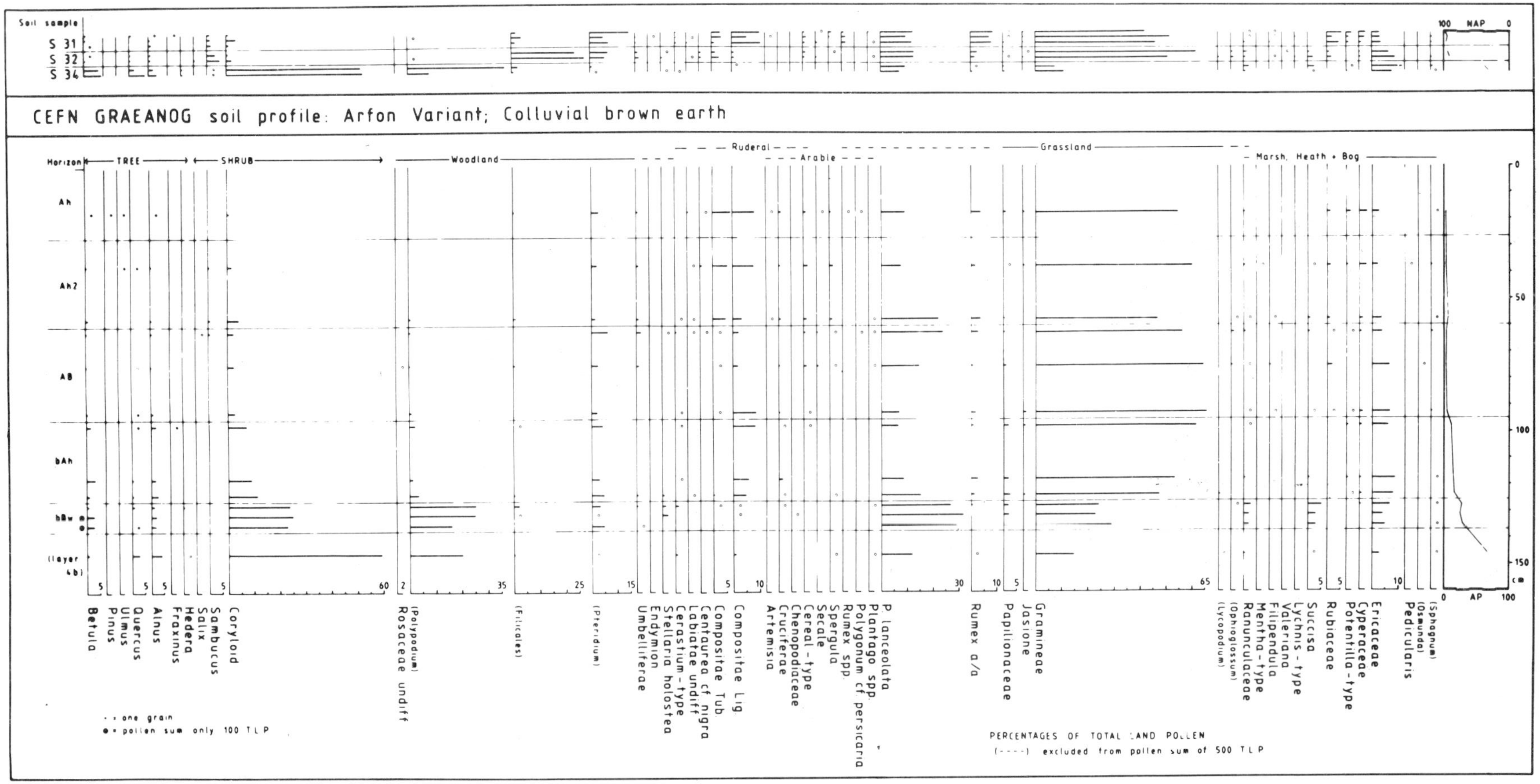

Fig. 5:2 Pollen diagrams through colluvium at the rear of the medieval farmstead (reproduced from Kelly 1982; Fig. 16). Soil samples S-34 are from continuation of layer 4b under farmstead; S-32 from beneath the wall of farmstead; S-31 from floor of farmstead.

RESULTS

Pollen Analysis of Buried Soils

Pollen samples were taken from soils buried beneath the recently excavated enclosed hut-group Cefn Graeanog II. The results of these analyses are presented and discussed more fully elsewhere (Chambers, forthcoming). In summary, analysis of pollen samples from soils beneath the hut-group walls confirmed that the earliest known hut (Hut J) had been built over a cultivated soil in which wheat had been grown. Hence the earliest period (Period I) of the Cefn Graeanog II enclosed hut-group is identified as a cultivated field (White, forthcoming).

The Colluvium Section

The four rectangular buildings which constituted the medieval farmstead lay side-by-side with their long axes north-west to south-east and their rear (south-east) ends cut back into the slope of the land. This slope, in part natural, is covered with hillwash (colluvium). A section to the rear of the farmstead was sampled for pollen analyses (Fig. 5:2). The lower horizons of this hillwash (bBw, bAh, AB) represent material which accumulated prior to construction of the medieval farmstead. The accumulation of horizon Ah2 is believed to be contemporary with occupation of that settlement; whilst horizon Ah accumulated more recently. The results of the pollen analyses have been discussed briefly elsewhere (Chambers 1982).

The pedological distinctions between the colluvium soil horizons (W.S.D.G. 1978; Keeley 1982) were confirmed by distinct differences between the pollen contents of the lower horizons (Fig. 5:2); only horizons Ah2 and Ah could not readily be separated on their pollen contents. The boundaries between the soil horizons were interpreted as stand-still episodes when little if any cultivation was practised. Indeed, these boundaries could well reflect breaks in occupation of the land which permitted revegetation and stabilization of the slope such that downslope soil movement was restricted.

The colluvium pollen spectra were then compared with the pollen spectra of soils buried beneath the archaeological sites (Fig. 5:2 and Chambers, forthcoming). By combining the palynological data with archaeological and documentary evidence (Kelly 1982; White, forthcoming), it has been possible to attempt a reconstruction of sequent occupance and agricultural history of the ridge (Table 5:1). Periods of arable activity would have stimulated downslope soil movement and the accumulation of colluvial material; hence, periods of little or no arable activity may be inferred for the boundaries between colluvial soil horizons. These boundaries appear to correlate with known or suspected breaks in settlement on the ridge.

It has been argued that horizon bBw (Fig. 5:2) accumulated during Period I of the Cefn Graeanog II hut-group in the first millennium BC (Chambers 1982: 889) but it may have accumulated earlier than this (Table 5:1) as pollen spectra in horizon bAh, above, seem to correspond most closely with pollen spectra in the

Table 5:1: Proposed correlation between buried soils, archaeology & colluvium horizons

Colluvium Horizon	Layers (Kelly 1982)	Corresponding Pollen Spectra	Horizon Accumulates	Archaeology
Ah	0a,1 Post-medieval	-	Post-medieval	Post-medieval & modern farms
			?Break in occupation	13th or 14th century AD
Ah2	2 Layer partially truncated by stone phase of medieval farmstead	Sample S-31 (floor of medieval farmstead) Sample S-32 beneath wall of medieval farmstead corresponds with AB/Ah2 boundary	Medieval Period	Medieval farmstead ? Early medieval occupation
			?Break in occupation	6th-7th century AD
AB	2a Layer truncated by medieval farmstead	*cf* some pollen spectra under Period IV hut-group walls	Roman & post-Roman	Periods III, IV of hut-group
			?Break in occupation	2nd-1st century BC
bAh	2c Layer truncated by medieval farmstead	Buried soil under walls of Period II hut (J)	First millenium BC (see Table 2)	Period I, II of hut-group
bBw	2c As above. Grave of inhumation cut down from this layer into layer 4, below	Lower horizons of lynchet under Period IV hut group walls	Second/first millennium BC (see Table 2)	?Bronze Age burial (Grave cut down from this layer)
4b	4b Charcoal in extension of this layer under farmstead produced date of 1390±90 bc (HAR-2554)	Extension of layer 4b beneath medieval farmstead - sample S.34	Before and/or during second millennium BC	?Bronze Age cairns ?Standing Stone

Period I soil (under Hut J) in the first millennium BC. The bAh/AB boundary is thought to correspond with an assumed break in occupation between Period II and Period III of the Cefn Graeanog II hut-group when the Period II hut (J) became a grass-grown ruin in the first two centuries BC. Hence horizon AB probably accumulated during cultivation associated with the later periods (III, IV) of this hut-group in Roman and post-Roman times.

Kelly (1982) argues that the upper part of horizon Ah2 (Fig. 5:2) is contemporary with occupation of the medieval farmstead. Hence, the lower part of horizon Ah2 corresponds either with cultivation associated with the preceding timber phase of the farmstead or with other, earlier settlement in the period since the abandonment of the Cefn Graeanog II hut-group in the 6th century AD. There were traces of "round-cornered rectangular structures, perhaps of early medieval date" (Kelly 1982: 887) to the north of the Cefn Graeanog II hut group (Fig. 5:1) but these were not excavated. Nevertheless, it became clear in analysing the palynological and palaeobotanical results from the two recent excavations that the accumulation of the whole of horizon Ah2 clearly post-dated the demise of the Cefn Graeanog II hut-group. This conclusion was reached after comparing pollen spectra in buried soils and colluvium horizons with records of charred weed seeds recovered from flotation samples taken from the two recently excavated sites. Hillman (1982) recorded abundant corn marigold (*Chrysanthemum segetum*) seeds in flots from the medieval farmstead, whilst this species was not apparently a feature of the segetal flora at the time of the Cefn Graeanog II hut-group. In the colluvium section, pollen of Compositae Tub. is conspicuous in horizon Ah2 but was not previously abundant; the increase in Compositae Tub. pollen records above the AB/Ah2 boundary was attributed to the importation of corn marigold into the area as an arable weed (Chambers 1982: 900). This would accord with the view that corn marigold is an introduction to Britain (Clapham, Tutin & Warbug 1981). According to Godwin (1975), seed records of *Chrysanthemum segetum* are confined to Roman and medieval occupation levels, though Iron Age records have since been reported from Hampshire (Monk and Fasham 1980; Jones, in press). At Graeanog, its arrival may be considered to have been in the medieval period.

A break in arable land use may be inferred for the Ah2/Ah boundary after the abandonment of the medieval farmstead in the first half of the 13th century AD. Post-medieval and modern tillage led to the accumulation of horizon Ah above.

This interpretation of the colluvium stratigraphy is summarised in Table 5:1.

The Valley Mire

A monolith of peat was extracted from the valley mire to the west of the excavations at a point 20 metres from the margin. Detailed results of the valley mire pollen and radiocarbon dating analyses will appear elsewhere (Chambers, forthcoming). The results of pollen analyses and humus determinations on the top 75 cm of the valley mire peat are summarised in Fig. 5:3. The base of this sequence apparently dates from *circa*. 850 bc.

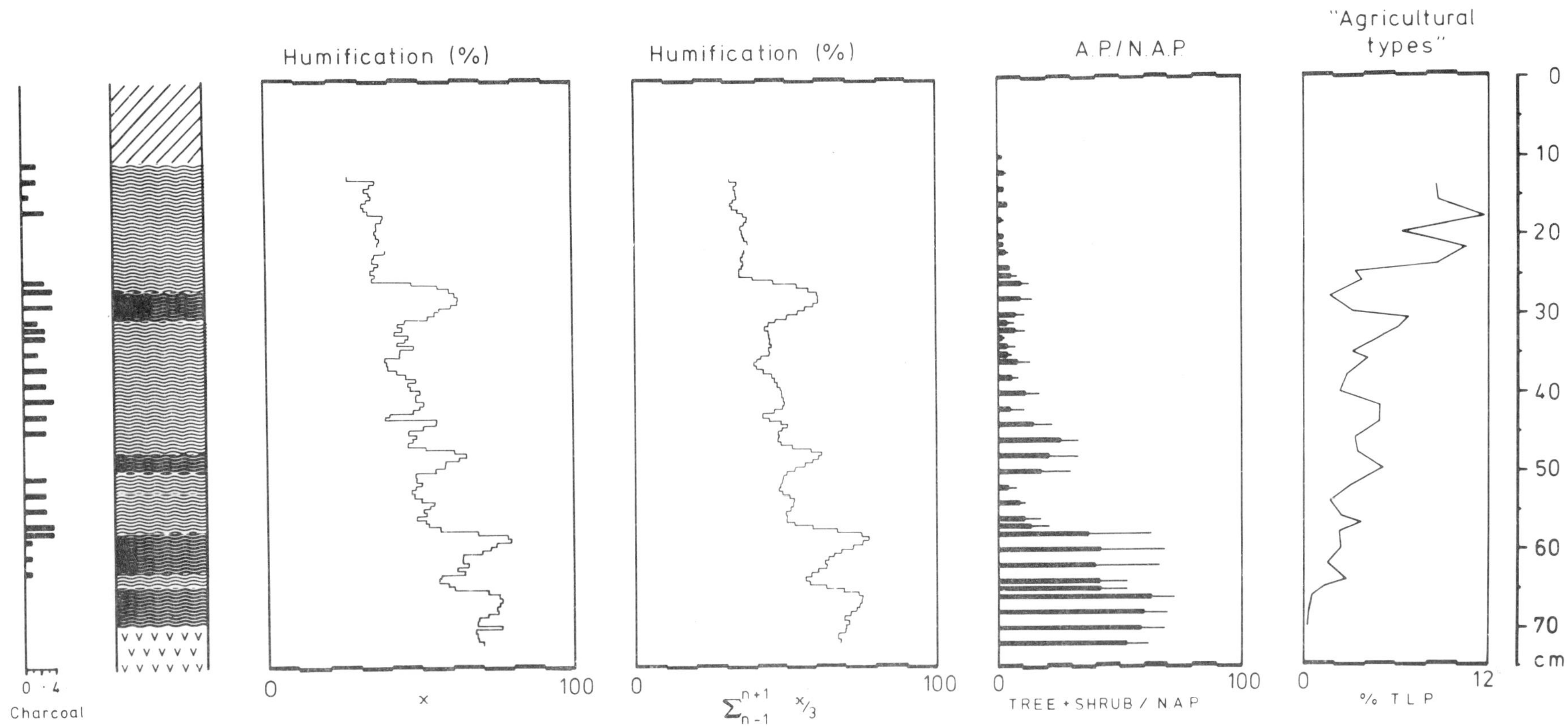

Fig. 5:3 Charcoal records in pollen preparation residues; peat stratigraphy; peat humification (raw data and 'smoothed' curve); A.P./N.A.P. ratios; and sum of pollen types indicative of agricultural activity in upper 75 cm of valley mire deposits, Cefn Graeanog (CG.B core). The base of this part of the core apparently dates from c.850 bc.

Fluctuations in the A.P./N.A.P. curve in the diagram are indicative of a series of forest clearances and regeneration episodes in the Graeanog area since the second millennium BC.

Correlation Between A.P. Levels and Degree of Humification

It is notable that both the percentage humification curve and the arboreal pollen (A.P.) curve in Fig. 5:3 exhibit an overall declining trend, but superimposed on this trend are similar marked fluctuations. While the magnitude of the fluctuations is not equivalent, the directions of the changes do appear so. The equivalence of peaks seems especially marked.

A Spearman's rank correlation of 36 pairs of A.P. and humification values gave a coefficient (r_s) of +0.83, significant at the 0.001 level. It must be acknowledged that this coefficient largely summarises the equivalence in the declining overall trends, but a coefficient of this magnitude indicates that the fluctuations in the curves may be closely correlated. Partial correlation coefficients were then calculated from Kendall's rank correlation coefficients (tau) of 0.65 (between A.P. levels and degree of humification), 0.68 (between A.P. levels and inverse depth), and 0.64 (between degree of humification and inverse depth). The corresponding partial correlation coefficients (with the third factor controlled in each case) were tau(ax,d)=0.38; tau(ad,x)=0.48; tau(xd,a)=0.36. The strongest correlation is therefore between decreasing depth (d) and arboreal pollen (a) values. However, a considerable degree of co-variance does exist between A.P. levels and degree of humification (x). (As a descriptive statistic, the significance of partial correlation coefficients cannot easily be assessed. An alternative approach would be to adapt established curve-matching techniques for these data.)

Peat Humification and Pollen Records of Agricultural Types

Fig. 5:3 contains a summation curve of pollen records of crop plants and probable weeds. A Spearman's rank correlation between percentage humification and recorded pollen of these taxa (excluding *Plantago lanceolata*) produced a negative correlation coefficient (r_s) of -0.74, significant at the 0.001 level (r = -0.83 when *P. lanceolata* included).

DISCUSSION

Significance of Changes in Humification

Aaby & Tauber (1975) showed that the degree of humification of raised-bog peat is closely related to humidity at the time of its formation and not due to secondary alteration during periods of dryness. Interpretation of raised-bog stratigraphy, therefore, has largely been in terms of climatically-induced changes. It has been suggested that horizons with high humification values in raised-bog peats can be correlated with periods of equable climate (warmer and drier), whereas horizons of low humification correspond with cooler and wetter conditions when a re-growth (recurrence) of peat is stimulated (cp. Granlund 1932; Godwin 1954; 1975). Weber's

Grenzhorizont was thought to mark a climatic deterioration of this type. This interpretation has been much modified by work demonstrating the occurrence of several similar boundary horizons (recurrence surfaces) of different ages in Sweden, and of recurrence surfaces with different ages in the same mire (Lundquist 1962; Godwin 1975).

Despite the conflicting evidence from continental Europe, some stratigraphic changes in raised mires have been solely attributed to climatic change. The 'flooding horizons' of the Somerset Levels were interpreted as climatically determined (Clapham & Godwin 1948), upon which prehistoric trackways had to be built to counter the rise in water-level: "The trackways are ... indicative of successional and climatic shift towards flooding.." (Godwin 1975: 42).

Apparently regular shifts from dark to light-coloured peat in Danish raised mires have been interpreted as recording climatic changes with a periodicity of 260 years (Aaby 1976). However, the hydrological relations between a raised mire found in a river valley and the surrounding landscape have been rather neglected: the effect of nearby forest clearance upon the drainage of such a raised mire is unclear.

It may be noted at this point that in certain instances, marked recurrence surfaces appear to be associated with periods of increased agricultural activity (cf. Turner 1964), but a clear correlation is not consistently borne out by the evidence. Current work in the Somerset Levels (Coles *et al.* 1975 *et* ff) and in the Humberhead Levels (B.M. Smith, in *prep.*) may allow comparison between climatic change and evidence of periods of increased agricultural activity. The difficulty here is that some of the evidence of past weather and climate is directly derived from stratigraphic studies in which a causal relationship between climate and peat humification was assumed (cf. Lamb 1977).

The matter is unlikely to be resolved without further detailed study.

Agricultural Activity and Humification of Valley-Bog Peat

In areas of marginal agriculture, Parry (1975) claimed a direct relationship between intensity of agricultural land-use and secular climatic change: one might expect, therefore, in marginal areas such as Cefn Graeanog, to find a correlation between degree of humification and level of agricultural activity, viz. agricultural activity is stimulated at times of more equable climate; a more equable climate increases the degree of humification.

Although in recent years the farming activity of Cefn Graeanog was mainly pastoral, arable use was made of land in the 1950's and 1960's with occasional crops of barley, roots and winter-sown oats; it is believed this may have been the practice for the preceding two centuries. Charred remains of crop plants (Hillman 1982) and soil pollen records indicate that the area has experienced arable activity (though intermittent) for millennia. There should therefore be a positive correlation between pollen evidence for agriculture and percentage humification if postulates A (humification relates directly to climate) and B (more equable climate stimulates marginal agriculture) are true. In fact, a strong negative correlation was found. In view of this, either (or both) postulate must be incorrect in this instance.

Humification in Valley-Bog Peat

Raised mires are believed to be sensitive to climatic change because they are ombrogenous: their existence is said to be directly determined by the precipitation and evaporation to which they are subject (Godwin 1975).

Valley mires are said to be topogenous and are directly influenced by the precipitation/evapotranspiration ratio of their hydrological catchment through the movement of groundwater and run-off. If forest clearance takes place within a drainage basin, then the amount of run-off may drastically increase. An intensification of arable agriculture has been shown to produce the same result: "Breaking up grassland and cutting woods results generally in a decreasing evapotranspiration and an increasing runoff. Consequently, a 'wetter' environment is created especially in the depressions of the landscape.." (Riezebos & Slotboom 1978: 81). Aaby & Tauber (1975) found a clear negative correlation between degree of peat humification and humidity at time of formation: it follows therefore that the creation of a wetter local environment will reduce the degree to which peat is humified. The wetter environment can be produced by forest clearance: there is thus a clear mechanism by which peat humification in valley mire deposits can be related to arboreal pollen levels. It is postulated that this explains the correlation observed at Cefn Graeanog.

In view of the high negative correlation between the pollen records of agricultural types and percentage humification, it seems appropriate to regard the changes in peat humification in the valley mire as being strongly influenced both by the clearance or regeneration of woodland and by the practice of arable agriculture on the ridge.

Dating Agricultural Activity at Graeanog

The colluvium section produced evidence of five episodes of increased agricultural activity, with four 'still-stands' possibly indicating breaks in land occupation. Tentative dates for these episodes were proposed in Table 5:1.

It was intended that these tentative dates might be confirmed (or otherwise) by comparing the pollen spectra in the colluvium section with the radiocarbon-dated mire pollen diagram. Unfortunately the radiocarbon dates obtained from the mire section were not in a chronological series; two dates were particularly suspect (Chambers, forthcoming). Further, although there was no stratigraphic evidence for hiatus or disturbance in that mire section (apart from the top 10 cm - see Fig. 5:3), two monoliths taken as 'replicates' from the valley mire showed clear signs of disturbance.

Table 5:2: Tentative correlation between periods of agricultural acitivity indicated in the mire pollen diagram and the accumulation of colluvium layers.

Horizons (valley mire)	[1]Age (yr bc/ad)	[2]Age (BC/AD)	Colluvium horizon	Archaeology
25 - 12 cm	*c.* 1430 ad to recent	AD 1420 to recent	Ah	Post-medieval
46.5 - 30 cm	*c.* 650 ad to *c.* 1330 ad	AD 680 to AD 1310/1385	Ah2	Medieval
58 - 50 cm	*c.* 40 ad to *c.* 520 ad	AD 75 to AD 620/640	AB Periods III, IV	Hut-group
64.5 - 60 cm	*c.* 350 bc to *c.* 120 bc		bAh Periods I, II	Hut-group
80 - 74 cm	*c.* 1050 bc to *c.* 800 bc		bBw	Bronze Age

1. The duration of agricultural activity was calculated from deposition-rates between pairs of radiocarbon-dated horizons, excluding the two dates apparently out of sequence (Chambers, forthcoming). The periods of agriculture are taken to correspond with reduced A. P. values and low peat humification.

2. The conventional radiocarbon years (ad) were converted to 'dendroyears' by interpolation from the calibration curve presented by Stuiver (1982). No attempt is made here to calibrate bc radiocarbon ages.

Hence there must be some doubt as to the date of clearance episodes in the mire pollen diagram.

However, these reservations apart, it may be no coincidence that humification peaks in Fig. 5:3 would correspond closely with the suggested ages of still-stands in the colluvium section if the two 'inverted' radiocarbon dates were excluded from calculations (the other six radiocarbon dates fit a smooth deposition-rate curve). Dates calculated on this selective basis are shown in Table 5:2.

SUMMARY AND CONCLUSIONS

Excavations of two archaeological sites on Cefn Graeanog, Gwynedd, Wales by G.A.T. Ltd., included a programme of sampling for environmental analysis. Pollen samples were taken from beneath structures on the sites, from a stratified accumulation of hillwash (colluvium) and from the margins of an extensive valley mire nearby.

The layered colluvium provided evidence of five episodes of intensified downslope soil movement since the Bronze Age, which appeared to be related to periods of increased local cultivation separated by still-stands of little or no arable activity, these possibly corresponding with breaks in land occupation.

Pollen analyses and humus determinations of the valley mire peat produced a positive correlation between arboreal pollen levels and degree of humification, and a negative correlation between the latter and records of pollen of crop plants and likely weeds. It is suggested that decreased evapotranspiration and increased run-off accompanied woodland clearance and arable activity. The local environment of the valley mire thus became wetter and hence reduced the degree of peat humification. Periods of reduced arable activity apparently allowed woodland regeneration, reducing run-off and producing a drier environment in the valley mire and hence increased peat humification. There may be implications in the results presented here for peat humification studies of raised mires in certain topographic situations.

Determination of valley mire peat humification <u>may</u> therefore be a useful technique for evaluating the impact of human activity in a catchment, and when combined with soil pollen studies, may allow reconstruction of both sequent occupance and periods of increased arable activity. However, there is evidence to suggest that some of the Graeanog valley mire deposits may be disturbed and radiocarbon dates on the peat monolith are not in a chronological series.

Acknowledgements

The author is indebted to Professor A. G. Smith for laboratory facilities and to Mrs. M. Patrick for drawing Fig. 1 and Table 1. I should like to thank R. B. White (Director) and R. S. Kelly (Deputy Director) of the Gwynedd Archaeological Trust for information on the archaeological sites and for helpful discussions and encouragement.

Bibliography

Aaby, B. (1976) Cyclic climatic variations in climate over the past 5,500 reflected in raised bogs. Nature 262 281-284

Aaby, B. & Tauber, H. (1975) Rates of peat formation in relation to degree of humification and local environment, as shown by studies of a raised bog in Denmark. Boreas 4 1-17

Barber, K.E. (1976) History of vegetation. In: Methods in Plant Ecology. (ed Chapman, S.B.) Oxford: Blackwell Scientific Publications pp. 5-83

Chambers, F.M. (1982) Appendix 3, Palynological studies. In: Kelly, R.S. The excavation of a medieval farmstead at Cefn Graeanog, Clynnog, Gwynedd. Bulletin of the Board of Celtic Studies 29 (4) pp. 898-900

Chambers, F.M. (forthcoming) Report on palynological investigations, Cefn Graeanog. In: White, R.B. Cefn Graeanog II - an Iron Age, Roman and post-Roman farmstead in Arfon Gwynedd. Cambrian Archaeological Association Monographs and Collection No. 2

Clapham, A.R. & Godwin, H. (1948) Studies of the post-glacial history of British vegetation. VIII. Swamping surfaces in peats of the Somerset levels. Philosophical Transactions of the Royal Society B 233 233-273

Clapham, A.R., Tutin, T.G. & Warburg, E.F. (1981) Excursion Flora of the British Isles, 3rd Ed. Cambridge: University Press

Coles, J.M. (1975 et ff) Somerset Levels Papers. Cambridge

Godwin, H. (1954) Recurrence-surfaces. Danmarks Geologiske Undersogelse RII (80) 22-30

Godwin, H. (1975) The History of the British Flora, 2nd Ed. Cambridge: University Press

Granlund, E. (1932) De Svenska hogmössarnos geologi. Sveriges Geologiska Undersokning Ser C 26 (373) 1-193

Hillman, G.C. (1982) Appendix 4, Crop husbandry at the medieval farmstead, Cefn Graeanog: reconstruction from charred remains of plants. In: Kelly, R.S. The excavation of a medieval farmstead at Cefn Graeanog, Clynnog, Gwynedd. Bulletin of the Board of Celtic Studies 29(4) pp. 901-906

Hogg, A.H.A. (1969) Cefn Graeanog: a native site of the Roman period. Transactions of the Caernarvonshire Historical Society 30 8-20

Jones, M.K. (in press) The Plant Remains In: Excavations of Daneburg, Hants (ed. Cunliffe, B.W.) London: CBA

Keeley, H. (1982) Appendix 2, The Soils. In: Kelly, R.S. The excavation of a medieval farmstead at Cefn Graeanog, Clynnog, Gwynedd. Bulletin of the Board of Celtic Studies 29(4) pp. 895-898

Kelly, R.S. (1982) The excavation of a medieval farmstead at Cefn Graeanog, Clynnog, Gwynedd. Bulletin of the Board of Celtic Studies 29(4) 859-908

Lamb, H.H. (1977) Climate; Present, Past and Future. Vol. 2 Climatic History and the Future. London: Methuen

Lundquist, G. (1962) Geological radiocarbon datings from the Stockholm station. Sveriges Geologiska Undersokning Ser C 56 (589) 8-12

Monk, M.A. and Fasham, P.J. (1980) Carbonised plant remains from two Iron Age sites in central Hampshire. Proceedings of the Prehistoric Society 46 321-344

Parry, M.L. (1975) Secular climatic change and marginal agriculture. Transactions of the Institute of British Geographers 64 1-13

Riezebos, P.A. & Slotboom, R.T. (1978) Pollen analysis of the Husterbaach peat (Luxembourg): its significance for the study of sub-recent geomorphological events. Boreas 7 75-82

Stuiver, M. (1982) A high-precision calibration of the AD radiocarbon time scale Radiocarbon 24(1) 1-26

Turner, J. (1964) The anthropogenic factor in vegetational history.

I. Tregaron and Whixall Mosses. New Phytologist 63 73-89

White, R.B. (1978) Cefn Graeanog. In: Roman Britain in 1977 (ed Goodburn, R.) Britannia 9 406-407

White, R.B. (forthcoming) Cefn Graeanog II - an Iron Age, Roman and post-Roman farmstead in Arfon, Gwynedd. In: Cambrian Archaeological Association Monographs and Collections No. 2

W.S.D.G. (1978) Welsh Soils Discussion Group Summer Field Meeting: Cefn Graeanog, Bryncir, Gwynedd (Field Guide). Aberystwyth: W.S.D.G.

6. POLLEN AND PLANT MACROFOSSIL ANALYSIS OF DEPOSITS FROM THE IRON AGE DITCHED ENCLOSURE AT SHIELS, GOVAN, GLASGOW

D. Robinson

INTRODUCTION

The ditched enclosure at Shiels, Govan, Glasgow (NS 523 667) is situated on the River Clyde floodplain, 0.5 km south of the present river course, 7 km west of Glasgow city centre and 6 km south of the Antonine fort at Bearsden (Fig. 6:1).

The site lies near the centre of the Clyde Valley, which at this point is ca. 6 km wide. The valley is a deep fluvial valley (to at least 50 m B.O.D.) infilled with glacial, marine and fluvial deposits. It is bounded to the north and south by drumlin fields (aligned approximately west to east) and beyond that by higher ground of subdued local relief rising to ca. 150 m A.O.D.

The monument was detected on aerial photographs taken by Dr. J. K. S. St. Joseph. No part of the structure was visible at ground level as it was covered by a shallow ploughed soil.

Excavation of the site took place in 1973 and 1974, under the direction of Mr. J. G. Scott of Kelvingrove Museum, Glasgow, in advance of factory construction. Mechanical stripping of the area revealed an oval enclosure elongated east-west and approximately 42 m by 36 m. There was a single entrance to the east (Fig. 6:2). A rectangular building, 5 m by 3 m, had been set across the entrance to the enclosure and traces of several circular structures, presumed to represent the remains of houses, were present as stains in the subsoil.

The ditch surrounding the enclosure was cut into the washed sand and gravel of the floodplain and the soil was used to construct a bank on its outer edge. The bank was later destroyed, presumably by natural erosion and repeated ploughing of the area. The instability of the sand and gravel presented problems for the modern excavators and also apparently, for the original builders. In antiquity various attempts had been made to stabilise the sides of the bank and ditch. Clay-lining, terracing and repeated recutting had all been employed and collectively they resulted in a ditch which varied in width from 4 to 6.5 m but which was at most only 2.5 to 3 m deep. As a consequence of these measures the stratigraphy of the ditch deposits is complex. Sections revealed many alternating layers of clay, silt, sand and gravel, each containing variable amounts of organic

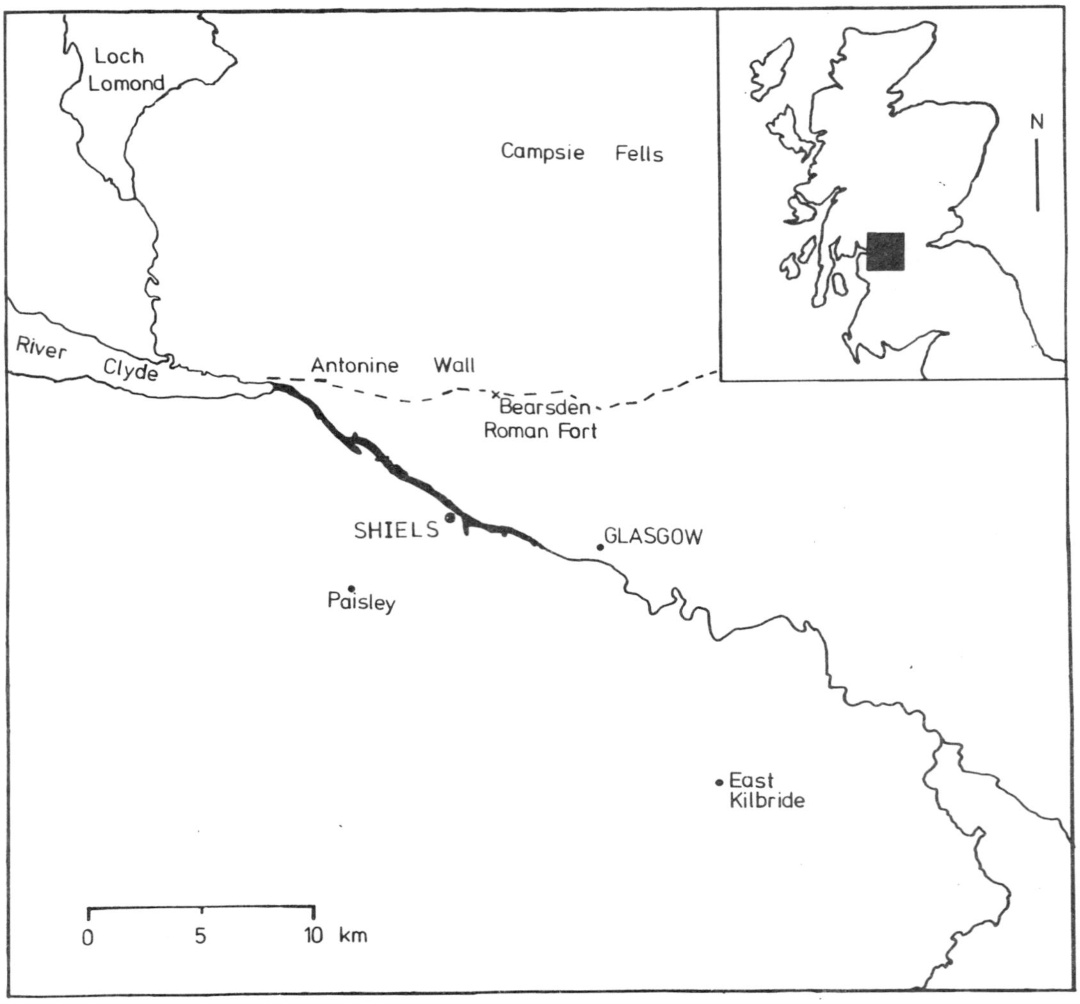

Fig. 6:1 (above) Location map for Shiels. Fig. 6:2 (below) Sketch plan of the ditched enclosure at Shiels showing the location of the section sampled for botanical analysis.

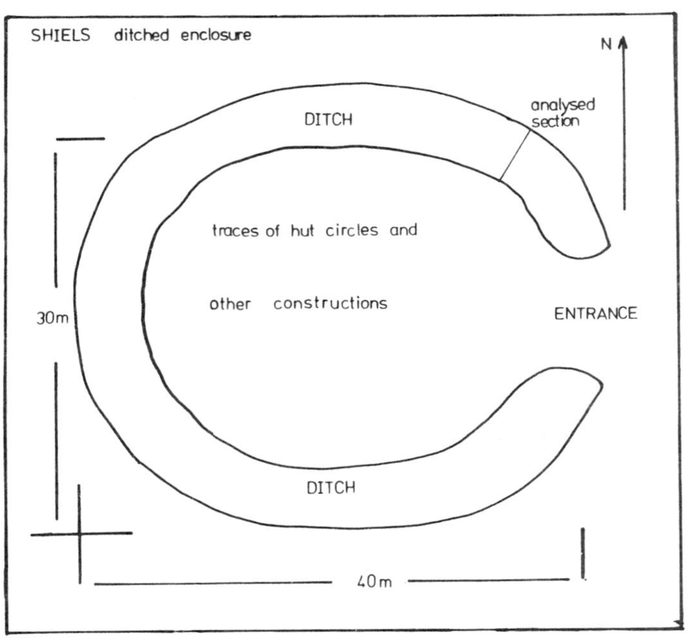

material. 'Vegetation, beetles, animal teeth and what appeared to be a mussel shell' were all noted from the excavation (J. G. Scott pers. comm.). Many wood chips, twigs and bark fragments were recovered along with several larger worked 'stakes'. Radiocarbon analysis of two of the 'stakes' produced dates of 1930 ± 140 bp (SRR-576) and 1640 ± 80 bp (SRR-577), suggesting the usage of the monument during the Roman Iron Age.

An excavation report is in preparation (Scott, forthcoming) and drawings and plans are not yet available for publication. Short summaries of the excavations are published in Discovery and Excavation in Scotland (Scott 1973, 1974).

SAMPLING AND ANALYSIS OF THE DEPOSITS

Sections were cut at several points through the ditch during the 1974 excavation. One section (Figs. 6:2 and 6:3) was sampled using metal monolith tins. The section at this point revealed a number of more or less organically-rich layers overlying the primary silt and sealed by several layers of sterile sand. The sediments sampled ranged from the primary silt at a depth of 170 cm (below the subsoil surface) to the lowest of the sterile sand layers at a depth of 90 cm. The sampled sediments, wood fragments and stakes were later submitted for botanical analysis.

In the analysis of the monolith samples 10 mm subsamples were removed where possible. Some layers within the section were narrower than this and smaller subsamples had to be used. Approximately 2 g of subsample was prepared for pollen analysis using standard techniques (Birks and Birks 1980). A pollen sum of 500 total land pollen was adopted and this was reached at most levels. The pollen data are expressed as a percentage of this sum. The remainder of the subsample (ideally 100 cm^3, but often less) was analysed for plant macrofossils. The material was broken down either in warm water or 1% NaOH solution before being wet sieved through coarse and fine sieves (1250 micron and 150 micron mesh respectively). The macrofossil data are calibrated to give values of macrofossils per 100 cm^3 of subsample.

The pollen data are presented in figure 6:4 and the macrofossil data in figure 6:5 and tables 6:1 and 6:2.

DATA AND INTERPRETATION

It was hoped by the excavator that the analysis of the ditch section would provide information about both the regional landscape and local events during the occupation of the enclosure, with particular interest being expressed in the economy of the inhabitants. To a great extent this has been possible despite the problems presented by the complex stratigraphy and the uncertainties concerning the origins and relative ages of the various layers and the time over which they accumulated. Much of the organic material in the sediments is obviously of local origin, having been washed in with soil from the ditch sides, the bank and the surrounding land surface. Other plant material would have been water- or air-borne. In addition, the presence of both carbonised and uncarbonised remains of heath and bog plants in many of the samples (Fig. 6:5) and the peat "pebbles" at 123 cm (Table 6:2), suggests that peat may have been brought to the site, perhaps for fuel. Peat falling into the ditch would be a rich source of non-local pollen and plant macrofossils.

The pollen input into a ditch is essentially of local origin with only a small contribution from the regional pollen rain. Ditch deposits are expected to yield only a limited number of pollen taxa. The deposits at Shiels were unusually rich, with 57 land pollen taxa being identified. It is possible that a combination of a species-rich local weed flora and an increased input of regional pollen due to the openness of the site were responsible for this. Despite the large number of taxa recorded and the complexity of the stratigraphy, the conditions reflected in the pollen diagram are remarkably uniform over the depth analysed. With regard to the effects of the human population two main features are apparent:-

1. The low levels of tree pollen. Apart from two isolated samples, tree pollen, mostly alder (Alnus), makes up less than 10% of the total. Deciduous forest trees, oak (Quercus), elm (Ulmus) and birch (Betula) are poorly represented.

2. The high incidence of pollen from open-habitat and agricultural weed species. Values of grass (Gramineae) pollen rarely fall below 50% of total pollen and the pollen of heather (Calluna vulgaris), rib-wort plantain (Plantago lanceolata), clover (Trifolium), buttercup (Ranunculus), docken and sorrel (Rumex), nettle (Urtica) and cereal type (Gramineae grains over 40 microns) is present in considerable amounts. There are also minor occurences of the pollen of other weed species such as spurrey (Spergula type) and mugwort (Artemisia). Weed species associated with both pastoral and arable agriculture are represented almost thoughout the diagram, with the rib-wort plantain curve being the most striking. There is some indication that pastoral agriculture, as represented by pollen of Plantago, Trifolium, Rumex and Ranunculus, is more prominent than arable agriculture in the lowermost and uppermost samples. Cereal type pollen and pollen of arable indicators such as Chenopodiaceae, Polygonum and Cruciferae tends not to be present in these samples.

Plant Macrofossil Data

Plant macrofossil remains preserved in a ditch provide a more local record of the vegetation than does the preserved pollen, however the deposits of Shiels are again unusually rich.

The overall picture revealed by the plant macrofossil analysis is similar to that revealed by the pollen analysis. In the category 'Plants Useful to Man', only a carbonised grain and seven carbonised rachis segments of barley (Hordeum) and a hazel (Corylus) nut fragment are present in a single subsample at 148 cm. Charcoal is, however, present throughout the section, indicating that human occupation of the site was probably continuous throughout the time that the deposits were accumulating.

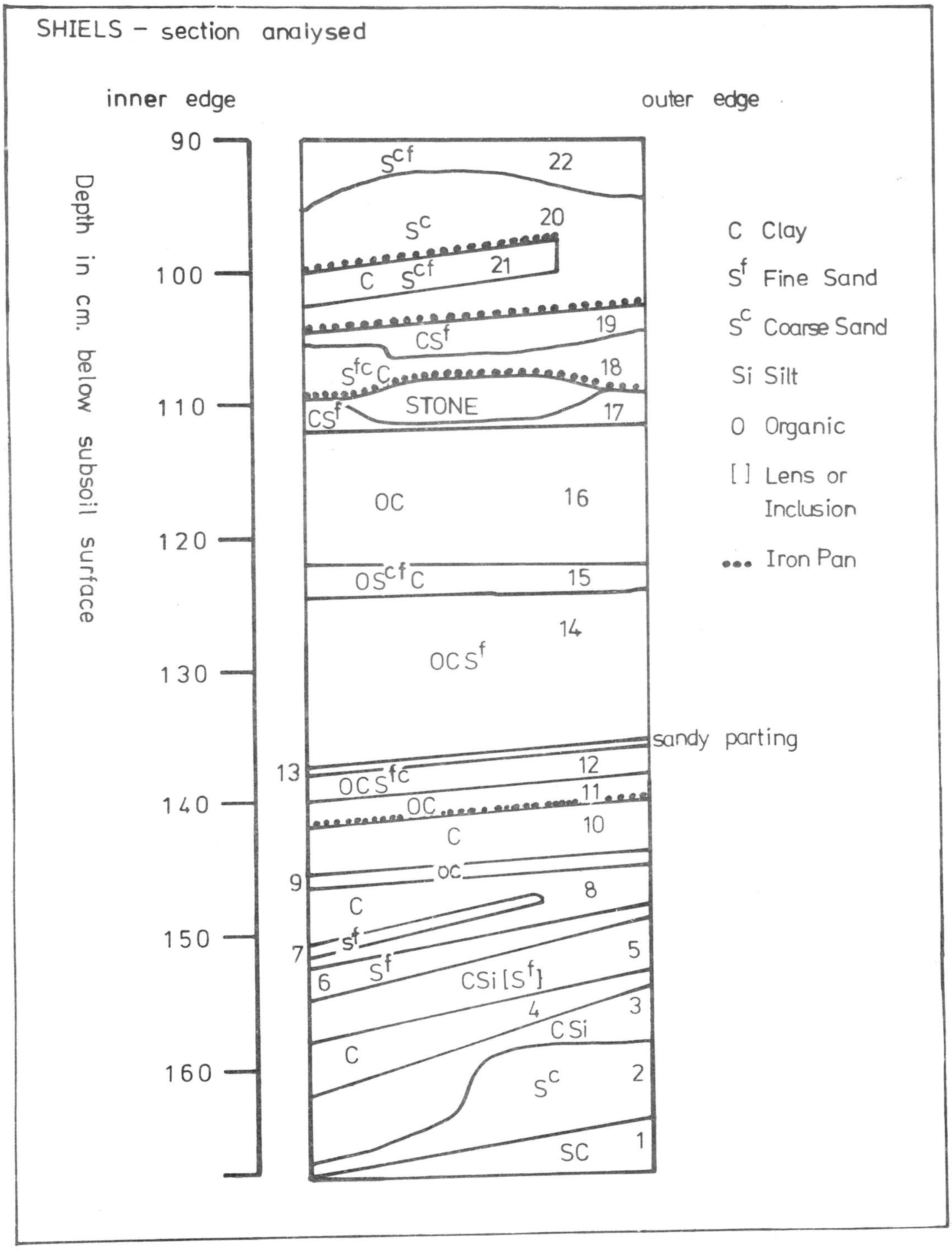

Fig. 6:3 Stratigraphy of the sediments sampled for botanical analysis.

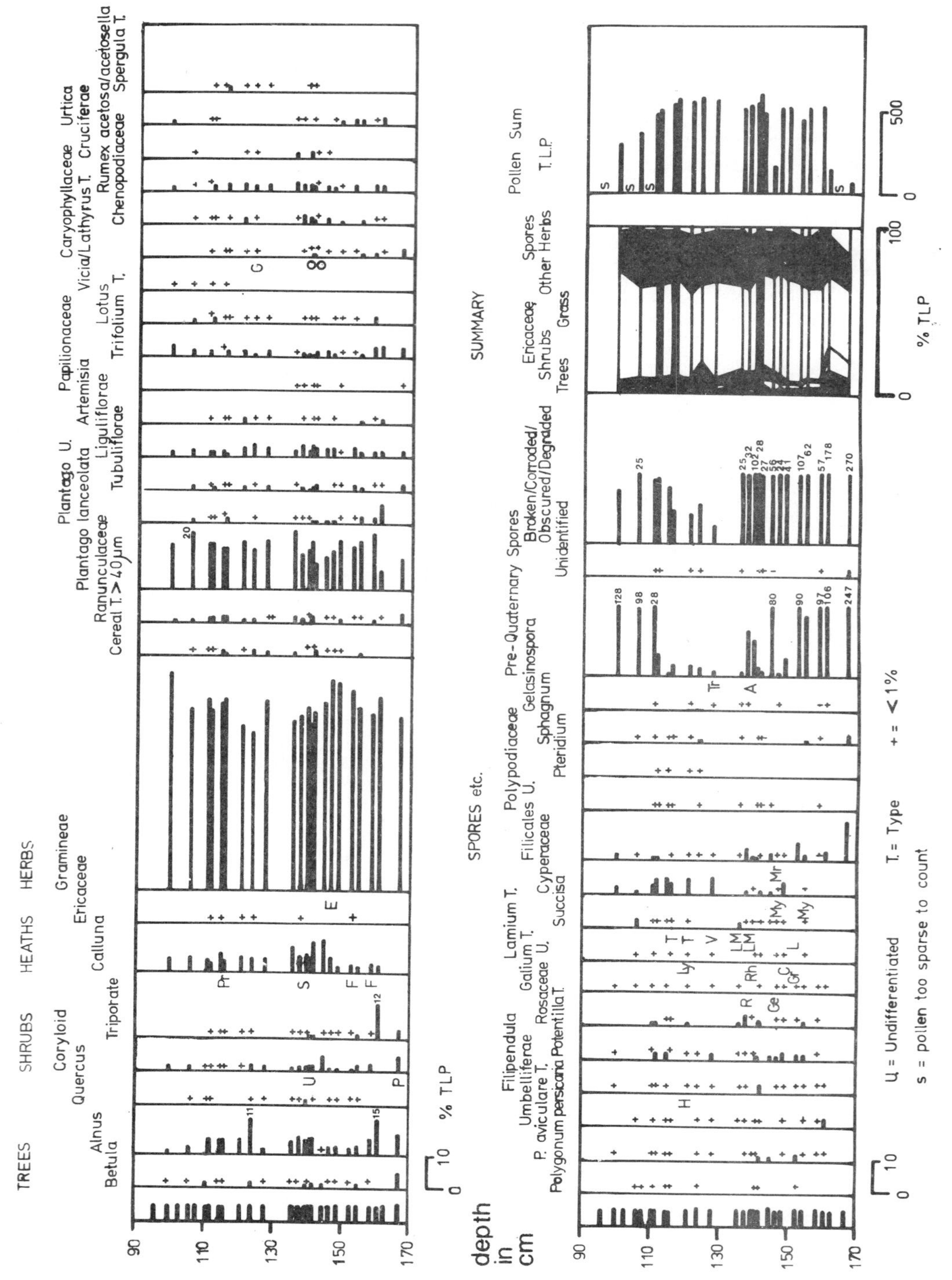

Fig. 6:4 Pollen data from Shiels, expressed as percentages of total land pollen. The presences of minor taxa are indicated within the bar diagrams by the following abbreviations: U = Ulmus, P = Pinus, Pr = Prunus T., F = c.f. Frangula, E = Empetrum, G = Genista T., O = Onobrychis T., R = Rubus T., Ge = Geum T., Ly = Lythrum, Rh = Rhinanthus T., T = Teucrium, L = Labiatae U., M = Mentha T., My = Myosotis T., Mr = Myriophyllum, La = Lycopodium annotinum, Lc = L. clavatum, Tr = Trichuris egg, A = Amphitrema (a rhizopod) (Pollen and spore nomenclature after Moore and Webb 1978).

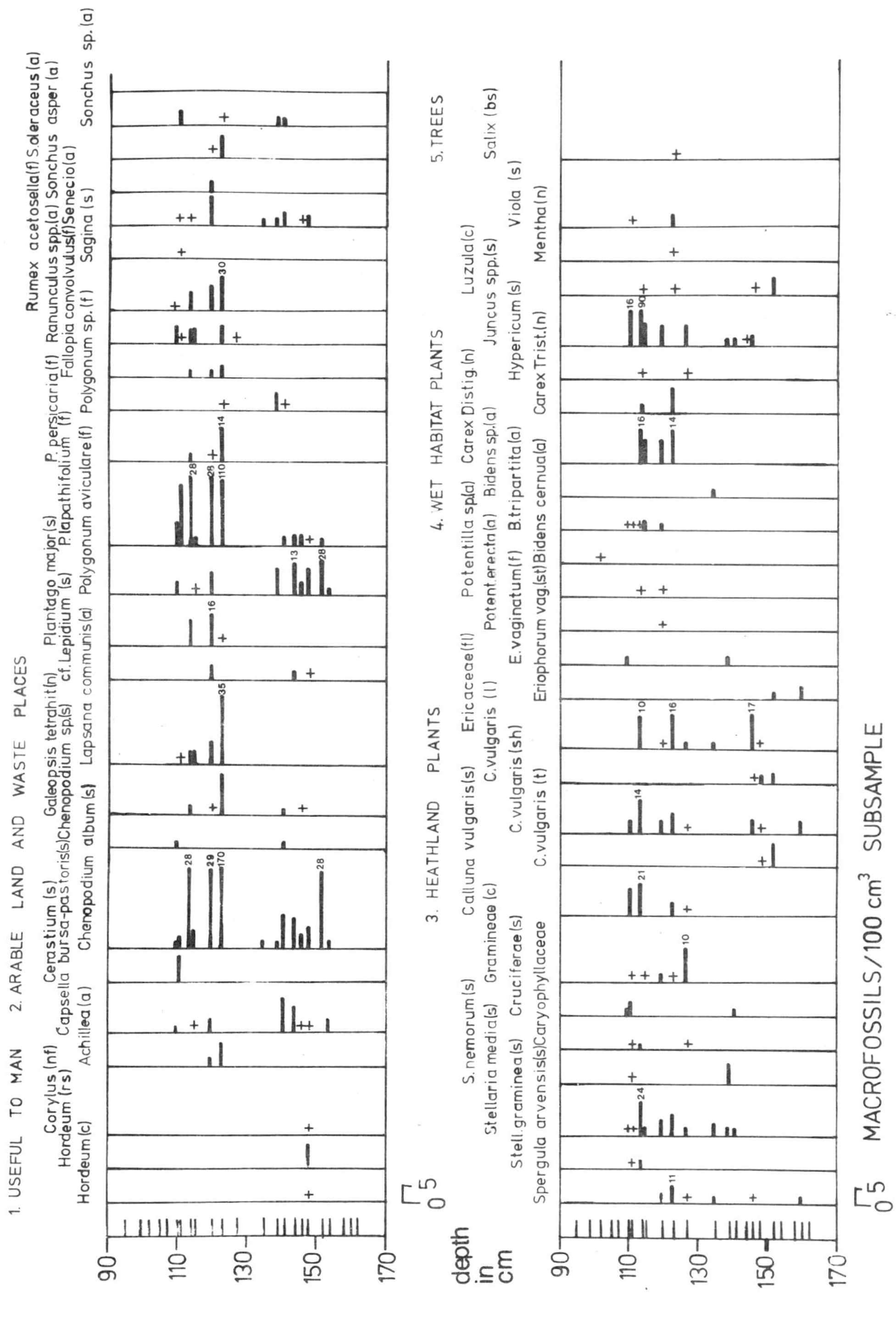

Fig. 6:5 Plant macrofossil data from Shiels, expressed as macrofossils/100 cm^3 of sediment. The following abbreviations have been employed: C = caryopsis, * = carbonized, rs = rachis segment, nf = nut fragment, a = achene, s = seed, n = nutlet, f = fruit, fl = flower, l = leaf, sh = shoot, t = twig, sl = leaf base spindle, bs = bud scale, Dist. = Distigmatae, Tris. = Tristigmatae (Nomenclature after Clapham, Tutin and Warburg 1981).

The fruits and seeds of weed species associated with arable and other disturbed ground such as *Polygonum lapathifolium*, *P. persicaria*, *Chenopodium album* and *Spergula arvensis* are common. Several weed species, including those mentioned above, are known to have been used as famine food. Most of the seeds and fruits from Shiels were recovered intact, however, and show few signs of having been ground or chewed.

A ditch such as this, surrounding a settlement site, might be expected to contain a certain amount of domestic debris and waste from the occupation, including sewage. With the exception of the charcoal, hazel and barley remains already discussed there are no other obvious examples of domestic waste. There is even less evidence for the presence of faecal material in the ditch. A single egg of the parasitic worm *Trichuris* was recorded from a pollen slide at a depth of 127 cm. It was not possible to determine whether the egg was from a species of *Trichuris* which infects humans.

Development of the Ditch Flora

The pollen and plant macrofossil diagrams both show an increase in the number of taxa present with time. This is probably accounted for by the progressive stabilisation of the ditch sides and surrounding soil and the development of a damp-ditch flora. It is interesting that of the taxa represented in the macrofossil diagram only *Viola*, *Hypericum*, *Juncus* and *Luzula* are not obviously represented in the pollen diagram. This is almost certainly due to the poor preservation of *Juncus* and *Luzula* pollen and the poor pollen dispersal of *Viola* and *Hypericum*.

IDENTIFICATION OF WOOD SAMPLES

The wood chips shavings, slivers and twigs recovered from the ditch during the 1974 excavation were of great interest. They were almost exclusively of hawthorn (*Crataegus*), the remainder being twigs of a species as yet unidentified. As the 'stakes' too were *Crataegus* (with the exception of one small worked piece of ash (*Fraxinus*) it would appear that the cut fragments found in the ditch resulted from the working of the 'stakes'.

It has proved difficult to relate the location of the wood specimens in the ditch to a specific level in the analysed section. The wood specimens are recorded as being recovered from the primary silt and clay layers of the ditch, however wood fragments recovered during the macrofossil analysis have not been positively identified as *Crataegus* below a depth of 141 cm. A concentration of *Crataegus* chips and shavings does, however, occur at 123 cm. *Crataegus* pollen was not identified at any level in the section but as identification is difficult, some grains may have been included in the Rosaceae Undifferentiated category.

There are a number of possible routes by which the *Crataegus* remains may have entered the ditch. The most obvious and most likely is that the wood was being used to produce stakes or stake-like artefacts in the absence of any other suitable wood in the vicinity. The possibilities that the wood was being used as firewood or came

Table 6:1

Bryophytes recovered from the ditch at Shiels (+ indicates presence)

Depth in cm.	162	160	158	154	152	148	146	144	141	139	135	123	120	115	114	111	110	107	105	102	99	95	90
Calliergon cuspidatum															+								
Ceratodon purpureus										+							+	+					
Eurhynchium sp.							+	+	+	+	+		+	+		+		+					
Hylocomium splendens					+		+	+	+	+	+	+	+	+	+	+	+	+					
Hypnum cupressiforme sl.							+		+			+		+		+	+						
Pleurozium schreberi												+		+			+						
Rhytidiadelphus sp.		+						+															
R. squarrosus					+					+	+	+	+	+				+					
Sphagnum sp.		+			+	+	+	+	+		+		+	+	+								

Table 6:2

Miscellaneous Macrofossils Recovered from the Ditch at Shiels

(a = abundant, + = present, numbers refer to number of macrofossils per 100 cm^3 of sample)

Depth in cm.	162	160	158	154	152	148	146	144	141	139	135	127	123	120	115	114	111	110	107	105	102	99	95	90
Charcoal	a	+			a	a		+	+		+	+	a	+	+	+							+	+
Wood	+	+							+	a			a	a	+	+								
Crataegus wood									+	+			+	+										
Coal and Shale	a	a		+	+	a		a	a	a	+		a	+	a	+	+	a	a	a	a	a	a	a
'Cinders'	+	+		+	+	+																		
Prequaternary megaspores		a	+		a	a		a	+					+	+	+	+	a	a	a	a	a	a	+
Daphnia egg cases		6		44	25	16	53	20		x	4	x	40	26	37	x	x	12 (x = more than 100)						
Insect remains		+		+	+	+	a	a	a	a	+	a		a		a	a	a		+				
Small mammal bone													1											
Peat 'pebbles'													a											

from a nearby hedge or thicket have also to be considered. The fact that the majority of the wood is worked and that no *Crataegus* charcoal was identified from the ditch argues against the former being the case. The apparent absence of *Crataegus* pollen and fruit stones tends to suggest that the latter situation is also unlikely. It may be that *Crataegus* branches were brought to the site as sources of stakes and brushwood, the two being combined to stabilise the ditch and bank sides and increase the defences of the enclosure. Whatever the route, it does appear that a human agency was involved and that the *Crataegus* did not arrive by natural means, for example as driftwood during a flood.

Conclusions

The pollen and plant macrofossil data show that the enclosure at Shiels was built and occupied in an open landscape with well established pastoral and arable agriculture. A similar open landscape of grassy heath and few trees is recorded for this period from Linwood Moss, 6 km to the west (Boyd 1982). It is obvious that this whole area south of the Clyde had been heavily used by human populations prior to the occupation of the site at Shiels. The fertility of the soil may have been maintained by regular flooding of the River Clyde, making the area a very attractive one for settlement. The Clyde does not flood the area at present, but the water level in the Clyde Estuary was higher during the Iron Age, possibly by as much as 2 m (W.E. Boyd pers. comm.). Occupation may, therefore, have been on a seasonal basis, the enclosure being abandoned during the winter due to flooding and/or waterlogging. Repairs and reoccupation would take place in the spring.

It has been impossible to estimate the time taken for the deposits to accumulate in the ditch due to the absence of suitable radiocarbon dates and information regarding the rate of accumulation. The presence of several well-developed iron pans suggests that sediment accumulation was not rapid. However, since clay-rich layers, responsible for impeding the movement of iron in solution, are not continuous across the ditch, the pans may have developed contemporaneously rather than successively. Therefore they cannot be taken to infer an extended period of sediment accumulation. The uniformity of the pollen and plant macrofossil contents of the ditch, despite the heterogeneous nature of the sediments, suggests that tens or hundreds, rather than thousands, of years may have been involved.

Acknowledgements

This work was supported by the Scottish Development Department (Ancient Monuments), Edinburgh.

Thanks are also due to Mrs. K. McKechnie who drew figures 6:1 and 6:5.

Bibliography

Birks, H.J.B. and Birks, H.H. (1980) Quaternary Palaeoecology. London: Edward Arnold

Boyd, W.E. (1982) The stratigraphy and chronology of Late Quaternary raised coastal deposits in Renfrewshire and Ayrshire, Western Scotland. Unpublished Ph.D. Thesis, University of Glasgow

Clapham, A.R., Tutin, T.G. and Warburg, E.F. (1981) Excursion Flora of the British Isles (3rd Edition). Cambridge: University Press

Moore, P.D. and Webb, D.A. (1978) An illstrated guide to pollen analysis. London: Hodder and Stoughton

Scott, J.G. (1973) Shiels. Discovery and Excavation in Scotland (1973) 66-67

Scott, J.G. (1974) Shiels. Discovery and Excavation in Scotland (1974) 82-83

Scott, J.G. (forthcoming) Report on the excavation at Shiels, Govan, Glasgow

7. THE ASSENDELVER POLDERS PROJECT:

INTEGRATED ECOLOGICAL RESEARCH

W. Groenman-van Waateringe and J. P. Pals

GENERAL ENVIRONMENTAL BACKGROUND (W.G.v.W.)

Introduction

The Assendelver Polders are situated about 25 km north west of Amsterdam (fig. 7:1) and about 20 km east of the present coastline of the North Sea. Amateur archaeologists have surveyed more than 75 settlement sites here (fig. 7:2). The occupation can be traced from the Early Iron Age (*circa* 6th century bc) until the Roman period and again in the medieval period. This paper deals with the Iron Age and Roman periods. Roman imports from the reign of Tiberius and Claudius-Nero point to contacts with the "Romans" (Brandt, in press), either with the advanced military outpost in Velsen (Morel & de Weerd 1980) about 2 km south-west of the Assendelver Polders, or with the Roman army stationed along the Rhine Limes.

The purpose of the Assendelver Polders Project was to study the spatial organisation and development of the settlement system from an economic, social and technological viewpoint (Brandt, Groenman-van Waateringe & van der Leeuw in preparation). Four levels of research have to be discerned:

1. the supra-regional level
2. the regional level
3. the community level
4. the household level.

This has all to be seen in the context both of environmental circumstances, which in the coastal area of the western part of the Netherlands are mainly determined by the changes in the relative height of the sea level, and of the possible influence of the more complex "Roman" society and organisation.

The excavations took place mainly in the years 1979-1981.

To understand the settlement history of an area which is as relatively young in geomorphological terms, a detailed knowledge of its genesis is absolutely necessary, because there is a direct interaction between this and habitability. Detailed pedological, palaeoethnobotanical and zoo-archaeological research is needed to determine, both qualitively and quantatively, the economic pattern

Fig. 7:1 (above) Location of the Assendelver Polders. Drawing IPP Fig. 7.2 (opposite) Location of the archaeologication sites in the Assendelver Polders. Drawing IPP.

Legend:

1. Sites surveyed
2. Sites excavated
3. Creek and natural levee deposits
4. Lagoonal and back-swamp deposits
5. Back-swamp deposits
6. Fen peat
7. Oligotrophic peat
8. Fen peat on top of DK I clayey deposits
9. Maximal extension of oligotrophic peat between the DK 0 and the DK I transgression phases.

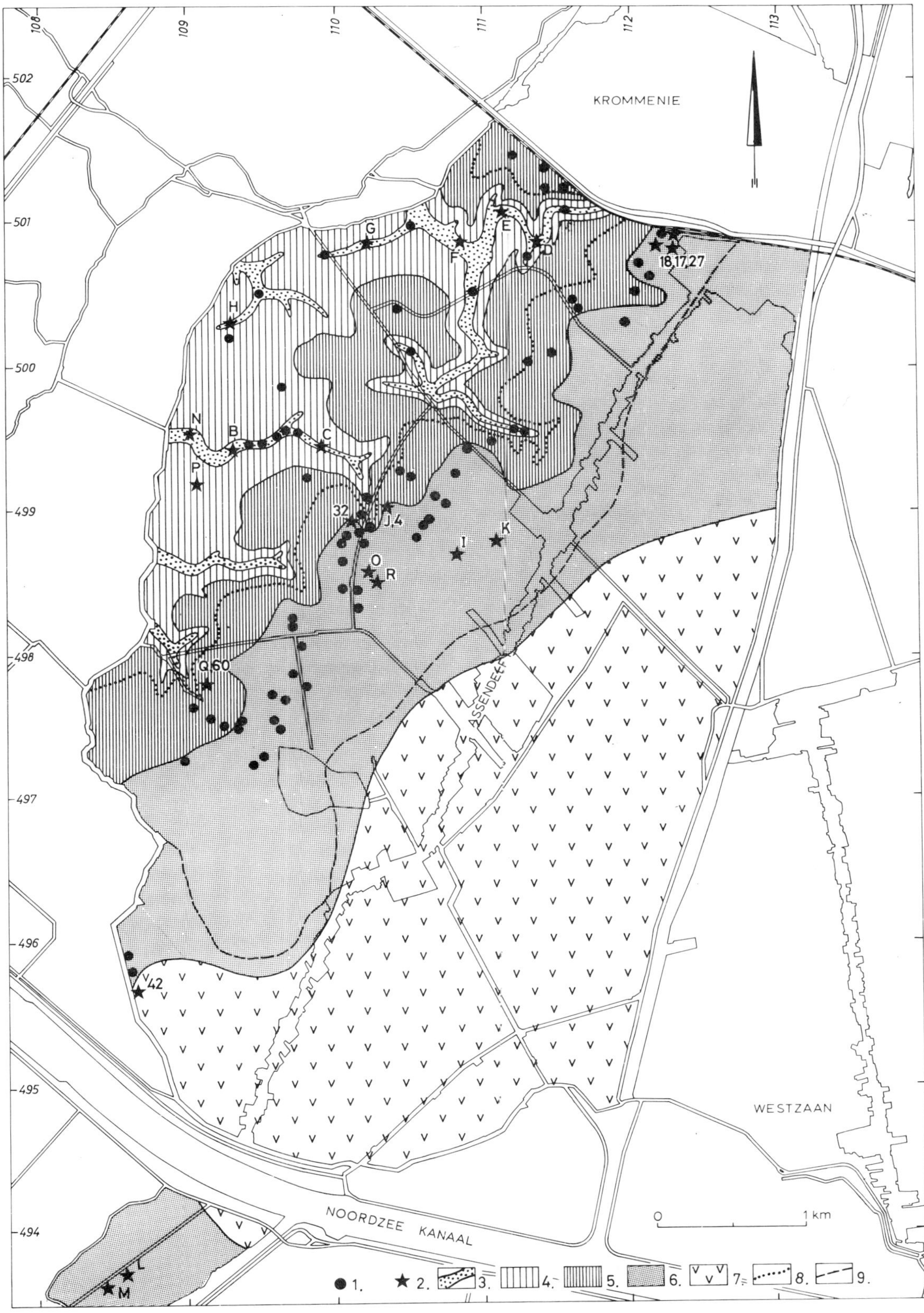

KROMMENIE
WESTZAAN
NOORDZEE KANAAL
ASSENDELFT
108
109
110
111
112
113
502
501
500
499
498
497
496
495
494
G
H
E
F
D
18,17,27
N
B
C
P
32
J,4
I
K
O
R
Q,60
42
L
M
0
1 km
1.
2.
3.
4.
5.
6.
7.
8.
9.

at every level of organisation and its variation in the time and space.

The (supra-) regional development of the landscape

The settlement history of the area of the Assendelver Polders is entirely determined by its positioning at the edge of a great estuary- the Oer-IJ. The changing picture of this estuary and the attendant possibilities for habitation will be briefly outlined here (Zagwijn 1971).

About 5000 bp. the hitherto strong rise in sea level, since the last glacial period, began to decrease. More or less parallel to the present coastline a series of low aeolian ridges, partly overlying coastal barriers were formed in two major phases, the eastern one between 4800-4100 bp, the western one between 4200 and 3500 bp (van Straaten 1965). Behind these coastal barriers a tidal flat area came into existence, its clayey and sandy sediments now lying circa 4-5 m below OD. A coastal inlet formed to the south-west of the Assendelver Polders.

In the next phase, from about 3800 bp, this coastal inlet moved more to the north-west, caused by south to north transport of material along the coast and the accretion of the coastal barriers at their northern end. The inlet was now much narrower. Several rivers discharged into the estuary. The land inbetween (i.e. the former beach plains) and behind the coastal barriers was now better protected against the sea, and a lagoon was formed supplied with fresh water from the rivers. The growth of fen peat began along the estuary and the Oer-IJ was still of a brackish character. Ombrogenous peat was formed as islets surrounded by belts of fen peat, representing the former drainage channels.

During the Dunkerque (DK) I transgression phase (circa 2500-2000 bp) the influence of the sea on the area increased. In the mouth of the estuary a pattern of sand bars and gulleys (ebb and flood channel) developed. Erosion cleared away the western edge of the peat, causing the formation of a lagoon. The situation in the remaining peat area changed drastically. Blocked drainage channels were re-opened and the excess amount of water was then able to discharge into the lagoon. The peat consequently dried out and became inhabitable. Sedimentation only took place in the north-west part of the Polders. Here clay was deposited in an aquatic milieu and via a tidal flat phase an area with salt-marsh character developed, intersected by creeks with sandy levees, in contrast to the finer, clayey material deposited in the back-swamp areas.

About 2000 bp the mouth of the estuary in the North had almost entirely silted up. There remained only a few rivulets for the drainage of the hinterland and along which transport to sea was possible. The southern part became a great fresh water lake - the IJ. The peat was subsided as a result of drying out. There was also subsidence in the clayey back-swamps more than in the silted up creek beds in their sandy levees , as a result of drying out in the salt-marsh area. This meant that the drainage of the hinterland could no longer take place via the creeks, but that run-off was instead discharged into the back-swamp area. This became increasingly wet leading to peat formation. The peat growth must have reached in the end at least the height of the levees, because the thickness of the overlying DK III sediments is nearly everywhere the same.

Possible sites for habitation

(a) In the area of the aeolian ridges, the so-called Older Dunes, from the Late Neolithic onwards. This type of landscape is not represented in the Assendelver Polders but its presence in the near neighbourhood was essential for the people living in the Polders as a source for many products not growing there (all kinds of wood, wild edible fruits) and as a dry north-south connection along the coast.

(b) In the peat area, when through natural (and/or man made) drainage the peat became dry and solid enough to build houses on. As far as we now know the oldest habitation in the Polders dates from the 6th century bc, when because of the DK I transgression the peat area was better drained. In the neighbourhood of the southern part of the north-west clay deck habitation was established in the 3rd century bc. This was made possible by drainage through peat channels connecting with creeks in the salt-marsh area that had not as yet silted up. Apparently as late as the 1st century AD year-round habitation was possible, the size of the farmsteads (circa 6 x 24 m) not pointing to a mere seasonal occupation (Groenman-van Waateringe 1961). Habitation had been established on top of a section consisting entirely of reed peat, with a reed-sedge peat overlying in turn a thin layer of reed peat, a meso-oligotrophic peat and an oligotrophic peat with a lot of Ericaceae pollen. At several sites *Myrica gale* strongly expanded as a result of the drying-out of the peat.

(c) In the clay area in the north-west part of the Assendelver Bolders on top of the natural levees of the salt-marsh creeks. The creeks initially provided good drainage. However, progressive silting-up of the main creeks and stronger subsidence of the clayey back-swamps relative to the sandy creek beds and levees led to further problems of water discharge from the hinterland. This led to the eventual drowning of the back-swamps and subsequently to renewed peat growth. The peat area then became too wet and hence uninhabitable, as did the clay area. Towards the end of the 2nd century A.D. even the artificial heightening of the settlement sites on the levees failed to remedy the situation. Moreover the entire subsistence strategy of the people was endangered when fields and pastures became simply too wet.

Stratigraphy and peat development in the Assendelver Polders

The basis of the so-called Holland peat are the Calais IV deposits at *circa* 4 m below OD. Depending on the geographical location there is either continuous peat growth with possible peaks in the Chenopodiaceae in the pollen diagram as indication for a halophilous vegetation, correlating with clay deposits elsewhere, or there is a discontinous peat growth, interrupted by clay sedimentation. It is, however, rather difficult to correlate the different phenomena when not enough radiocarbon datings are available. Peaks in the Chenopodiaceae curve can occur because of brackish conditions, but also as a phenomenon accompanying human occupation. Younger clay bands can be intercalated in older peat beds, and a single transgression phase clay sedimentation may alternate with peat growth. It is necessary to study the genesis of the Polders in great detail to be able to understand the entire settlement history in space and time. Because of the difference in drainage in the peat area, in time as in space, generalisations are not possible and each site must be studied. As we have seen habitation on the peat is not so much dependant on the stage reached in the peat development as on the drainage possibilities. In any case the top of the peat may be eroded or removed as a result of human interference (e.g. deliberate burning for cultivation purposes).

The peat development in the Polders, if not interrupted by man or nature is as follows. On the clayey subsoil a reed or reed-sedge peat develops, turning via a reed-fern peat to a _Sphagnum_-reed peat and then where the circumstances were favourable for oligotrophic peat growth to a true _Sphagnum_ bog with Ericaceae. Elsewhere the succession is arrested at an intermediate stage. In the back-swamps there is no peat available for analysis, partly because of erosion, but mainly because of the heavy compaction by the Dk III sediments, when only a thin black layer without recognisable plant remains persists. As we nevertheless wanted information from the immediate environment of the settlement on the levees in the clay area, samples from the settlements itself, in creeks, drainage ditches around fields, house mounds (_terpen_) and houses, and (refuse) pits were analysed.

All these samples are far from ideal for pollen analysis since even the best are composed of markedly humic clay which is not the ideal medium for pollen preservation. Moreover the pollen contained in the sandy-clayey samples may have arrived from various sources, which comprise the following categories:-

(a) Pollen already present in the sediment at the time of deposition. It is our experience that the amount of this can be considerable. Of this pollen of _Abies_ and _Picea_ can easily be recognised as such because these genera are not native to our region. However, it is more difficult to define the extent of the remaining secondary pollen.

(b) Pollen from the vegetation present in the region at the time of sedimentation. This is the pollen produced in the Holland peat as described above.

(c) Long-distance pollen blown in by the wind both at the time of sedimentation and later with the filling-up of the creeks etc. Palynological analysis of the peat deposits in the coastal belt indicates that the potential habitats for trees such as _Quercus_, _Tilia_, _Ulmus_, _Fraxinus_, _Fagus_ and _Carpinus_ should be sought in the area of the Older Dunes. From the salt-marshes west and north of the Polders pollen may have blown in and the same holds true for the eastern peat area.

(d) Pollen of the local vegetation in and around the settlement. It is this pollen that is of the greatest interest in relation to the economy of the settlements excavated. It is therefore especially important to determine the component of this pollen present in the analyses and to isolate it from the other three sources.

Sources a and b can be eliminated by analysing the subsoil. Source c cannot be isolated, but will only differ from the pollen included in the subsoil in so far as the vegetation in the area of origin has changed throughout time. Sources a-c can be recognised on the basis of six criteria. These are:

1. the presence of <u>Abies</u> and <u>Picea</u> with occasionally relatively higher values for <u>Pinus</u>,

2. the presence of, or rather higher values for, <u>Pediastrum</u>,

3. the occurrence of spore types 114-116, restricted to clay laid down under marine conditions (see Pals, van Geel & Delfos 1980: 403),

4. the occurrence of Hystrichosphaeridae,

5. percentages of arboreal pollen higher than 50%

6. regular appearance, if in low quantities, of spore type 127, which seems to be indicative of a dry, more acidic environment (Pals, van Geel & Delfos 1980: 401, 408).

Applying these criteria to the analyses (cf. fig. 7:3) allows us to conclude that

1. the local vegetation was treeless but for some willow and alder (straight curves for all other trees),

2. fresh water must have stood in some of the ditches during the period of habitation (presence of spore type 128, indicative for shallow eutrophic fresh water (Pals, van Geel & Delfos 1980: 407)),

3. arable and pastoral farming took place near the settlement (presence of Cerealia and pasture plants),

4. there was a considerable washing in from the sides and erosion near the bottom of the ditches, and several pits were deliberately backfilled (presence of characteristic of subsoil material),

5. in the upper part of the ditches the pollen represents the local vegetation (absence of subsoil characteristics),

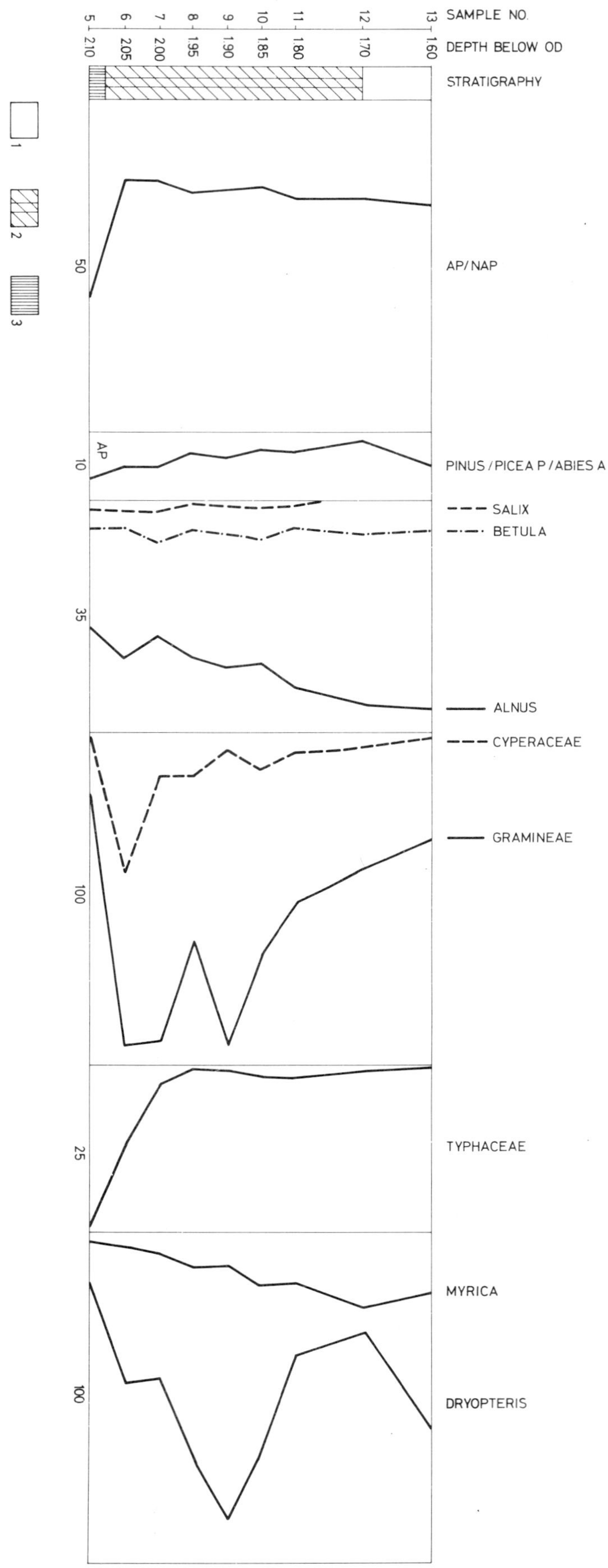
SAMPLE NO.
5 6 7 8 9 10 11 12 13
DEPTH BELOW OD
210 205 200 195 190 185 180 170 160
STRATIGRAPHY
1
2
3
AP/NAP
50
AP
10
PINUS / PICEA P / ABIES A
SALIX
BETULA
35
ALNUS
CYPERACEAE
GRAMINEAE
100
TYPHACEAE
25
MYRICA
DRYOPTERIS
100

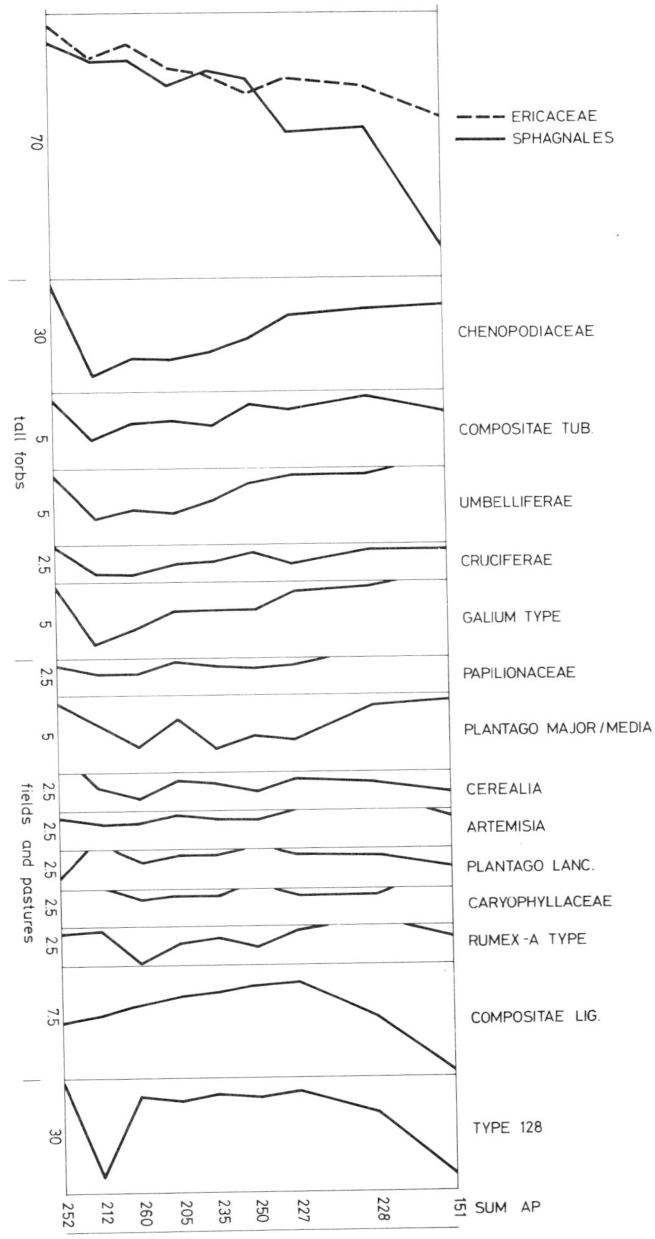

Fig. 7:3 Pollen analysis of a section at site B, selection of curves. Drawing IPP.

Legend

1. DK III clay
2. Peaty clay
3. Sandy subsoil

6. the succession in the back-swamp area is comparable with those in the peat area, i.e. the peat growth started with a reed-sedge peat and developed via a reed-fern peat into a *Sphagnum*-reed bog. This succession is not necessarily contemporaneous across the area; its development is entirely dependent on local conditions,

7. rather early in the whole development tall forbs appear. This may reflect a natural phenomenon of washing up of organic material, but could also be caused by human interference with the natural vegetation, e.g. reed cutting, accumulation of waste from man and his animals.

Pollen analytical evidence for cereal cultivation

Several sites, especially in the clay area, have given high percentages for cereal pollen, most probably pointing to processing of the cereals quite near the sampling site (Vuorela 1973). Because of the high number of cereal type pollen per site we thought it worthwhile to present the data concerning length of pollen, annulus measurements etc. in separate diagrams for the different sites (Groenman-van Waateringe & Troostheide in prep.) (fig. 7:4, from a section through a channel and a *terp*-ditch cut into it).

The point of segregation between wild grass pollen and Cerealia was established as 40 microns. In addition, the Gramineae pollen larger than 40 microns was, where feasible, examined under phase contrast for grouping under *Triticum* type, *Hordeum* type, *Avena* type, *Secale* type and definitely wild grass type. The guiding principles for identification were the criteria established by Faegri & Iversen (1964: 195-197), Beug (1961: 21-42, and especially 31 for the criteria which must be met by Cerealia pollen), Grohne (1954), and also Andersen (1979). In a district such as the Assendelver Polders, with periodic marine influence it is still, however, possible that wild grass pollen may have been included under the cereal type pollen.

In most of the samples, if cereal type pollen did occur, *Triticum* type pollen was dominant, followed by *Hordeum* type, while *Avena* type was only found once. *Secale* type is missing altogether. The dominance of *Triticum* type means that there is a serious discrepancy between the data provided by the seed and fruit analysis and the pollen data. Thus we must consider the possibility that all (or part) of our as *Triticum* type identified pollen belongs to (one of) the wild grasses, mentioned by Beug (1961: 40) under *Triticum* type that is, certain species of *Bromus*, *Hordeum* and *Agropyron*, and some pollen of *Elymus arenarius*, although the pollen of these wild grasses belongs only in part to this type. Comparing our measurements for the *Triticum* type pollen with those given by Beug (1961: 31-33 and fig. 10) is is clear that at least part of our *Triticum* type pollen can be nothing other than *Triticum*, especially in view of the length of the pollen grains, but also of the pore diameters and widths of the annulus. For example the range of pore diameters in fig. 7:4 lies between 3-7 microns and peaks at 4 and 5 microns. Such high values in combination with equally high values for the width of the annulus and pollen length are only to be found

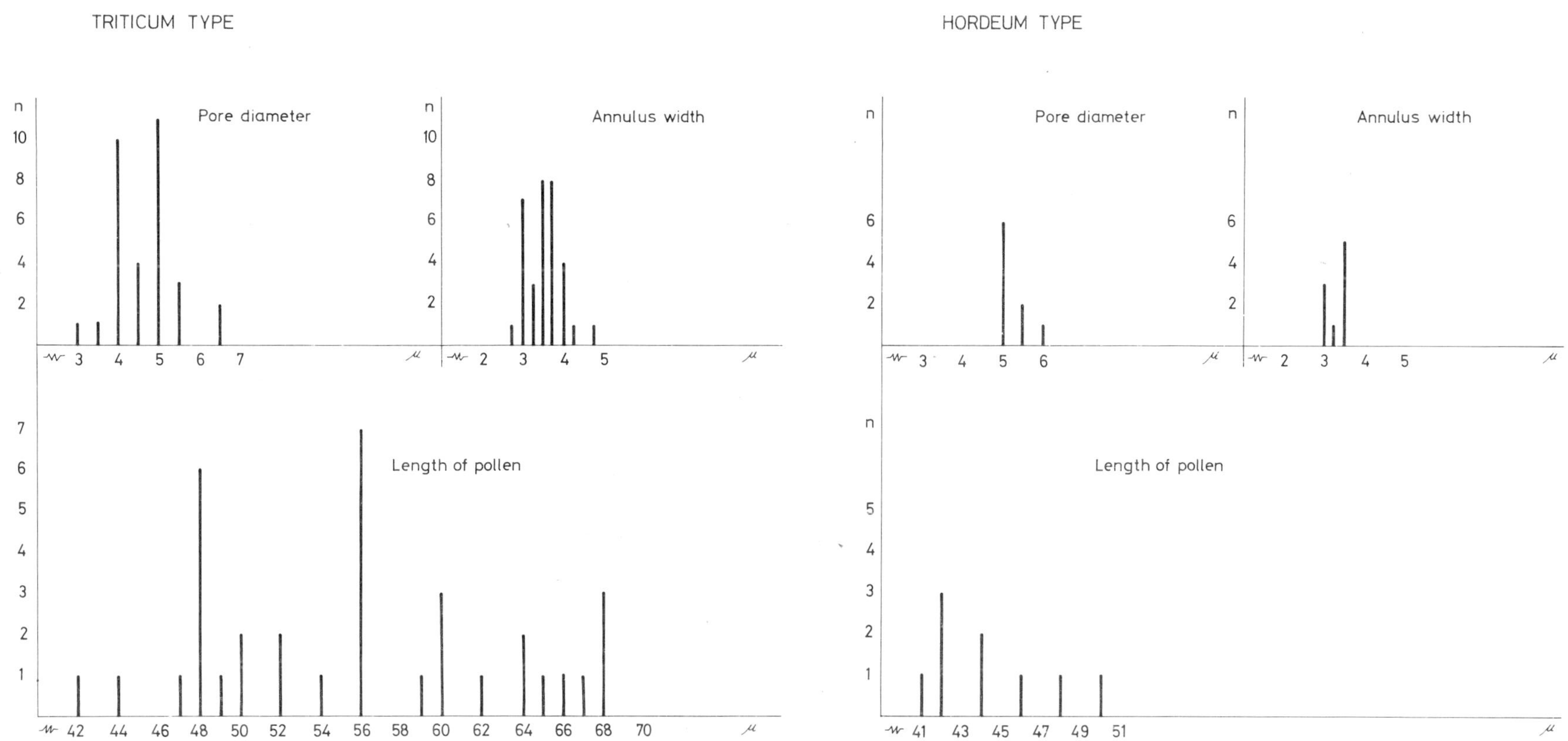

Fig. 7:4 Measurements of <u>Triticum</u> type and <u>Hordeum</u> type pollen in a section at site C. Drawing IPP.

within the Triticum type in Triticum species, i.e. Triticum compactum Host, T. aestivum L. and T. spelta L. These high values cannot be the result of excessive swelling of the pollen because of the sizes for Corylus pollen in our samples were between 25 and 30 microns and the measurements for the Hordeum type pollen fall neatly in the ranges given by Beug, whereas the same mounting medium was used.

All the diagrams give more or less the same picture: the measurements for Hordeum type restricted to a rather narrow range, those for Triticum type showing a much larger variation. Is it possible that different Triticum species are represented, e.g. Triticum dicoccum Schubl., T. compactum Host. and T. spelta L.?

Diatom analysis

Two methods have so far been used: a rough inventory of several sections of the Polders conducted by Vos (1982) and a quantitative analysis of a tidal channel/creek near one of the larger sites in the clayey area (site F) conducted by Fontaine (1981). Only clayey sediments could be used, peat being very poor in silicon and an unfavourable medium for diatoms. Besides the analysis executed for gaining more insight into the genesis of the landscape the diatom content of potsherds were analysed, to determine if the ceramic was made from locally available clay or not.

The rough inventory

Vos distinguished among the diatoms four main groups

(a) Coastal allochthones. A group of marine to marine - brackish species, always found in coastal sediments, but without a sufficiently clear indication of a marine milieu to be regarded as allochthonous.

(b) A highly varied diatom group, characteristic of a meso- to oligihalobous lagoon with waterplants.

(c) Scoliopleura tumida flora, typical of tidal flats.

(d) Diploneis interrupta flora, occurring in humic reed-clays with lots of organic material, e.g. in salt-marshes.

In the area with DK I sediments Vos found indications of the following succession: a Scoliopleura tumida flora, followed by a varied diatom flora capped by a Diploneis interrupta flora, thus pointing to a tidal flat phase (the first deposition of marine sediment on or in the place of the Holland peat), followed by a lagoonal phase (drowning of the tidal flats because of compaction of the underlying sediments) and then gradual silting-up to reed marsh level.

The quantative analysis

The analysis by Fontaine of the creek filling at site F (fig. 7:5) is difficult to interpret. One of the main problems with diatom

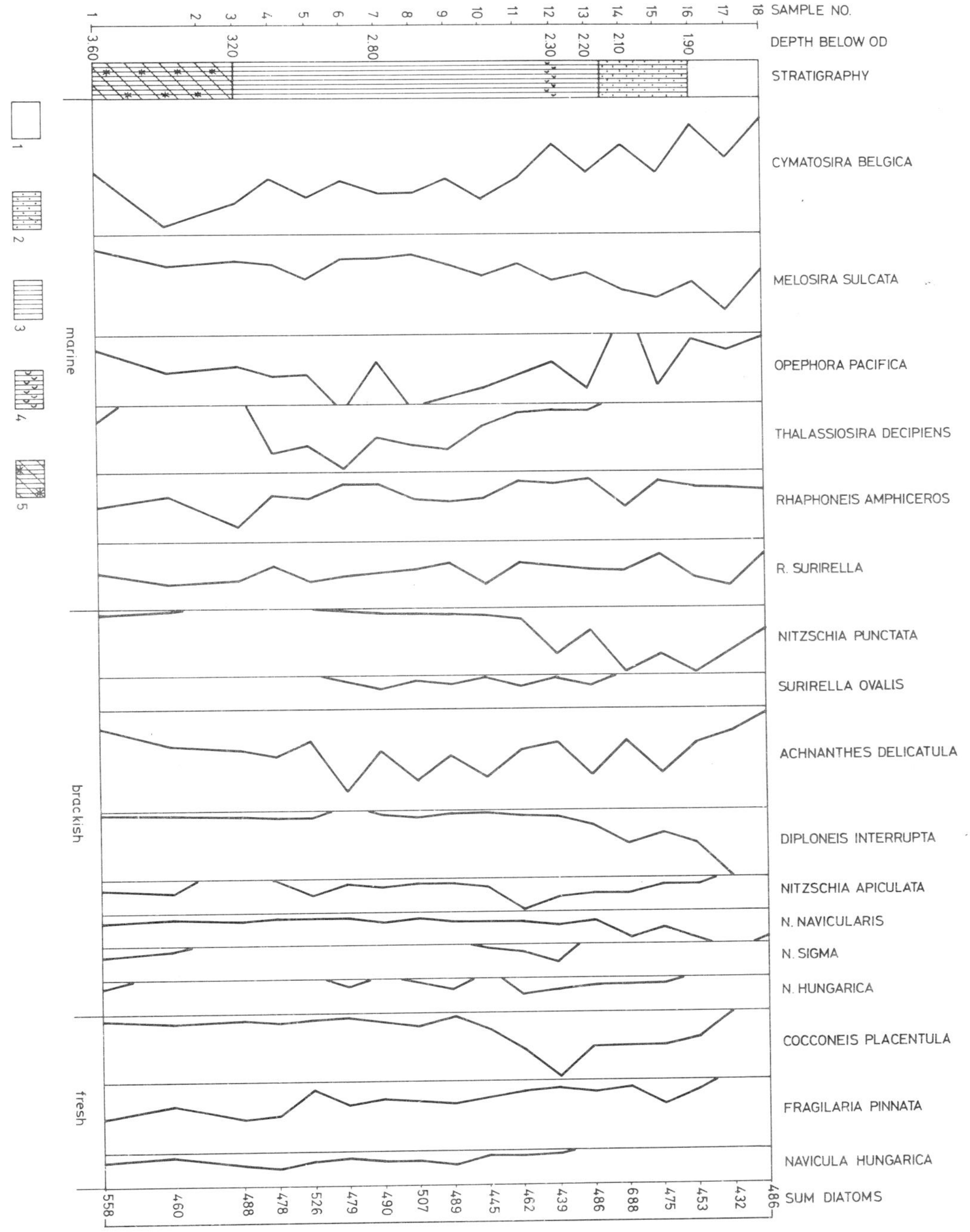

Fig. 7:5 Diatom analysis of a section through a tidal channel at site F, selection of curves. Drawing IPP.

Legend

1. DK III clay
2. Brown-grey clay
3. Grey clay
4. Marine molluscs
5. Subsoil, grey clay with reed and wood remains.

analysis is that there is always a high number of secondary diatoms, i.e. diatoms that did not live where they were found, but were deposited there after their death. One of the criteria for autochthonous versus allochthonous is unbroken against broken, but this is highly dependent on the size and shape of the diatoms, a small round species having much better chance of staying undamaged during transport than a long thin one.

The same section has been analysed for pollen (fig. 7:6) and the combination of both data may give us an answer to our problem. The stratigraphy of the section from top to bottom is

samples 1- 2: grey clay with reed and wood remains, subsoil material
samples 3-13: light grey clay with shells, homogenous finely levigated material, distinct layer of marine shells in living position at 2.30 m minus OD (sample 12)
samples 14-16: brown-grey sandy clay
samples 17-18: DK III clay (not analysed for pollen).

In both diagrams species (genera) characteristic of marine and brackish conditions on the one hand and fresh water conditions on the other hand are found. It is clear that both environments contributed to the results, whereby the marine sediments brought in all the pollen and diatoms characteristic of this kind of sediment and the fresh water influx from the hinterland provided the fresh water species.

In the pollen diagram the differences between the subsoil samples and samples 3-9 are not very great, amounting to slightly higher AP percentages and minor differences in the other curves. Around sample 9 there are fluctuations in most curves. The Chenopodiaceae peak in sample 12 ties in with the level of marine shells in the section. Between the subsoil and the samples 3-11 there are also slight changes in the diatom diagram. For example Thalassiosira decipiens has high values in samples 4-9 and Surirella ovata occur only in samples 5-13. From sample 11 onwards changes occur, for example in the Nitzschia punctata, the Diploneis interrupta curve and the Cocconeis placentula curve. From both pollen and diatom analysis one gets the impression that the filling-in of the lowermost part of the channel was an event of short duration (very homogenous material, rather straight curves in pollen and diatom diagrams). Around the layer of marine shells there is probably a standstill or at least a decrease in the silting-up of the channel, which shows clearly in a change of sediment from sample 13 onwards. The presence in the pollen diagram right to the top of Abies, Picea, types 114-116 points to a steady unflux of marine sediment. The rise in the Diploneis interrupta curve may indicate salt-marsh conditions, the rise in Cocconeis placentula the presence of waterplants and thus much quieter water movements

THE ECONOMY (J.P.P.)

The geological evidence outlined above is here considered in the context of other forms of information.

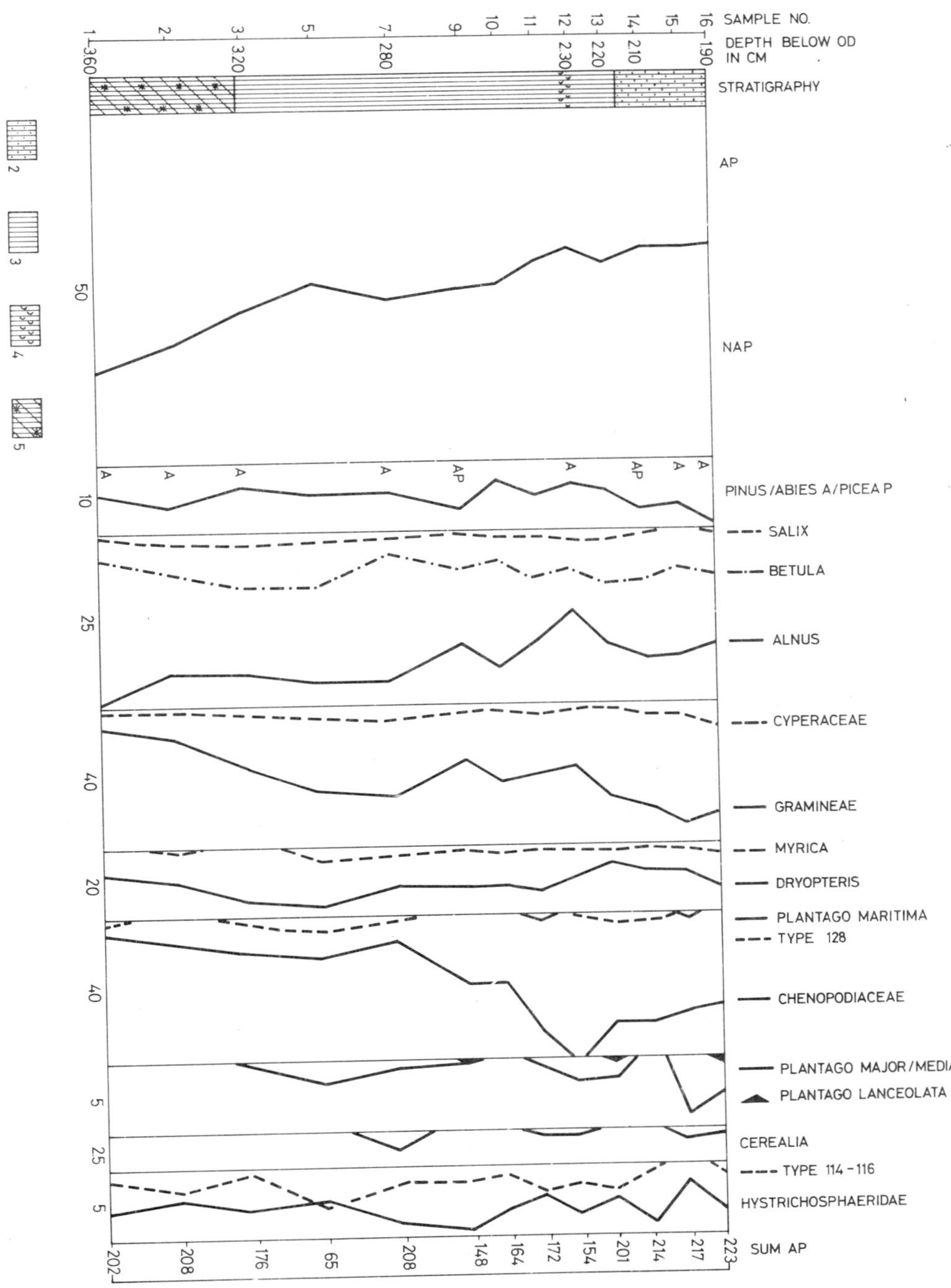

Fig. 7:6 Pollen analysis of the same section as in Fig. 7:5 minus the two topmost samples, selection of curves. Drawing IPP. Legend: see Fig. 7:5.

The archaeobotanical analysis

The sampling programme for the archaeobotanical analysis was designed to provide information about environment and plant economy, and in relation to the latter the following questions were asked:

- which crops were cultivated, and what was their distribution in space and time?

- is it possible to establish the crop processing activities involved?

It was considered unrealistic to pose questions of a more sophisticated nature. Apart from the samples for pollen and wood analysis, the following material was available for the examination of macroscopic plant remains:

(1) In the clay area those creek or ditch fills below the water table were expected to be contemporaneous with the settlements. These contexts yielded waterlogged and occasionally carbonized plant remains.

(2) fills of ditches and postholes above the water table (being the only identifiable features). The intention was to take samples from all these features in order to extract carbonized plant remains by flotation, but the extreme scarcity of carbonized plant remains (in the first and second excavation campaign, only 30% of the samples yielded one or more seeds and the difficulty of processing the stiff clay, made us shift the sampling strategy towards an ad hoc basis, concentrating on contexts appearing to be rich in carbonized material. Even so, the efficiency could be raised to no more than 70%, only 11 samples out of over 200 'fertile' ones yielded more than 100 cereal grains, whereas chaff was extremely scarce. Hand-sieving of the flotation residues resulted in occasional addition of cereals, but failed to produce chaff remains.

(3) From the peat area: various features within house plans, such as infilling layers, floors and hearths. Most samples contained both waterlogged and carbonized plant remains.

(4) From both clay and peat area: contents of pots.

It is obvious that a straightforward comparison of the plant remains from all these types of samples would not be justifiable. Each type of sample has its own interpretive value: Type 1 would reflect the local stand of vegetation in and around the creek or ditch, and possibly contain elements of the regional vegetation, transported by wind or water. In addition, waste from the settlement may be present. The carbonized plants remains from the type 2 samples are considered to be of heterogenous origin, a 'noise' present throughout the settlement, consisting of remains of plants that have had the chance to become carbonized: crop plants with associated weeds, especially the cereals, being parched prior to threshing or grinding, and the waste from the processing activities, which is used to tinder. The peat samples (type 3) contain all kinds of domestic refuse, possible mixed with litter and/or dung, which provides a reflection of the area grazed by the animals (at least that part of the grazing area near the farmstead). The find of a pot containing analysable charred plant remains (type 4 samples) is obviously a chancy matter: some cases will be discussed individually.

The size of type 2 samples circa 12 litres of type 1 and type 3 samples circa 1 litre, whereas type 4 samples did not exceed 500 cc. Only a small part of the samples has been analysed as yet, so that the conclusions drawn are necessarily of a tentative nature.

Type 1 samples

A synopsis of the vegetation units represented in type 1 samples is shown in table 7:1.

site + dating	F (1st bc - 2nd AD)	C (1st AD)	D (2nd-3rd AD)
water plants	2 (x)	3 (x)	4 (xxx)
drying, mineral-rich mud	4 (xx)	3 (xxx)	5 (xxx)
annual ruderals & field weeds	1 (x)	4 (x)	3 (x)
grassland	1 (x)	7 (xx)	4 (x)
reedland	3 (xxx)	4 (xx)	8 (xxx)
mesotrophic peat	-	2 (x)	2 (x)
salt marsh	2 (xx)	4 (xx)	-

Table 7:1:synopsis of the types of vegetation represented in type 1 samples: number of species with indication of seed quantity (x = 1-10, xx = 10-100, xxx = 100-500)

This record would suggest an increasingly wet environment and diminishing influence of the sea during the first centuries AD, which is completely in agreement with the geological evidence. This material does not provide information about land use.

Type 2 samples

Table 7:2 demonstrates that a great number of samples is needed to produce an acceptable level of presence for each crop species; sites such as D and H were too small to produce a reasonable number of features which could be sampled, and therefore, species with a small chance of becoming carbonized, such s Linum, may be easily missed in the analysis.

It appears that Hordeum vulgare L. emend (hulled six-row barley) was the main crop; Linum usitatissimum (flax), Triticum dicoccum Schubl. (emmer wheat), Panicum miliaceum (broomcorn millet), Camelina sativa (gold-of-pleasure), and possibly Avena sativa L. (oats) were additional crops, oats and emmer possibly only admixtures. This species combination is quite familiar in Iron Age sites in the coastal area (cf Korber-Grohne 1967, Behre 1976) except for the absence of Vicia faba (horsebean). This species' absence in type 2 samples is not very surprising, but it was also lacking in the extremely well preserved material from site Q, where the items especially would be expected to appear (cf. Behre 1976).

Site + dating	K (1st BC)	H (1st AD)	C (1st AD)	D (2nd-3rd AD)	F (1st BC - 2nd AD)
number of samples	15	9	25	5	115
Crop Plants					
Hordeum vulgare	435(15)	61(8)	24(12)	3(3)	1269(70)
id., internodes	1	-	-	1	3(1)
Avena sp.	1	1	-	-	48(12)
Cerealia indet.	-	-	6(2)	1	73(15)
id., stem frags.	-	-	-	-	present(6)
Panicum miliaceum	-	-	-	-	15(2)
Triticum dicoccum	1	-	-	-	3(3)
Camelina sativa	2(1)	-	-	-	1
Linum usitatissimum	-	-	-	-	15(5)
Reed lands					
Eleocharis palustris	76(11)	84(4)	32(12)	17(3)	418(27)
Scirpus sp.	51(13)	80(4)	40(11)	4(2)	1162(58)
Others, nr. of spp.	2	2	1	-	8
nr. of items	3	2	1	-	30
Annual ruderals & field weeds					
nr. of spp.	6	4	6	3	12
nr. of items	92	17	8	5	272
Damp grasslands					
nr. of spp.	8	-	6	-	12
nr. of items	23	-	10	-	140
Drying, mineral-rich mud					
nr. of spp.	1	-	1	-	3
nr. of items	1	-	1	-	24
Others					
nr. of spp.	1	1	6	1	8
nr. of items	-	4	44*	4	13

* many corroded, unidentified seeds

Table 7:2 Synopsis of the carbonized plant remains from type 2 samples. Numbers in brackets indicate sample frequency.

Crop plants		
Avena sp.	2	oats
Hordeum vulgare	572	hulled six-row barley
Cerealia indet.	43	unidentified cereals
id., stem fragments	x	
Annual ruderals & field weeds		
Chenopodium album	4	fat hen
Ch. ficifolium	1	'fig-leaved goosefoot'
Hyoscyamus niger	5	henbane
Polygonum aviculare	2	knot grass
Stellaria media	2	chickweed
Perennial ruderals		
Sambucus ebulus	6	danewort
Urtica dioica	1	stinging nettle
Damp grasslands		
Carex otrubae	11	false foxsedge
Galium pulustre	1	marsh bedstraw
Mentha cf. arvensis	1	mint
Potentilla anserina	4	silverweed
Rannunculus repens	1	creeping buttercup
R. sardous	2	hairy buttercup
Rumex sp.	4	the docks
Poaceae	4	grasses
Fabaceae	7	leguminous plants
Mesotrophic peat		
Ranunculus flammula	3	lesser spearwort
Carex flava agg.	7	yellow sedge
C. rostrata	2	beaked sedge
Drying, mineral-rich mud		
Atriplex sp.	13	orache
Polygonum lapathifolium	1	pale persicaria
Ranunculus scleratus	1	celery-leaved crowfoot
Reed lands		
Carex paniculata	4	panickled sedge
C. sp.	1	sedge
Eleocharis palustris	219	common spike rush
Myosotis scorpioides	1	water forget-me-not
Montia chondrosperma	3	blinks
Rumex hydrolapathum	1	great water dock
Scirpus sp.	242	bulrush/club rush
Others		
Calluna vulgaris	2	heather
Juncus gerardii	2	mud rush

Table 7:3

Carbonized plant remains from six samples from ditch 50800/2.2 at site F.

In six samples from a house ditch at site F, stem fragments of cereals were found together with many weed and cereal seeds (see table 7:3). This might well reflect the waste of the processing of straw for thatching (cf. Reynolds 1981: 119). Table 7:3 also demonstrates the weed assemblage: not only those plants known today as weeds are present, but also species from damp, and even marshy habitats; the abundance of *Eleocharis* and *Scirpus* is especially striking.

Type 3 samples

The only site where samples of this type have been analysed is site Q (dated *circa* 600 bc). The results of these samples are published by Beemster (forthcoming) and Therkorn et al. (forthcoming) and will only be discussed briefly here.

The following conclusions were arrived at:

(a) concerning the environment: on the spot the peat was covered with heath; in the vicinity of the site, it was intersected by one or more tidal creeks, and stood in contact with a fresh-water marsh. In the gradient zones the vegetation was dominated by bog myrtle.

(b) concerning plant husbandry: the most important crop plants were barley and gold-of-pleasure; emmer wheat, broom corn millet, oats and flax were occasionally found, orache (*Atriplex*) seeds may have been gathered for consumption.

The abundance of gold-of-pleasure pods and straw indicated that this crop was grown near the farmstead and, consequently, on the peat.

Evidence for the cultivation of cereals in the immediate vicinity of the farmstead was not found: barley internodes were numerous, even more so than the grains (grain: internode ratio = 0.51), but stems and pollen were absent, the latter may even indicate that the processing of barley took place somewhere else. Remarkably, *Triticum*-pollen was found regularly, although emmer grains and spikelet forks were scarce. This discrepancy between *Triticum* pollen and seeds also occurs at other sites, and has not yet been satisfactorily explained. (See above discussion.)

Type 4 samples

Five pots are discussed, with very different contents. At site B, a pot contained 58 carbonized *Stellaria media* (chickweed) seeds. Since no evidence for the gathering of wild fruits was found, it was suggested that certain plants such as *Stellaria* and *Rumex* (the docks) were consumed as vegetables to supply the need of vitamin C. These plants would have been cooked, in contrast to, e.g. *Salicornia* (glasswort), which is presently highly appreciated (Mabey 1975) and must have been common and easy to gather, even for the inhabitants of the peat sites.

At site H, a pot contained 121 Hordeum, 25 Stellaria, 7 Scirpus and 1 Eleocharis, all carbonized. This indicates:

- once more the association of Scirpus and Eleocharis with cereals (see above)

- that barley was prepared in pots, and consequently used for human consumption, and not only as animal fodder.

- that the case for Stellaria being a vegetable is not very strong.

A pot at site Q, located near the entrance of the farmstead and interpreted as an offer pot, contained three carbonized Hordeum grains and was further backfilled with peat. If the interpretation as offer pot is correct, then the original contents must have disappeared.

The most interesting pot samples are from site P: here two pots were found with charred cereals immediately visible to the naked eye. The archaeological context, a concentration of sherds and bones in a house-ditch, near the supposed entrance virtually excluded the interpretation as offer pots, and the contents affirmed this: more than 70% of the crop appeared to be under-developed or to consist of empty glumes, and the grains (barley) were mixed with weed seeds and chaff (fragments of rachises, glumes and awns, and an occasional basal node or stem fragment). Full results are presented in table 7:4.

Thus the material appears to be the waste from stage 5 or 10 of the crop processing sequence described by Hillman (1981). The pots have apparently been used as dustbins and disposed of into the house ditch together with other rubbish.

Wood and charcoal analysis

The analysed material has been sampled rather hapharzardly, due to the unfavourable conditions of preservation. Four house plans have been sampled systematically: two at site F, and two in the peat area, at sites L and Q, of which not only the posts, but also the wattle walls were preserved. Sites F and L were analysed by Drs. I. Stuyts, and site Q by E. van Hoorn-van Berkel. The overall wood spectra do not show spectacular differences (see table 5) but there is a differentiation in the wood used for the posts: at site F they are made of alder, whereas the peat sites show oak as the main constructional timber, at site Q supplemented by ash for the stall end.

During the occupation of sites L and Q, the peat area apparently was not able to supply large stems of any species, so the inhabitants of these sites had to rely on the dunes for their supply of building wood.

The use of alder at site F may have been made possible by an extension of this species over the levees (a strong increase of Alnus pollen in the area during the first centuries AD is recorded by De

Find number	16202	16207	
Hordeum vulgare	560	2300	hulled six-row barley
incl. empty spikelets	(3.9 gm.)	(12.6 gm.)	
id., chaff fragments	21.2 gm.	56.2 gm.	
Avena sativa	45	350	cultivated oats
Avena fatua	32	59	wild oats
Bidens tripartitus	11	75	tripartite bur-marigold
Atriplex sp.	17	26	orache
Polygonum lapathifolium	65	169	pale persicaria
Sonchus asper	22	48	spiny sow-thistle
S. arvensis	-	7	corn sow-thistle
Chenopodium album	2	5	fat hen
Ch. ficifolium	2	6	fig-leaved goosefoot
Ch. rubrum/glaucum	-	4	red/glaucous goosefoot
Bromus mollis/secalinus	-	13	chess
Stellaria media	1	3	chickweed
Eleocharis palustris	-	1	common spike-rush
Scirpus lacustris	1	2	bulrush
Rumex sp.	1	-	the docks
Carex paniculata	1	-	panickled sedge
C. sp.	1	-	sedge
Poaceae	3	9	grasses

Table 7:4. Carbonized plant remains found in two pots at site P.

Site	F	L	Q	
Fraxinus	+	-	+	ash
Alnus	+	+	+	alder
Salix	+	+	+	willow
Quercus	+	+	+	oak
Pyrus/Malus	+	+	-	pear/apple
Betula	+	-	+	birch
Corylus	+	+	-	hazel
Picea	+	-	-	spruce
Juniperus	+	+	-	juniper
Cornus	-	+	-	cornel
Rhamnus	-	-	+	buckthorn

Table 7:5. Presence of wood species in three sites in the Assendelver Polders

Jong, in press), but also have been induced by large scale chopping of oak by the Romans for the construction of their castellum and harbour works near Velsen (Morel & de Weerd 1980).

Zooarchaeological analysis

Zooarchaeological data are scarce, due to the unfavourable conditions for the preservation of bones. The finds from four sites in the clay area and one from the peat, listed in table 7:6, give an idea of the meat consumption patter. Site Q was analysed by Dr. L. H. van Wijngaarden-Bakker and the other sites by Seeman (1980). Cattle was obviously the most important domestic animal, even if over representation (bias towards larger bones caused by bad preservation) is accounted for. The animals were small (shoulderheight 1.05-1.15 m) and slender, and Roman influence in the younger sites in terms of better breeding/selection (increase of size) could not be traced. Pig was of minor importance, which is easily explained by the scarcity of forest in the area. Sheep/goat is relatively well represented in the peat site, which is also characterized by the absence of horse. Perhaps the environment was too damp for this animal to be of much use, although the land must have been dry enough to allow transportation by wagon, in view of the disc-wheels found at site Q (cf. Therkorn et al., forthcoming). The only zooarchaeological indication for possible contact with the Romans is the find of chicken bones at site N. (See footnote.)

	clay				peat
site	B	C	D	F	Q
cattle	77	71	78	60	44
sheep/goat	18	23	6	28	51
pig	2	-	5	4	3
horse	2	1	6	3	-
dog	+	1	-	+	-
other	+	4	5	5	2
Total number of bones	139	105	18	166	70

Table 7:6. Percentages of animal bones found at five sites in the Assendelver Polders

Integrating the results; conclusions

If we want to make observations on the exploitation of the landscape in the Assendelver Polders by the Iron Age inhabitants, the evidence is not very helpful. In both peat and clay area, houses with stalls or cattle boxes occur, whereas pits, raised granaries or other structures associated with crop storage (cf. Buurman 1979) were lacking. There is, however, evidence for crop fields on the levees in the clay area, in the form of shallow ditches enclosing plots of land; in the peat area, ditches were not found.

The evidence presented in the preceding sections allows us to outline tentatively some aspects of the subsistence of the Assendelver Polders settlement system.

At site Q, considered with respect to subsistence to be the type site for the peat area, in which it was the best investigated site, the first inhabitants, having traversed the peat with one or more wagons drawn by oxen, settled down around 600 bc on a heath-covered hillock. The wood for the construction of a farmstead was brought from the coastal dunes, which were about 7 km away. Most probably the basis of subsistence was stock raising: the peat was primarily used for grazing goats (Therkorn _et al_. forthcoming) and cattle, the latter probably covering a more extended area, including the banks of a tidal channel. Agriculture on the peat was restricted to the cultivation of gold-of-pleasure; cereals (predominantly barley) were consumed, but not grown locally. It is impossible to determine whether the inhabitants of site Q had their own cereal fields at some distance from the site, e.g. in the dune area or on a sandy creek levee, or that the cereals were being traded.

From the second century bc onwards, the clay area became inhabitable, and a rather dense occupation developed. The local environment was able to provide wood for construction, although there is evidence that the dune area was still being exploited. Both agriculture and stock raising were practised here, the fields being situated on the sandy creek levees. The many plant species of marshy habitats that were associated with the crops suggest that the fields must have been rather damp, and in view of the rising water table and consequent peat growth during the first centuries AD, increasingly restricted in area. Probably the high water table was prohibitive for storage of crops on or under the soil surface.

Apart from pottery, there is only one indication for possible contact with the Romans: the find of chicken bones at site N (dated around the beginning of our era).

Notes

Results of site N are not listed, since botanical and zooarchaeological material are still under investigation. After completion of the manuscript, two carbonized hilums of _Vicia faba_ were found in a sample from a ditch.

Acknowlegements

This publication is only an interim report. More than 15 people are working on the environmental part of the Assendelver Polders project and all specialized studies are still in progress. Although the main outline of the environmental background is known many details have yet to be filled in. The following is based partly on published work, but is mainly composed on unpublished data by Dr. M. J. Jansma, J. M. Fontaine and Drs. P. Vos (diatom analysis) Drs. D. P. Hallewas (Rijksdienst voor het Oudheidkundig Bodemonderzoek, Amersfoort) and Drs. P. Vos (geology, pedology), Dr. J. P. Pals and D. C. M. Beemster (Palaeoethnobotany), Dr. B. van Geel (Hugo de Vries laboratorium, University of Amsterdam), Professor W. Groenman-van

Waateringe, J. de Jong (Rijks Geologische Dienst, Haarlem, C.D. Troostheide and H. van Veluwen (palynology), E. van Hoorn-van Berkel and Drs. I. Stuyts (Biologisch-Archaeologisch Instituut, State University, Groningen) (wood identification), Dr. L. H. van Wijngaarden-Bakker, F. Laarman and M. Seeman (zooarchaeology). Unless otherwise stated, all are staff or students of the IPP. The excavations were conducted under the direction of Drs. R. W. Brandt, Dr. S. E. van der Leeuw and Drs. A. Voorrips were supported in 1979-1980 by the Netherlands organisation for the advancement of pure research (Z.W.O.), in 1981 by the Z.W.O. foundation for archaeological research (ARCHON).

Bibliography

Andersen, S.Th. (1979) Identification of wild grass and cereal pollen. Danmarks Geologiske Undersøgelse Arbog 1978 69-92

Beemster, D.C.M. (forthcoming) Huis Q van A tot Z: botanische analyse van een boerderij uit de vroege Ijzertijd in de Assendelver Polders. Internal Report IPP. Amsterdam

Behre, K.-E. (1976) Die Pflanzenreste ans der frühgeschichtlichen Wurt Elisenhof. Studien zur Küstenärchäologie Schleswig-Holsteins, serie A, band 2. Bern: Frankfurt/Main

Brandt, R.W. (in press) A short-lived relation along the northern frontier. In: Roman and Native in the Low Countries (eds Slofstra, J. and Brandt, R.W.) Oxford: BARS

Brandt, R.W., W. Groenman-van Waateringe & S.E. van der Leeuw, eds (forthcoming) The Assendelver Polders Papers I

Buurman, J. (1979) Cereals in circles: crop processing activities in Bronze Age Bovenkarspel (NL). Archaeo-Physika 8 21-37

Faegri, K. & J. Iversen (1964) Textbook of Pollen Analysis. Copenhagen.

Fontaine, J.M. (1981) Diatomeenonderzoek van grondmonsters en aardewerk-scherven afkomstig uit Assendelft. Internal Report IPP. Amsterdam

Groenman-van Waateringe, W. (1961) Een boerderij uit de eerste eeuw na Chr. te Krommenie (N.H.). De boerderij. In: In het Voetspoor van A.E. van Giffen. (eds Glasbergen, W. & W. Groenman-van Waateringe) Groningen. pp. 110-115

Grohne, U. (1957) Die bedeutung des Phasenkontrastverfahrens fur die Pollen-analyse, dargelegt am Beispiel der Gramineenpollen vom Getreidetyp. Photographie und Forschung 7 237-248

Jong, J. de (forthcoming) Pollenanalytical investigation and radiocarbon dating of two cored sections at Assendelft (NL). In: The Assendelver Polders Papers I (eds Brandt, R.W. et al.)

Hillman, G. (1981) Reconstructuring crop husbandry practices from charred remains of crops. In: Farming practice in British prehistory (ed Mercer, R.) Edinburgh. pp. 123-162

Körber-Grohne, U. (1967) Geobotanische Untersuchungen auf der Feddersen Wierde. Wiesbaden

Mabey, R. (1975) Food for free. Fontana

Morel, J.M.A.W. & M.D. de Weerd (1980) Early Roman harbours in Velsen. In: Roman frontier studies 1979. (eds Hanson, W.S. and Keppie, L.J.F.) Oxford: BARS 71 pp. 475-494

Pals, J.P., B. van Geel & A. Delfos (1980) Palaeoecological studies in the Klokkeweel bog near Hoogkarspel (prov. of Noord-Holland). Review of Palaeobotany and Palynology 30 371-418

Reynolds, P. (1981) Deadstock and livestock. In: Farming practice in British prehistory. (ed Mercer, R.) Edinburgh. pp. 97-122

Seeman, M. (1980) Onderzoek aan dierlijke resten uit opgravingen in de Assendelver Polder gedurende de jaren 1978 en 1979. Internal Report IPP Amsterdam

Straaten, L.M.J.U. van (1965) Coastal barrier deposits in South- and North-Holland, in particular in the areas around Scheveningen and IJmuiden. Mededelingen van de Geologische Stichting, new series 17 41-75

Therkorn, L., R.W. Brandt & J.P. Pals (forthcoming) An Early Iron Age farmstead: site Q of the Assendelver Polders Project. Antiquity

Vos, P. (1982) Een diatomeeninventarisatie in de Assendelver Polder (sites N, P en Q). Internal Report IPP Amsterdam

Vuorela, I. (1973) Relative pollen rain around cultivated fields. Acta Botanica Fennica 102

Zagwijn, W.H. (1971) De ontwikkeling van het 'Oer-IJ' estuarium en zijn omgeving. Westerheem 20 11-18

8. MAN AND THE ENVIRONMENT AT FRESWICK LINKS:

PRELIMINARY OBSERVATIONS FROM A LATE NORSE SETTLEMENT IN CAITHNESS

A. K. G. Jones, C. E. Batey, C. D. Morris and D. J. Rackham

INTRODUCTION

Preliminary analyses of animal and plant remains extracted from Late Norse deposits at Freswick Links (NGR ND 3765 6760), Caithness, have highlighted a number of interesting problems concerning the accumulation of archaeological deposits within a developing sand dune system. In addition to outlining the history of Late Norse archaeological research in Caithness, this paper attempts to describe the development of a sampling strategy for the site; to explore which kinds of biological remains were deposited as a consequence of human occupation and which formed part of the natural background biota; and, finally, to demonstrate differences in the remains found in two depositional environments - a midden deposit and an area surrounding domestic buildings.

THE CAITHNESS CONTEXT

Caithness is formed mainly of Old Red Sandstone, a formation which includes mudstones, sandstones and flags. Much of the country is overlain by Pleistocene drift deposits with ice and ice/water deposition. All but the southern and western parts of the country are covered by a shelly till upon which soils have developed. Glacial acitivity is largely responsible for the gently undulating landscape (mostly below 100 m O.D.) cut by deeply incising rivers and streams. The heart of Caithness is clothed by a deep bed of peat, up to 6 m deep in places, which is still the most common source of domestic fuel (Sharpe and Saxon 1972; Omand 1972a; Omand 1982).

The spectacular coastline comprises high sandstone cliffs and rocky inlets with characteristic flagstone layering, and a few large sandy bays backed by extensive dunes, e.g. Sandside Bay, Sinclair's Bay and Freswick Bay. Caithness is characterized by a lack of tree-cover which appears to date back to the last period of glaciation (Peglar 1979: 245-6). Although rather bleak and windswept, the north-east corner of the country, particularly the parish of Canisbay, now includes rich farmland supporting cereals and beef cattle.

Settlement, probably from prehistoric times, appears to have been concentrated along the coastal zone with some inland

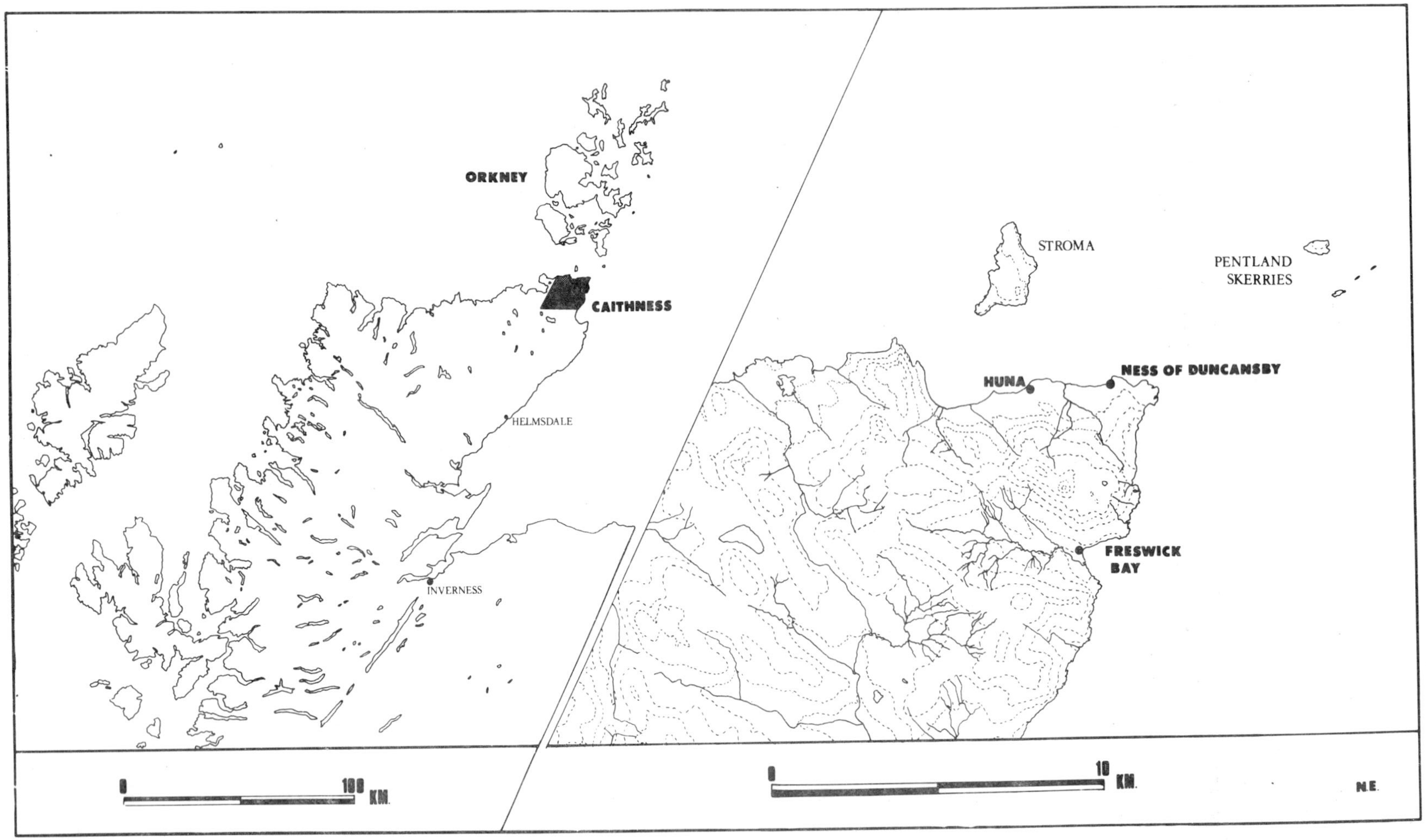

Fig. 8:1 Location map of Freswick Links, Caithness, Scotland (N. Emery). Crown copyright.

colonization around lochs and along rivers. However, it should be stressed that field work has been focussed along the coast and many traces of early occupation may lie undiscovered. Many place names in the county contain Old Norse elements (Omand 1972b) and suggest considerable settlement in the Viking and Late Norse periods. Nicolaisen (1982) distinguished between the Norse names of the north and east of the county and Gaelic names in the south and west. In the parish of Canisbay, all the surviving name elements, where not modern, are Norse or Norse-derived (D. Waugh pers. comm.).

Archaeological evidence for Viking and Late Norse activity is relatively sparse. The site of Freswick, 5 km south of John o' Groats, has been known as a Late Norse occupation site from field walking and excavation since the early part of this century. Sites of similar period are well known on the islands of Orkney and Shetland (Wainwright 1962: 117-62; Small 1968; Morris forthcoming), but until recently Freswick was the only settlement site in Caithness producing Late Norse material. However, two sites on the north coast, Robertshaven (NGR ND 3880 7360) and Huna (NGR ND 3995 7358) have recently yielded pottery which is very similar to that found at Freswick.

Broadly contemporaneous archaeological evidence in the country include pagan Viking graves at Westerseat (Anderson 1874: 551-2) Castletown (Anderson 1874: 549-51), and Reay, Sandside Bay (Grieg 1940: 21-2) and a fine Runic cross-slab from Thurso (Romilly Allen and Anderson, 1903: 36-7). The towns of Thurso and Halkirk may have been settlements since the eleventh century; however, attestable archaeological evidence of this period is remarkably elusive.

FRESWICK LINKS: THE LATE NORSE SITE

Work by A. O. Curle in 1937-8 (1939: 71-108) and V. G. Childe in 1941 (1943: 5-17) concentrated on recovering artefacts and structural information from the site. Childe's excavations were conducted in advance of extensive sand-quarrying while Curle's work contained more of a research component. Both excavators encountered midden deposits associated with the structures, but these were not subjected to detailed scrutiny. Margery Platt simply reported that bones of short-horned cattle were the most abundant taxon, with smaller quantities of pig, sheep, dog, horse, red deer, seal, gannet and cod (Curle 1939: 109). This phase of investigation at the site is re-assessed by Batey (1982).

In more recent years, visits by members of Glasgow and Durham Universities produced evidence that Freswick Links is an extensive site where archaeological deposits are being subjected to considerable erosion. Two major kinds of erosion are recognized. Firstly, the sand dune terrace fronting the sea is receding as a result of sand-quarrying and consequent exposure and destabilization; and secondly, areas to the landward are being blown-out as a consequence of the activities of a large rabbit colony. Detailed examination of the eroding cliff faces has revealed large numbers of artefacts in addition to animal and plant remains associated with human occupation. The Inspectorate of Ancient Monuments of the Scottish Development Department agreed to fund a small-scale survey of the site in September 1979. This work included the collection of column samples for laboratory analysis, recording the extent of erosion and the collection of artefacts. In addition, limited phosphate and auger surveys were undertaken. Results of these investigations indicated considerable spatial differences in extent and thickness of midden deposits across the links; midden deposits were traced for 150 m along the edge of the dunes and up to 100 m inland. As a consequence, the Inspectorate supported further survey work in the spring and excavation in the summer of 1980. Two further seasons excavation, designed to sample the midden deposits both to the north and south of the area excavated by Curle, have been financed by the Inspectorate and the University of Durham. Results of parts of this work have been published (Batey et al. 1981; 1982) while other aspects are currently being prepared for publication (Batey et al. forthcoming a, b).

THE EVOLUTION OF A STRATEGY FOR SAMPLING THE DEPOSITS

As a major component of the preliminary investigations of 1979, two small column samples (approximtely 25 cm square) were collected and processed in the Biological Laboratory of the Department of Archeaology, University of Durham; summary results of the first have been published (Batey et al. 1981: 9-10). Detailed results of analyses of the animal and plant remains from both columns awaits publication (Rackham et al. forthcoming). This work demonstrated that a large number of different kinds of animal and plant remains (including charred grain and seaweed, fish, bird and mammal bones, and invertebrate shells and carapace fragments) were present in the eroding cliff edge deposits. Particularly abundant were fish remains, mainly from the cod family (Gadidae). However, the number of bones collected from these small samples did not provide sufficient data for the fishery to be understood. In order to obtain such data, it was agreed to sieve large quantities of archaeological deposits from a number of small trenches known as 'environmental trenches' located along the cliff face to the south and north of the structures excavated by Curle (see figure 8.2). In addition to producing interpretable assemblages of fish remains, it was thought that sieving large quantities of the deposits would produce representative samples of shells, bones and other small remains which are often overlooked by excavators.

In 1980 an attempt was made to sieve, using 1 mm sieves, the entire fills of a trench 4 x 1 x 1.25 m (Area 4). The trench was divided into 4 metre squares in order that spatial variation in the density and kinds of find may be observed. The individual midden deposits were usually 5-10 cm thick and could be seen to run for the length of the trench. Layers of midden were separated by thin (usually 2-5 cm) layers of clean wind blown sand. Unfortunately, severe practical problems were encountered because of the instability of the sand which made it necessary to batter the sides of the trench. Thus, at its deepest the trench had decreased in area to measure 3.5 x .75 m. In the light of this experience, trenches 4 x 2 m were excavated in later seasons. Sieving was restricted to a central strip 50 cm wide which was divided into 4 metre long sections. The remainder of the deposits within the environmental trenches was excavated by hand. In addition, small column samples

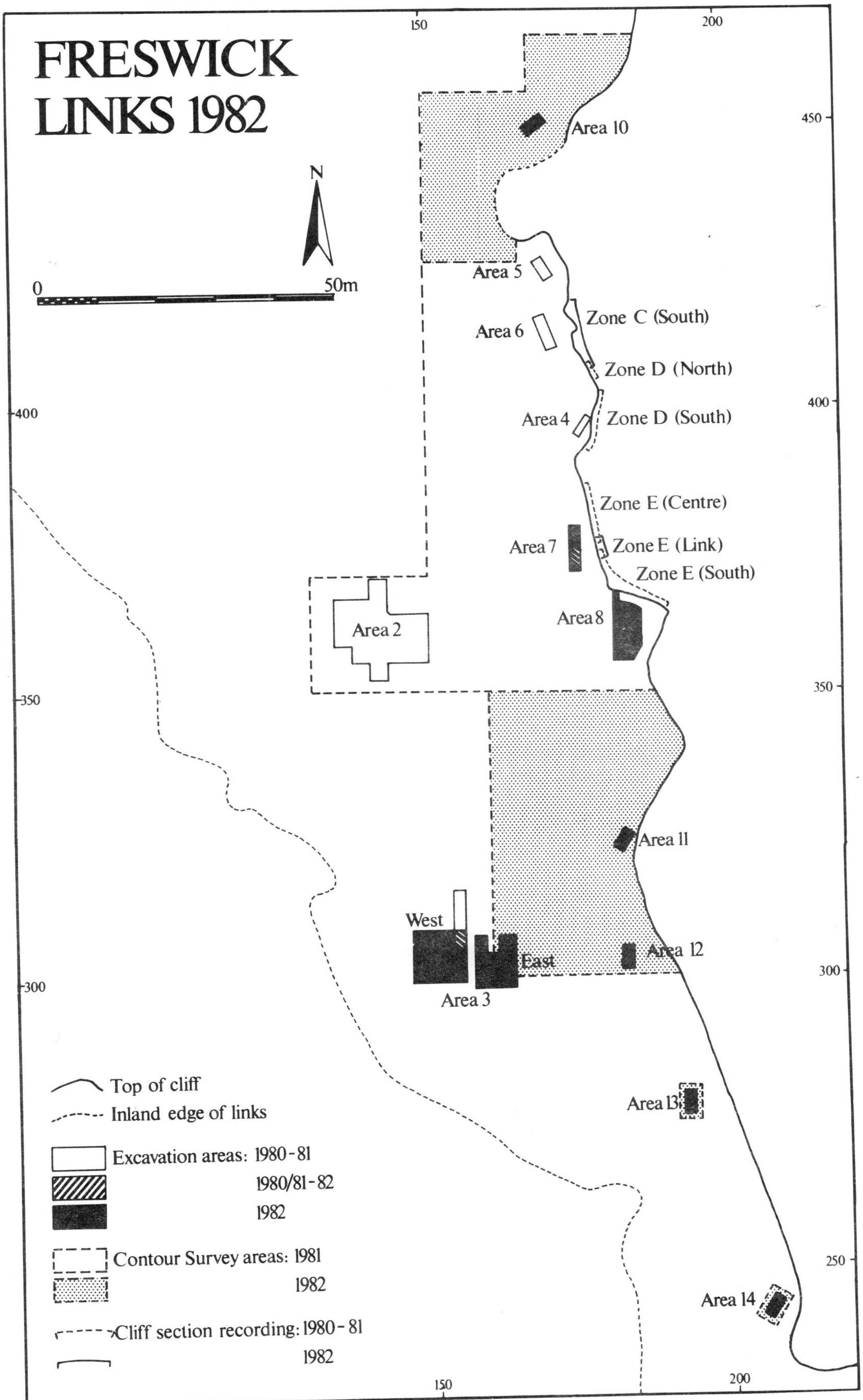

Fig. 8:2 Plan of Freswick Bay showing location of trenches (M. J. Rains). Crown copyright.

(approximately 25 cm square) were collected from the section of each trench and sieved in the laboratory using 500 micron sieves to monitor biological remains not collected on 1 mm sieves and to provide a preliminary indication of the range of material incorporated into the deposits.

In Area 2, which included a building excavated by Curle (op. cit) and dark midden layers immediately surrounding the structure, a fairly large area was excavated around the buildings and four metre squares on the site grid were selected and all deposits bulk-sieved. In situ deposits within the structure were not investigated.

Most samples weighed between 50 and 100 kg and generally comprised an in situ volume of 1 metre square by 2-5 cm thick. The volume of each sample was recorded prior to sieving and is the excavated volume. Under certain circumstances both the areas sampled and the meshes used were altered. Where charred grain were observed at frequencies greater than approximately 15 grains/15 kg floating materials were collected on a 500 micron rather than a 1 mm sieve in order to collect small charred seeds. In two of the environmental trenches structural remains were recovered. To continue collecting samples from a central 50 cm strip from these trenches ignoring the stratigraphy and walls was judged to be inappropriate. So large samples, usually 50 - 100 kg, of each context were collected and their extent carefully plotted. Where possible samples were taken on a metre square grid but frequently the size of the trenches restricted the sample area.

In summary, a three-tiered strategy has been developed for the recovery of animal and plant remains. Firstly, a series of small columns collected for detailed laboratory analysis using 500 micron meshed sieves. Secondly, large amounts of deposit from standard sized trenches have been sieved to 1 mm where the deposits appear to be simply layers of midden; and large quantities of soil have been sieved from areas of the site which contain the remains of buildings. Thirdly, all unsieved deposits were carefully trowelled and finds recovered by hand. This approach has allowed relatively large areas to be excavated while objectively testing the levels of recovery and assessing variation across the site.

During the 1980 excavation, clear differences were observed in the nature and ingredients of samples from midden deposits and around the structures. The remainder of the paper consists of a comparison of preliminary results from these two different archaeological contexts.

AREAS 2 AND 4

One trench, Area 4, was excavated behind the cliff edges in Zone D (see figure 8:2) and excavated to a depth of approximately 1.25 m where clean wind-blown sand was encountered. The stratigraphy of this trench consisted of alternate bands of midden deposit and wind blown sand. No structural remains were recovered. All the midden deposits from Area 4 were sieved using a modified 'Siraf' sieving tank fitted with 1 mm mesh and flot sieve as described by Kenward, Hall and Jones (1981). The same method was used to sieve samples, 1 metre square, taken on regular grid from areas around a building recognized as Building VI, probably a house, excavated by Curle. This part of the site is known as Area 2 (see figure 8:2).

The soil samples were washed using water pumped from Freswick Burn. The weight and volume of each sample was recorded. In order to reduce the amount of material stored, stones larger than 2.5 cm diameter were weighed and classified by type before they were discarded. Table 8:1 shows the basic data for the samples here considered.

Area	Number of Samples	Total wt. (kg)	Mean wt. (kg)	Mean vol. excavated deposit (1)	Mean % stone by weight
2	17	879	52	39	12
4	20	975	49	37	14

Table 8:1: The number of samples, total weight sieved, mean weight and percentage of discarded stone from soil samples from Areas 2 and 4, Freswick Links, Caithness (information abstracted by Mr. J. Maytom).

While the deposits from Areas 2 and 4 contain similar pottery dating them to the Late Norse period there is no evidence to suggest that the materials from the two areas were laid down synchronously. Regrettably precise dating evidence is not yet available. Furthermore, it should be noted that only a quarter of the samples collected from in Area 4 and approximately half of the samples from Area 2 are considered. Work on the remainder of the samples from the two areas may alter some of the general observations presented here. Definitive catalogues are not yet available for all the animal and plant remains, neither are detailed descriptions of the sediments. Nevertheless, this paper provides an opportunity to illustrate one approach to sampling large coastal midden deposits. In addition, an attempt is made to present a general picture of the nature of the deposits and their contents, and to give preliminary observations in advance of detailed anaylses. Figures 8:3 shows the percentage of samples which contained the different kinds of material from Areas 2 and 4.

THE MINERAL COMPONENT

The deposits at the cliff edge (Area 4) and near Building VI (Area 2) were mainly blown sand; a large component of the sand was made up of shell particles with some quartz, presumably mainly derived from the local sandstone rocks. Around the building the deposits tended to be black, containing clay and other fine particles which made the soil damp and sticky. At the cliff edge the sediment tended to be lighter in colour and rather more friable. Although the proportion of discarded stone in the two sets of samples was similar (see table 8:2), the type of stone was markedly different. Large amounts of flagstone were present around the house. This is not surprising since the floors and walls of the buildings contained large amounts of rather crumbly flagstone. By contrast the midden

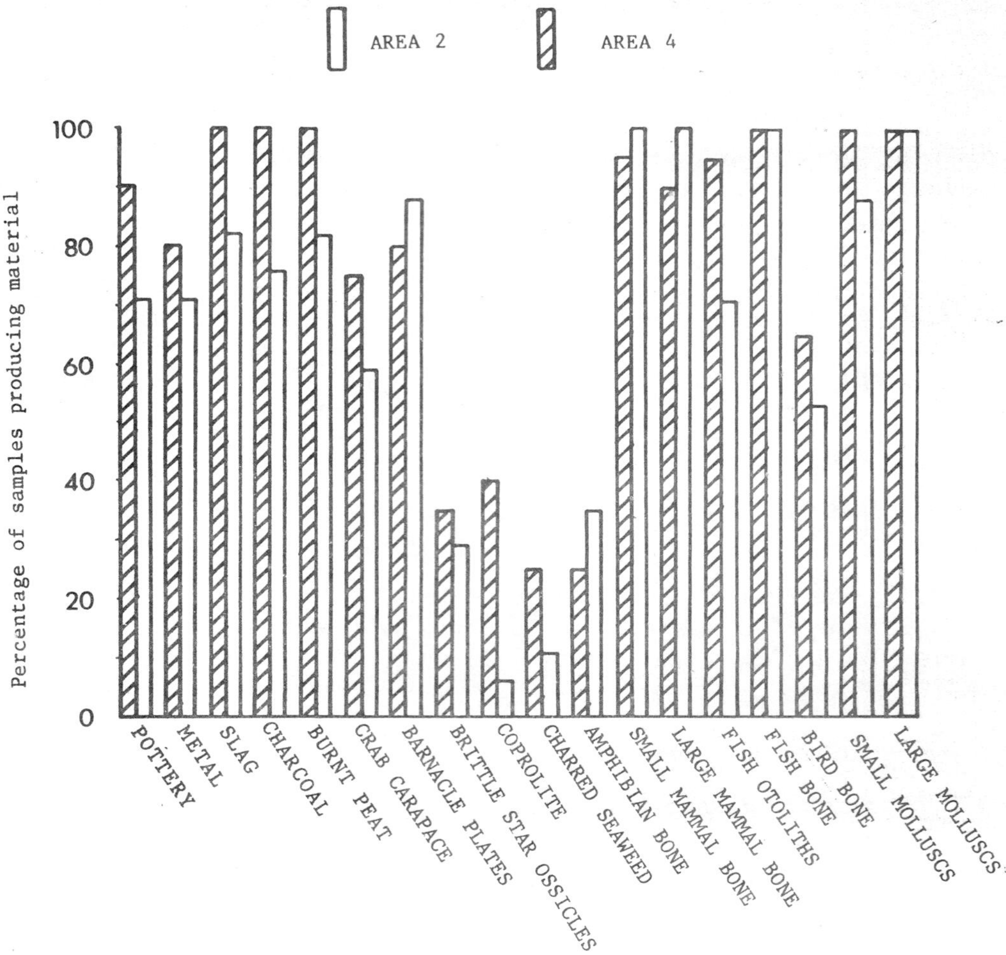

Fig. 8:3 (above) Histogram showing percentage of samples from Areas 2 and 4 containing various kinds of finds (J. Maytom). Crown copyright.
Fig. 8:4 (below) Histogram showing the maximum breadth measurements of saithe Pollachius virens (L.) otoliths from square 4 Area 4 (J. Maytom). Crown copyright.

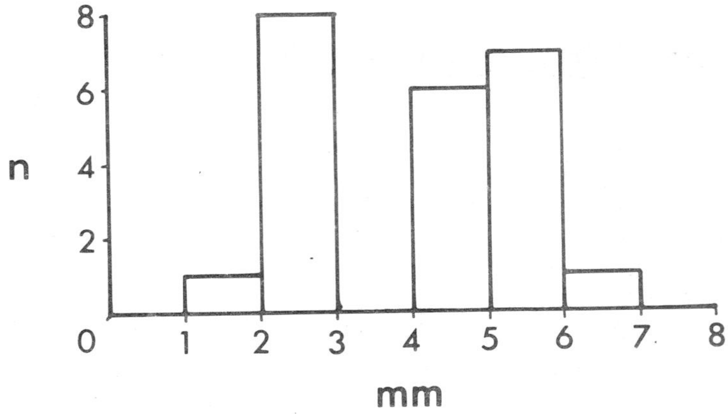

deposits contained predominately fire-cracked sandstone beach pebbles. How these stones were used is in some doubt: possibilities include that they were hearth-stones, pot-boilers used in cooking, or in preparation of grain prior to milling.

Although excavators were instructed to remove all pottery from the deposits at the time samples were taken, fragments of hand-made grass-tempered pottery were nevertheless recovered from the sieved samples in both areas. Around the building, sherds tended to be small and numerous while in the midden deposit they were large and less abundant. Dr. D. F. Williams (pers. comm.) has noted that this pottery is different from ware from Orkney of similar form and date and it is therefore probably of local origin. Pockets of pure clay were found in the samples taken from around the building, clay being also present in deposits within the structure, but absent from the midden.

ANIMAL REMAINS

Bones

Area	Total Wt. mammal bone (g)	Mean wt. mammal bone (g)	Total wt. fish bone (g)	Mean wt. fish bone (g)
2	1120	66	331	20
4	647	32	1600	80

Table 8:2: The mammal and fish bone from Areas 2 and 4, Freswick Links, Caithness (information abstracted by Mr. J. Maytom).

Large mammal bones, mainly cattle, were a major component of the samples from Area 2 but were much less common in the midden of Area 4. Bones of amphibia, mice, shrews and voles were fairly evenly distributed in the two sets of samples. Bird bone (identified by Ms. E. P. Allison), like large mammal bone, was found in larger quantities in the midden and consisted of bones of domestic fowl *Gallus gallus* L., and various seabirds: gannet *Sula bassana* (L.), manx shearwater *Puffinus puffinus* (Brunnich), cormorant *Phalacrocorax carbo* (L.), shag *P. aristotelis* (L.), guillemot *Uria aalge* (Pontoppidan) razorbill *Alca torda* L., puffin *Fratercula artica* (L.) and gulls *Larus* spp. - the last being the most abundant wild species. Bird bones were generally very fragmentary and approximately only half of the specimens from the midden were identifiable. Substantial quantities of immature birds and species which only come to land to breed in summer months e.g. guillemot, razorbill, puffin and manx shearwater suggest that fowling activities were concentrated in the summer.

Fish bones were abundant in most of the layers examined. Three species, cod *Gadus morhua* L., saithe *Pollachius virens* (L.) and ling *Molva* sp., appear to have been the most important species as their bones were abundant in both Areas 2 and 4. Relatively large numbers of otoliths were present and these have been measured using vernier calipers. Similar measurements on modern saithe otoliths from fish of known size shows that a fish of 20 cm total length will have otoliths of approximately 3 mm breadth, fish 50 cm total length approximately 5 mm, and fish 115 cm over 8 mm. The histogram (figure 8:4) of breadth measurements for saithe otoliths from Area 4 suggests that two size classes were caught: small individuals usually less than 20 cm total length and larger specimens ranging from 35 to 90 cm total length. Generally the fish bones from Area 2 were more fragmentary than those from the midden. Remains of gurnards (Triglidae), haddock Melanogrammus aeglefinus (L.) and cartilaginous fish (Selachii) tended to be more common around the building

Shells and other invertebrate remains

Shells were very common in both areas. They have been divided into three groups: edible marine, inedible marine, and terrestial molluscs (identified by Ms. E. P. Allison and Dr. T. P. O'Connor). Limpets Patella sp. were the most abundant edible molluscs and were probably used as bait for fishing or as human food. A smaller number of common winkles Littorina littorea (L.) were also present.

Taxon	Area 2 Total number	Area 2 Sample mean	Area 4 Total number	Area 4 Sample mean
Patella sp.	787	46	1829	91
Littorina littorea	81	5	299	15

Table 8:3: The total number of limpet and winkle apices and the mean number for each sample recovered from Areas 2 and 4 (information abstracted by Mr. J. Maytom).

Apart from these very abundant species, occasional fragments of mussel shell Mytilis edulis L. and specimens of common whelk Buccinium undatum (L.), were recovered.

Several species of inedible marine molluscs were represented in the deposits. Small numbers of specimens of Littorina littoralis (L.), L. saxatilis (Olivi), Gibbula cineraria (L.), Acmaea virginea (Muller) and Trivia artica (Montagu) formed a group of large inedible molluscs. In addition, small marine prosobranch molluscs were also present. This group presented special problems of identification as the shells have been subject to surface abrasion which had in most cases, removed the characteristic sculpturing typical of the different species. A few individuals could be identified to species and comprised specimens of Rissoa parva (da Costa), R. guerini Recluz, Cingula semicostata (Montagu), Chrysallida obtusa Brown and Odostomia eulimoides Hanley. These species must have arrived in the deposits as a component of the beach sand blown into the dunes.

The land molluscs are rather more informative although a restricted range of taxa were present. Aegopinella nitidula (Draparnaud), Cochlicopa lubrica (Muller) and Succinea putric (L.) were the most widespread species, while a group fo slug shells (Limacidae spp.) and other species from Area 2 were typical of an assemblage from stable, rather damp grass-land. Those from Area 4 suggested that the midden sediments were laid down fairly slowly allowing a few pioneer species to colonize the area but never slowed to a point where species intolerant of disturbance could become established.

Crab (Crustacea, Decapoda) carapace fragments were present in small quantities in most samples. Barnacles (Cirripedia) occured in large numbers, particularly in Area 4 and, like the small marine prosobranchs, were mainly blown into the dunes by the wind. In addition, numerous brittle star (Ophiuroidea) ossicles and a few sea-urchin (Echinoidea) shell fragments have been recognized.

THE PLANT REMAINS

Burnt peat fragments were abundant around the building and small quantities were recovered from the midden. Charcoal fragments were present in most samples but in very small quantities while charred shoots of ling *Calluna vulgaris* (L.) Hull, were a rare find in Area 4. Seaweed fragments were not very common in any deposits. Charred grains, mainly barley *Hordeum* sp. and oats *Avena* sp., occurred in small numbers, totals of 298 and 363 charred grains were recovered from Areas 2 and 4 respectively. This is a density of between 0.3 and 0.4 seeds/kg for the two areas. Modern roots, moss fragments and some uncharred seeds of weed species growing on the links were recovered from most samples indicating that modern contamination may be a problem when assessing the smaller animal and plant remains.

DISCUSSION

It is clear that the deposits contain a mixture of animal and plant remains which have been transported by a number of different agencies. Some have been blown in by the wind, while others were brought onto the site by Man. A number are the remains of organisms which once colonized the site, while others may have been introduced by seabirds. Nevertheless, differences between the finds in Areas 2 and 4 suggest that these deposits were laid down under contrasting depositional regimes. The remains from Area 2 are generally rather poorly preserved, consistent with trampling. However most finds from the midden samples were well preserved. In brief, the midden (Area 4) contains domestic refuse including bulk fish waste, while around the building (Area 2) fragments of building materials are present in addition to domestic debris.

SUMMARY OF THE RESOURCES USED

It is probable that the site was selected because of the wide variety of natural resources in the vicinity. Freshwater was available from Freswick Burn. Areas of light shell sand landward of the main dune-hills provided an ideal substratum for cereal cultivation, with seaweed collected from the beach as a readily available fertilizer. Evidence for cultivation of the links was not revealed in the excavation of Area 4 but was noticed in the cliff section in Zones D and E and found in subsequent excavations in Areas 6, 11 and 14.

Flagstones from the south of Freswick Bay and the John o'Groats sandstone from the north of the Bay and water-rolled beach pebbles provided building materials. In addition, large quantities of beach pebbles were also brought onto the Links and heated. Clay for pottery manufacture may well have been taken from the shally till exposed in the Burn bank to the west of the site. Fish were caught from the now famous fishing grounds of north-east Caithness, using limpets as bait from the rocks exposed at low tide in the Bay. Birds were also available locally on the cliffs to the north and south of the site. The beach provided a convenient base for small boats while the extensive peat banks were available for fuel.

With such a range of resources available to settlers it is perhaps surprising that the site should be abandoned. As yet no clear indications have emerged from documentary research or archaeological investigation. However, the later growth of fishing port at Staxigoe and Wick may not be without significance as fishing appears to be a major activity at the site. Clearly the site contains a wealth of animal and plant remains, many of which hold clues to the lifestyle and activities of people living on the site. It is hoped that with detailed analyses of the material collected to date a clear picture will emerge of the complex relationship between Man and nature in Caithness during the Late Norse period.

Acknowledgements

This work could not have taken place without the assistance of a large number of people. During the excavation most of the site staff took samples, labelled bags, transported sacks across the links and helped wash the soil samples. However, we are particularly endebted to Ms. Carol Garbutt for her conscientious work on-site washing and recording the samples.

After the excavation, the samples have been sorted at the Environmental Archaeology Unit, University of York, and particular thanks are due to Ms. E. P. Allison and Mr. J. Maytom for their painstaking work, in addition to their contributions to the text. We are also indebted to Drs. T. P. O'Connor and D. F. Williams for their specialist work.

Finally we wish to thank the Scottish Development Department (Ancient Monuments Branch) through Mr. N. Reynolds who funded both the excavation and analysis of soil samples collected in 1980.

Bibliography

Anderson, J. (1874) Notes on the relics of the Viking Period of the Northmen in Scotland. Proceedings of the Society of Antiquaries of Scotland 10 (1872-4) 536-94

Batey, C.E. (1982) The Late Norse site of Freswick, Caithness. In: Caithness. A Cultural Crossroads (ed Baldwin, J.) Edinburgh: Edina Press pp. 45-59

Batey, C.E., Jones, A.K.G., Morris, C.D. and Rackham, D.J. (1981 Freswick, Caithness. Excavations and Survey at Freswick Links and Freswick Castle, 1979-80. Summary Report. Durham: Department of Archaeology

Batey, C.E., Jones, A.K.G. and Morris, C.D. (1982) Freswick Links, Caithness: Progress report on survey and excavation, 1981. Archaeological Reports for 1981. Universities of Durham and Newcastle Upon Tyne pp. 54-8

Batey, C.E., Jones, A.K.G., Morris, C.D. and Rackham, D.J. (forthcoming a) Freswick Links, Caithness: Progress report on survey and excavation, 1982. Archaeological Reports for 1982. Universities of Durham and Newcastle Upon Tyne

Batey, C.E. Jones, A.K.G., Morris, C.D. and Rackham, D.J. (forthcoming b) Freswick Links, Caithness. Excavation and survey at Freswick Links, 1981-2. Summary Report. Durham: Department of Archaeology

Childe, V.G. (1943) Another Late Viking House at Freswick, Caithness. Proceedings of the Society of Antiquaries of Scotland 77 (1942-3) 5-17

Curle, A.O. (1939) The Viking Settlement at Freswick, Caithness. Report on Excavations carried out in 1937 and 1938. Proceedings of the Society of Antiquaries of Scotland 73 (1938-9) 71-110

Grieg, S. (1940) Viking Antiquities in Scotland (=Viking Antiquities in Great Britain and Ireland, ed. H. Shetelig, Vol II), Oslo: Aschehous & Co.

Kenward, H.K., Hall, A.R. and Jones, A.K.G., (1980) A tested set of techniques for the extraction of plant and animal macrofossils from waterlogged archaeological deposits. Science and Archaeology 22 3-15

Morris, C.D. (forthcoming) Viking Orkney. In Prehistoric Orkney. (ed. Renfrew, C.) Edinburgh: Edinburgh University Press

Nicolaisen, W.F.H. (1982) Scandinavians and Celts in Caithness: the place name evidence. In: Caithness. A Cultural Crossroads (ed Baldwin, J.) Edinburgh: Edina Press pp. 75-85

Omand, D. (1972a) The Ice Age. In: The Caithness Book (ed Omand, D.) Inverness: Highland Printers pp. 26-32

Omand, D. (1972b) Some Caithness Place-Names. In: The Caithness Book (ed Omand, D.) Inverness: Highland Printers pp. 222-8

Omand, D. (1982) The making of the Caithness landscape. In: Caithness. A Cultural Crossroads. (ed Baldwin, J.) Edinburgh: Edina Press pp. 6-23

Peglar, S. (1979) A radiocarbon-dated pollen diagram from Loch of Winless, Caithness, north-east Scotland. New Phytologist 82 245-63

Rackham, D.J., Batey, C.E., Jones, A.K.G. and Morris, C.D. (forthcoming) Freswick Links, Caithness: Report on Environmental Survey 1979 submitted to Proceedings of the Society of Antiquaries of Scotland

Romilly Allen, J. and Anderson, J. (1903) The Early Christian Monuments of Scotland. Edinburgh: Society of Antiquaries of Scotland

Small, A. (1968) A historical geography of the Norse Viking colonisation of the Scottish Islands. Norsk Geografisk Tidsskrift 22 1-16

Sharpe, R.M. and Saxon, J. (1972) Geology. In: The Caithness Book (ed Omand, D.) Inverness: Highland Printers pp. 3-16

Wainwright, F.T. (ed) (1962) The Northern Isles. Edinburgh: Nelson

9. IRON AGE TO LATE SAXON LAND USE IN THE BRECKLAND

Peter Murphy

INTRODUCTION

Drawing together the diverse types of information available on ancient farming so as to propose some overall reconstruction of former subsistence economies is perhaps most easily attempted in areas of marginal land where there are environmental constraints on the range of possible forms of land use. In this respect the East Anglian Breckland is an ideal study area. The very term 'Breckland' reflects the marginal nature of the area: it refers to the large sandy open fields or 'brecks' which in historical times were reclaimed from heathland and allowed to revert to heath when prices fell or when soil nutrients were exhausted.

The soils of the area have been described by Corbett (1973) and some relevant aspects will be discussed further below, but in general they are sandy and very well drained. There is evidence indicating that soil degradation had already occurred in parts of the Breckland before the period considered here: radiocarbon dating of humus in the B_h horizon of Breckland podzols by Perrin et al (1964) indicates a minimum age of formation of 3000 years. Other soils have clearly undergone changes over the last three millennia, notably as a result of windblowing and marling with cultivation (Corbett 1973: 94-5) but considered in broad terms the present-day distribution of soil-types is thought to be similar to that obtaining between the Iron Age and the Late Saxon period. The climate of this part of central East Anglia is semi-continental with a low average annual rainfall (23-25 in, or 584-634mm in the southern Breckland) and high potential soil moisture deficit in the spring and summer. Temperatures tend to extremes with a high incidence of ground frosts and high summer temperatures (Corbett 1973: 17; Duffey 1976: 62). The possible significance for farming of climatic change is considered below.

The semi-natural vegetation has been strongly influenced by these features of soil and climate and also by grazing pressure from sheep and rabbits. It includes some distinctive types of heathland and grassland marked by the presence of a number of plant species more common in coastal dune habitats (e.g. Carex arenaria, sand sedge) and others which are typical of continental climatic conditions or characteristic of the steppe grasslands of central Europe (e.g. Scleranthus perennis, perennial knawel) (Clarke, W. G. 1925: 51; Petch and Swann 1968: 31). The factors influencing vegetation have also had effects on land use in historical and modern times. In the later Middle Ages much land in the Breckland was

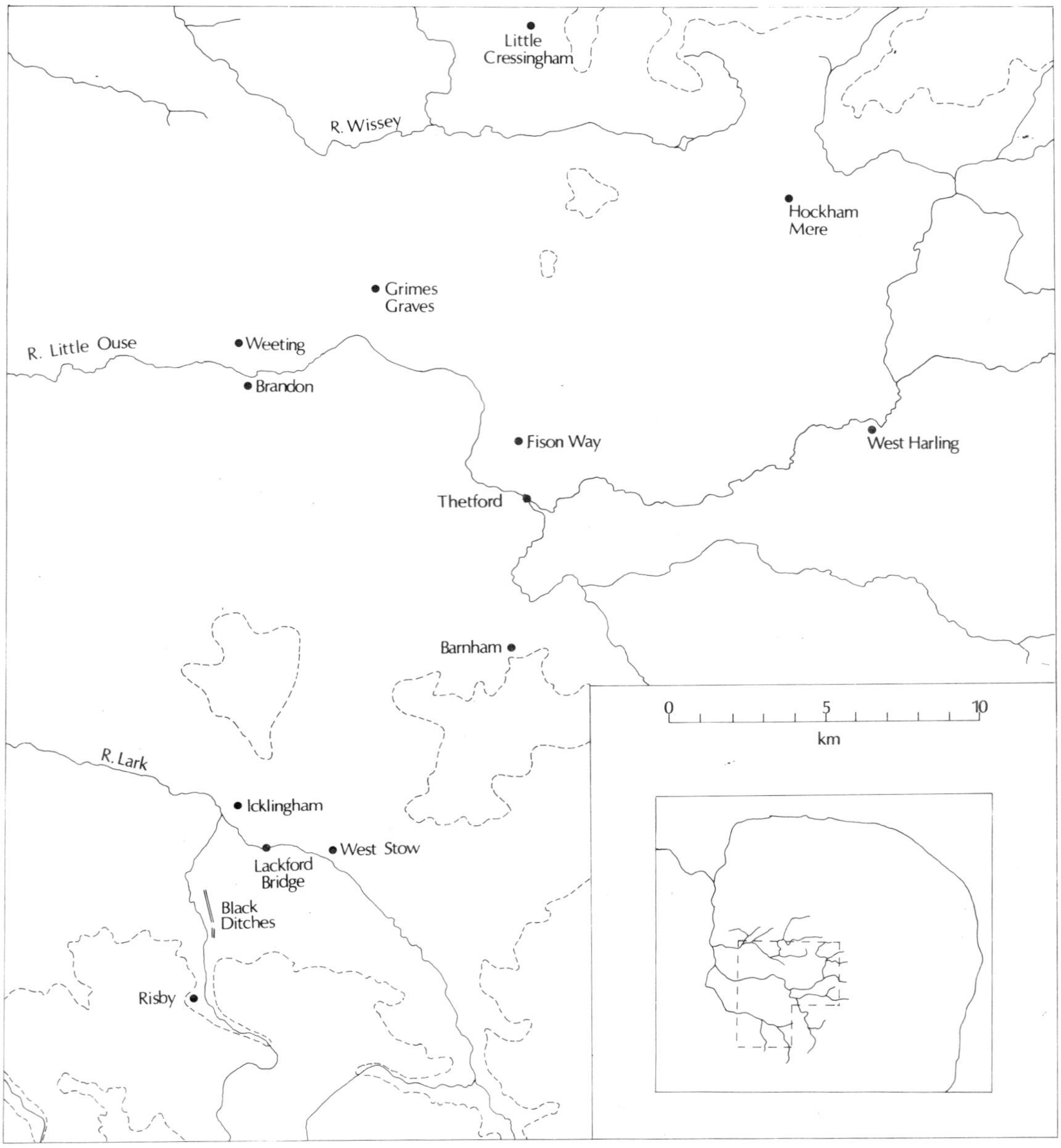

Fig. 9:1 Map showing principle Breckland rivers, 50 m contour and main sites mentioned in text.

Table 9:1: THE MAIN SITES DISCUSSED IN THE TEXT

Site	Grid Reference	County Site Number	Type of Site	Period	Ancient Monuments Laboratory Reports Series No.
Lackford Bridge, Suffolk		-	Peat section	-	2515
Stanch Meadow, Brandon, Suffolk	TL 778864	BRD 018	Peat section; settlement	Iron Age - Middle Saxon	3524, 3567
West Stow, Suffolk	TL 791713* TL 797713**	WSW 030	Settlements	Iron Age - Middle Saxon	3325, 3491, 3493, 3494, 3495 3496, 3498, 3507, 2944
Icklingham, Suffolk	TL 783719	IKL 063	Settlement	Roman	2521
Barnham, Suffolk	TL 866777	BNH 009	Enclosed settlement	Iron Age	2837
Fengate Farm, Weeting, Norfolk	TL 777878	WWB 5636	Settlement	Roman	forthcoming
St. Barnabas' Hospital, Thetford, Norfolk	TL 869822	THD 1092	Urban site	Late Saxon	2516
Little Cressingham, Norfolk	TL 863988	CRL 5053	Ring-ditch	Iron Age Roman (upper fill)	2888
Fison Way, Thetford, Norfolk	TL 864848	THD 5853	Settlements	Beaker - Late Roman	3566 and forthcoming
Gallows Hill, Thetford, Norfolk	TL 864846	THD 5744	Turf-stack	Roman	2945

Definitive reports on these sites will be published in the journal East Anglican Archaeology. Interim reports, subject to alteration and addition, are included in the Ancient Monuments Laboratory Report Series.

* Iron Age Site
** Saxon Site

sheep-walk or used for infield-outfield farming maintaining soil fertility by sheep and by marling (Clarke, W.G. 1925: 20-1; Darby 1935). Rabbit warrens were established during the Middle Ages and rabbit farming was economically important until the 19th century (Clarke, W.G. 1925: 131-145). The introduction of large rabbit populations was a new factor supplementing the effects of sheep on vegetation (Crompton and Sheail 1975). Wind-blowing of soils is known to have occurred in the Middle Ages: at West Stow in Suffolk a wind-blown sand deposit, perhaps of late 13th century date, overlay 12th-13th century ridge-and-furrow (West 1969: 3). From at least the 16th century onwards overgrazing resulted in the widespread occurrence of unstable surface sand with occasional catastrophic sand-storms, notably in 1668 when the River Little Ouse was blocked and the village of Santon Downham partly buried. During the 18th and early 19th centuries there were extensive enclosures associated with planting of Scots Pine shelter-belts, which are now a very familiar feature of the landscape, and the introduction of a four-course farming rotation. Later 19th and 20th century agricultural depressions resulted in a long-term population decline. During the 1920s the Forestry Commission began planting conifers and today about 21,000 ha of the Breckland consists of plantations, though more than two thirds of the region is agricultural land, predominantly arable, producing mainly spring barley and sugar beet (Clarke, W.G. 1925: 20-4; Corbett 1973: 23-30).

Our present understanding of pre-medieval land-use in the Breckland is based upon Godwin's (1944) interpretation of his results from pollen analysis at Hockham Mere. He proposed the then-controversial view that the change from forest to heath indicated by palynology was a consequence of neolithic and later deforestation by man rather than being of climatic origin. More recent work at Hockham Mere by Sims (1978) has built upon this foundation, and analysis of biological remains from excavations at Grimes Graves by Legge (1981) has provided specific information on the economy of a Bronze Age farming settlement. In the present paper land use between the Iron Age and Late Saxon periods will be considered. The selection of this period has to a large extent been determined by the exigencies of Rescue Archaeology, but it does seem to comprise a real and distinct phase in the economic development of the Breckland, characterised by a substantial and, at times, predominant arable element in the agrarian economy and by repeated and, in some cases, continuous occupation and farming at sites on the terraces of the main rivers. Within this period Sims (1978) has distinguished several phases of land use. Between 2750-1860 b.p., a period corresponding to the Iron Age, there is evidence for a markedly increased scale of forest clearance and for increased dependence on cereals as part of a mixed economy. Subsequently between 1860-1610 b.p. there was little change in forest cover, but a further increase in cereal cultivation. This trend was reversed during the Late Roman to early Saxon period (1610-1300 b.p.) when cereal cultivation declined being replaced by pastoralism. During the subsequent Middle Saxon period, however, (1300-1140 b.p.) Sims considers that cereals became for the first time "the major product". These results provide an overall picture of land-use changes within the Hockham Mere pollen catchment. More detailed consideration will be given below to the diffentiation of land use between the main landscape sub-regions of the Breckland, as defined by Corbett (1973). The sub-regions discussed will be the river valleys, river terraces, slopes and upland areas at elevations of over 30m. O.D.

THE RIVER VALLEYS

The rivers Lark, Little Ouse, Wissey and their tributaries drain westward from the Breckland to the Fen Basin. Their broad floodplains are flanked by gravel terraces, and they contain up to about 3m of peats overlying coarse sands and gravels. Organic soils (Adventurers series) are formed on the peats with sandy gley soils at the floodplain margins (Corbett 1973). The river valleys and to a lesser extent the Meres were of crucial importance for the economies of Breckland settlements. Clarke W.G. (1925: 18-19) first drew attention to the regular spacing of modern villages along the river valleys and to some unusual features of Breckland parish boundaries laid out to ensure access to water.

Two valley floor sites have been investigated in recent years, and both provide some limited information about vegetation and land use. The first of these, at Lackford Bridge in the Lark Valley, was a section exposed in a gravel pit, and was unfortunately recorded and sampled in hurried and adverse circumstances (Fig. 9.2a). The section showed a 23cm thick layer of heat-shattered flint and charcoal, mainly of alder (*Alnus*), but with some oak (*Quercus*) and hazel (*Corylus* sp.), overlying a thin (5cm) brushwood peat with abundant wood and fruits of alder and some oak wood. The peat overlay coarse grey fluviatile sand. Above the archaeological deposit was a 7cm thick alder brushwood peat merging upwards into dark fen peat showing sandy layers and laminations, with an estimated thickness of 1.5m. Charcoal from the archaeological deposit gave a date of 3940 ± 70 bp or 1990 b.c. (HAR-2484) and this site thus pre-dates the main period considered in this paper. The section represents a local clearance and burning of valley-floor alder woodland. The brushwood peat over the archaeological layer evidently indicates regeneration, and this was not, therefore, a permanent clearance. There is no clear evidence for the site's function.

A second peat section was exposed at the northern edge of the excavation of Iron Age and Middle Saxon settlement sites located on a sand hummock over river gravels in the Little Ouse valley at Stanch Meadow, Brandon, Suffolk (Fig. 9.2b). This showed 63cm of peats at the flood plain margin. Overlying a greyish-brown sand at the base was 8cm of dark reddish brown to black very sandy peat with irregular sandy patches. This was covered by 24cm of dark reddish brown somewhat less sandy peat. Above this was a thin (1.5cm) band of *Calluna* charcoal with charred shoots, capsules and seeds and some fired clay fragments. Further peats above this charcoal layer were humified and disturbed by earthworms. A sample from the lowest 5cm of peat gave a date of 1810 ± 80 b.p. or ad 140 (HAR-4087), and a date of 1350 ± 70 b.p. or a.d. 600 (HAR-4086) was obtained for the charcoal band. Pollen samples were taken but analysis is not yet complete. The results of macrofossil analysis are listed in Table 9.2 and summarised in Fig. 9.3. More delicate types of macrofossils such as grass fruits and *Carex* utricles are rare in these samples, and there was a relatively high proportion of poorly-preserved partly

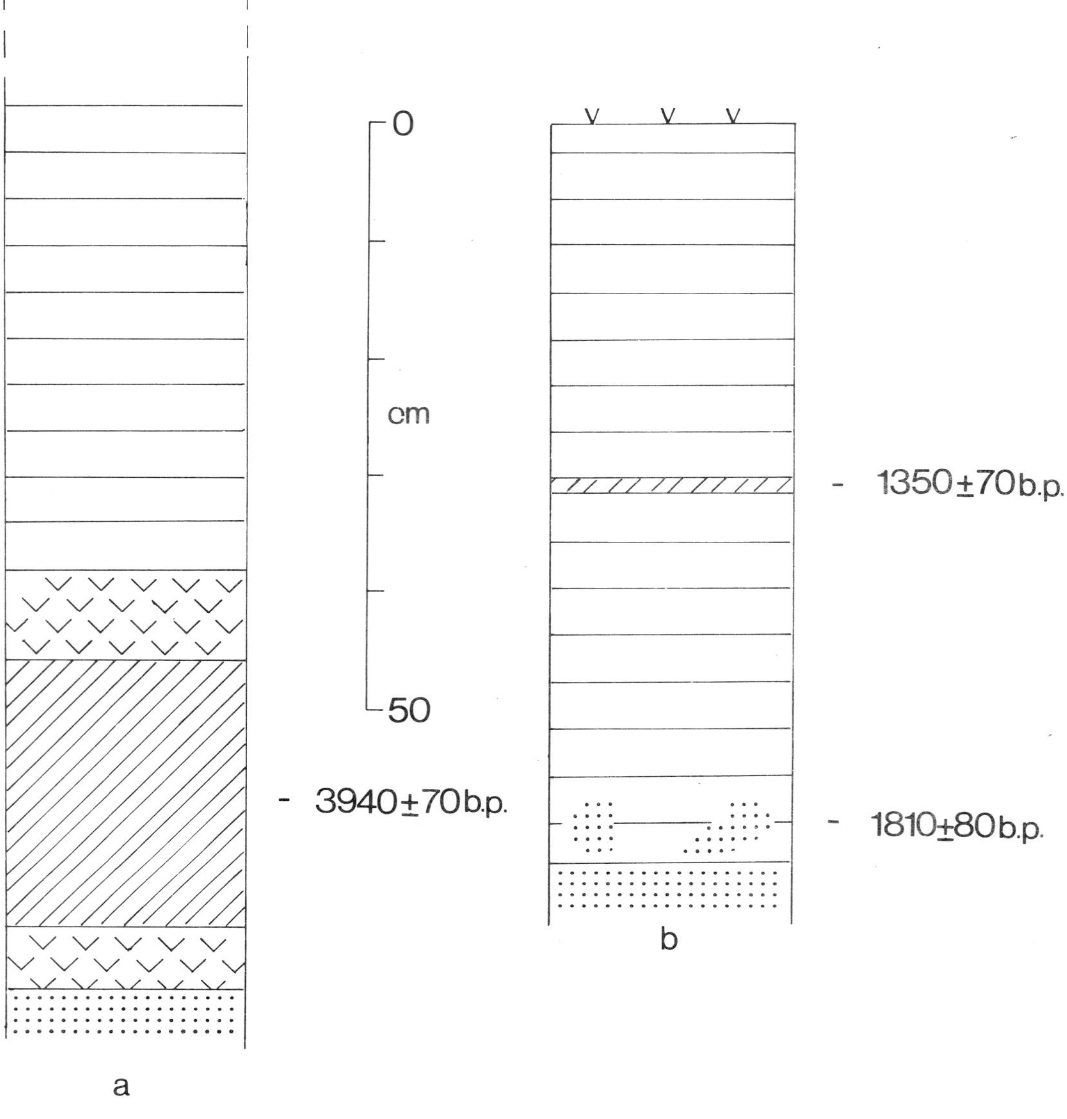

Fig. 9:2 Two valley-floor peat sections. a. Lackford Bridge; b. Stanch Meadow, Brandon. See text for details.

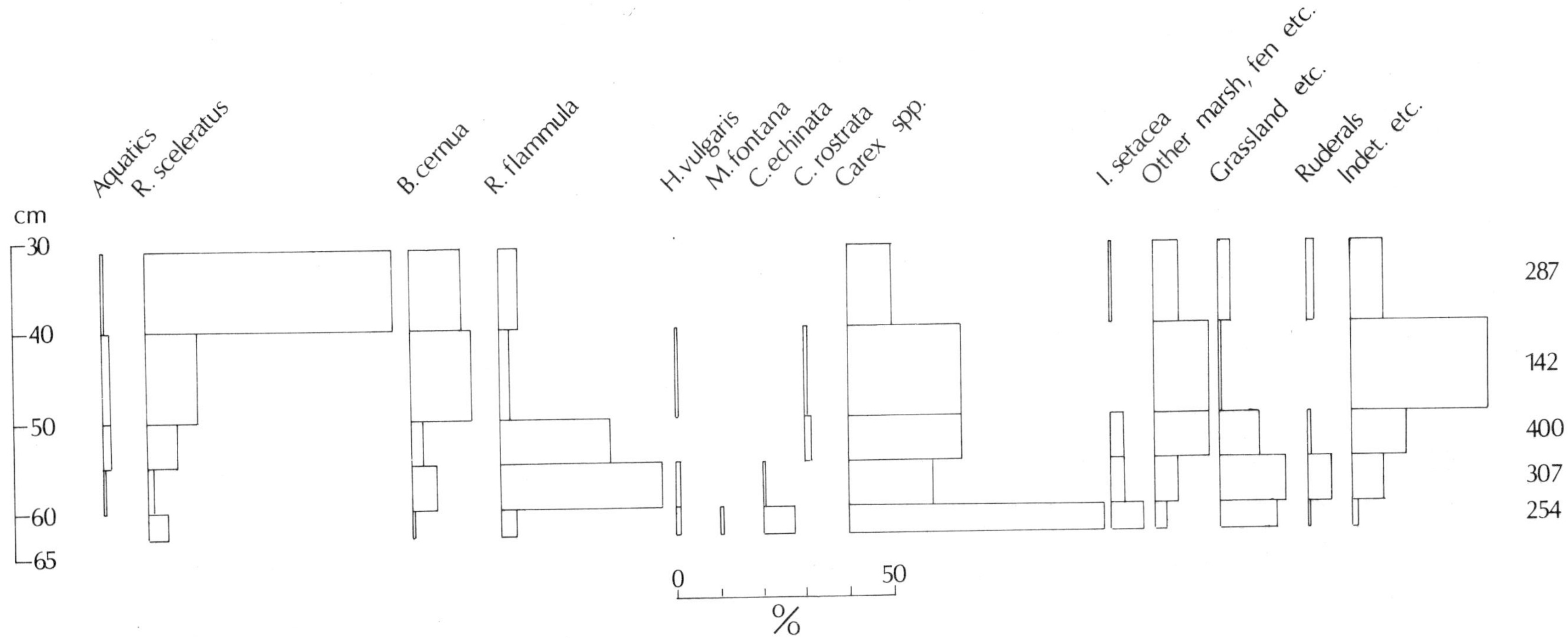

Fig. 9:3 Summary diagram of macrofossil frequencies in the peat section at Stanch Meadow, Brandon. Not included in this diagram are vegetative remains of Phragmites, seed fragments of Iris pseudacorus and Juncus seeds.

Table 9:2: Plant remains from the Brandon peat column. (1kg. samples). All taxa represented by fruits or seeds unless otherwise indicated.

Depth (cm)	30-31	31-40	40-50	50-55	55-60	60-63
Ranunculus acris/repens	-	-	1	6	4	3
Ranunculus flammula L.	-	12	3	99	113	9
Ranunculus sceleratus L.	-	161	16	27	4	11
Ranunculus sp.	-	-	-	-	1	4
Reseda sp.	-	-	-	1	-	-
Hypericum sp.	-	1	-	3	-	-
Lychnis flos-cuculi L.	-	2	10	36	1	1
c.f. *Cerastium* sp.	-	9	13	18	-	-
Caryophyllaceae indet.	-	1	1	-	3	-
Montia fontana L. subsp *chondrosperma*	-	-	-	-	-	2
Chenopodium album L.	-	1	-	c.f.1	-	-
Chenopodiaceae indet.	-	1	-	-	-	-
Potentilla sp.	-	6	-	27	41	26
c.f. *Fragaria vesca* L. (a)	-	-	1	-	-	-
Cicuta virosa L.	-	-	-	4	1	-
Hydrocotyle vulgaris L.	-	-	1	-	3	2
Berula erecta (Hudson) Colville	-	7	1	1	1	6
Rumex acetosella agg.	-	2	-	3	-	-
Polygonum aviculare agg.	-	-	-	-	1	-
Polygonum c.f. *hydropiper*	-	-	-	1	-	-
Urtica dioica L.	-	1	-	1	12	1
Calluna vulgaris (L) Hull (b)	+	-	-	-	-	-
Hyoscyamus niger L.	-	1	-	-	-	-
Mentha arvensis/aquatica	-	-	-	-	1	-
Lycopus europaeus L.	-	4	2	6	9	-
Galeopsis tetrahit/speciosa	-	-	-	-	1	-
Scutellaria c.f. *galericulata* L. (c)	-	-	5	2	1	-
Bidens cernua L.	-	33	19	11	17	1
Cirsium sp.	-	-	-	-	1	-

Depth (cm)	30-31	31-40	40-50	50-55	55-60	60-63
Sonchus arvensis L.	-	1	-	-	-	-
Sonchus asper (L) Hill	-	-	-	-	1	-
Alismataceae indet.	-	2	1	1	-	-
Juncus spp.	-	+	+	+	+	+
Iris pseudacorus L. (d)	-	-	+	+	+	+
Lemna sp.	-	-	1	2	-	-
Typha sp.	-	2	-	-	-	-
Eleocharis sp.	-	1	-	3	3	-
Carex rostrata Stokes	-	-	1	5	-	-
Carex echinata Murray	-	-	-	-	2	18
Carex sp. (e)	-	-	-	23	36	128
Carex spp.	-	26	36	79	22	20
Isolepis setacea (L.) R.Br.	-	1	-	10	9	18
Gramineae indet.	-	-	-	4	15	-
Phragmites australis (Cav) Steudel (f)	-	-	+	+	+	-
Charcoal fragments	+	-	-	-	+	-
Indet.	-	12	30	26	4	4
Totals	-	287	142	400	307	254

Notes (a) Form + size appropriate for this species, but pattern of ridges obscure.
(b) Charcoal, charred leaves, capsules + seeds.
(c) Intermediate in size between *S. galericulata* and *S. minor*. No reference specimens of the rare Breckland plant *Scutellaria hastifolia* were available.
(d) Seed fragments showing 'spongy' internal tissue.
(e) All one species. Attempts to identify these nutlets using the key of Nillson + Hjelmquist (1967) were unsuccessful + they could not be matched with available reference specimens.
(f) Culm nodes + fragments.

degraded fruits and seeds which have not been identified. Also present but not included in this diagram were *Juncus* seeds, fragments of *Iris pseudacorus* and, between 40 and 60cm rare vegetative remains of *Phragmites*.

The irregular sandy patches in the peat below 55cm do not appear to indicate fluviatile deposition of sand, but rather some artificial disturbance of sand on the adjacent terrace. Since there is no evidence for settlement at a period corresponding to the radiocarbon date for this lower peat it seems that this disturbance may have resulted from trampling by cattle or, less probably, ploughing related to a minor peak in frequencies of seeds from ruderal plants at 55-60cm (*Urtica dioica*, *Galeopsis tetrahit/speciosa*, *Cirsium* sp., *Sonchus asper*). The majority of seeds from this level are, however, from grassland, marsh and fen plants. Three species from this peat are particularly characteristic of wet or damp acid soil conditions (*Montia fontana* subsp. *chondrosperma*, *Hydrocotyle vulgaris*, *Carex echinata*), and a fourth, *Isolepis setacea* is common on damp, bare sandy ground (Clapham *et al.* 1962; Petch and Swann 1968; Walters 1953). All four decline in frequency higher in the peat. In the less sandy peat between 31-50cm *Ranunculus sceleratus* and *Bidens cernua* show marked increases in frequency. *R. sceleratus*, which is by far the most abundant species at 31-40cm. occurs in shallow-water and on mud typically in mineral-rich conditions. *Bidens cernua* is found by streams, particularly where water stands only in the winter. (Clapham *et al.* 1962.) Fruits of aquatics (*Lemna*, *Cicuta virosa* and indeterminate Alismataceae) are present at low frequencies in this peat.

The sequence is interpreted as follows. From 140 ad peat formation began over the terrace sands at the present floodplain margin. The presence of seeds of calcifuges seems to indicate that wet acid soil conditions were initially maintained by an influx of acidic groundwater from the decalcified sandy terrace soils. Subsequently the seed assemblages appear to represent more eutrophic conditions, perhaps related to increased frequencies of seasonal river flooding. Increasing wetness may perhaps be correlated with deteriorating rates of freshwater drainage associated with freshwater flooding in the southern Fen basin from the later Roman period onwards (Churchill 1960: 142). Alternatively some local factor could be involved. It may, for example, be relevant to note that bones of beaver were identified in an Iron Age context at Micklemoor Hill, West Harling adjacent to the River Thet (Clark and Fell 1953: 36). Throughout the sequence herbaceous plants predominate and there is no sign of woody species such as *Frangula alnus*, *Rhamnus catharticus*, *Salix* or *Viburnum opulus* which are commonly the primary colonists of open fen (cf. Tansley 1939: 657) nor any alder fruits. This, together with the evidence for disturbance of the lowest peat is thought to indicate suppression of scrub development by grazing at the floodplain margins. The thin band of charred *Calluna* remains dated to a.d. 600 which seals these peats is in-washed material thought to have been produced during clearance by fire of heath on the adjacent sand hummock prior to construction of the Middle Saxon settlement.

Results from two isolated peat sections clearly do not provide a firm basis for any generalisations about land use in the valley floors. By analogy with modern East Anglian valley-floor wetlands some zonation of vegetation across the river valleys is to be expected. Much more information will be required before it can be determined when large-scale use of the valley floors for pasture began. However two types of activity are distinguishable: an early temporary clearance of alder woods and a later maintenance of open vegetation apparently by grazing.

Settlement sites on the river terraces have provided some further evidence for the exploitation of valley-floor resources. Charred culm fragments probably of Phragmites were present in Middle Saxon features at Brandon, and this seems to indicate some reed-cutting for use as thatch. Besides the beaver remains from West Harling remains of crane were also identified. The crane breeds on the ground in wet bogs, lightly-wooded swamps and reed-beds (Peterson et al. 1954: 109). Evidence for river fishing is at present scanty: Late Saxon fishbone from St. Barnabas' Hospital, Thetford consisted largely of imported marine fish, although remains of perch, pike, eel and possibly tench have been identified by Andrew Jones (forthcoming). Fish-bone recovered by bulk-sieving at Brandon has yet to be examined. Finally, there is reason to suspect that in the Middle and Late Saxon periods flax and hemp cultivation extended from the terrace margins onto the organic floodplain soils. It is to farming on the terraces that I now turn.

RIVER TERRACE SITES

Between the main river floodplains and the upland slopes are broad flat terraces consisting of fluvial gravels overlain, in some areas (e.g. at Brandon and West Stow) by blown sand deposits. The soils of these terraces are well-drained stony brownearths (Freckenham and Croxton series), humus podzols (Redlodge) and raw soils on blown sand (Broomhouse series). At the floodplain margins are damper gley soils (Corbett 1973: 49, 65-72). In uncultivated areas the present-day 'natural' vegetation is Calluna-heath with bracken (Watt 1955) and grass-heath, whilst on unstable blown sand the sand-sedge (Carex arenaria) is common. These plant communities are generally well-represented on West Stow heath (Darrah et al. 1980), although pure stands of bracken are very rare.

Charred crop plant remains have been recovered from several settlement sites on terraces of the River Lark at West Stow and Icklingham and the Little Ouse at Brandon, Fengate Farm, Weeting and St. Barnabas' Hospital Thetford. In addition, cereal impressions on Saxon pottery from West Stow have been identified, some uncharred crop plant remains from wet deposits have been recovered at Brandon and Thetford, and samples from Icklingham have produced some mineralised seeds. Crops identified from these terrace sites are shown in Table 9:3. So far as cereals are concerned, the principle crops represented in Iron Age and Roman samples were spelt (Triticum spelta L.) and six-row hulled barley (Hordeum vulgare L.Emend). Emmer (Triticum dicoccum Schübl.) and wild or cultivated oats (Avena sp.) occurred at low frequencies in the Iron Age samples, and in

Table 9:3: SYNOPSIS OF THE DISTRIBUTION OF CROP PLANT REMAINS FROM SOME BRECKLAND TERRACE SITES

	IRON AGE		ROMAN			EARLY SAXON	MIDDLE SAXON		LATE SAXON
	West Stow	Brandon	West Stow	Weeting	Icklingham	West Stow	West Stow	Brandon	Thetford
Triticum c.f _dicoccum_ (emmer)	+	+	0	+	0	0	0	0	0
Triticum spelta (spelt)	+++	+++	+++	+++	+	++	0	0	0
Triticum aestivum s.1. (free-threshing wheat)	0	0	+?	+	0	++	+	+	+
Triticum sp (indeterminate wheat)	++	++	++	++	+	++	0	0	0
Hordeum sp. (barley)	+++	++	++	++	+	++	++	++	+
Secale cereale (rye)	0	0	+	0	+?	++	+++	++	+
Avena sp. (oats)	+	+	+	+	0	+	+	+	+
Vicia faba var _minor_ (horse-bean)	0	0	0	0	0	0	0	+	0
Vicia/Pisum sp. (vetch or small pea)	0	0	0	0	0	+	0	0	0
Linum usitatissimum (flax)	0	0	0	0	0	0	0	+	0
Cannabis sativa (hemp)	0	0	0	0	0	0	0	0	+
Prunus domestica (plum)	0	0	0	0	0	0	0	+	0
Prunus c.f. _avium_ (cherry)	0	0	0	0	+	0	0	0	0
Malus-type (apple)	0	0	0	0	+	0	0	0	
No. of contexts or samples producing crop plant remains	12	5	3	12	18	1	5	30	9

0 - absent
\+ - rare
++ - common. Frequently the predominant crop species of small samples, or secondary crop species in large samples
+++ - abundant. Predominant crop species in large samples.

Tentative identifications of cereals refer to grains not associated with spikelet fragments, to damaged/fragmentary spikelet remains or (particularly in the case of _Avena_) to awn fragments only.

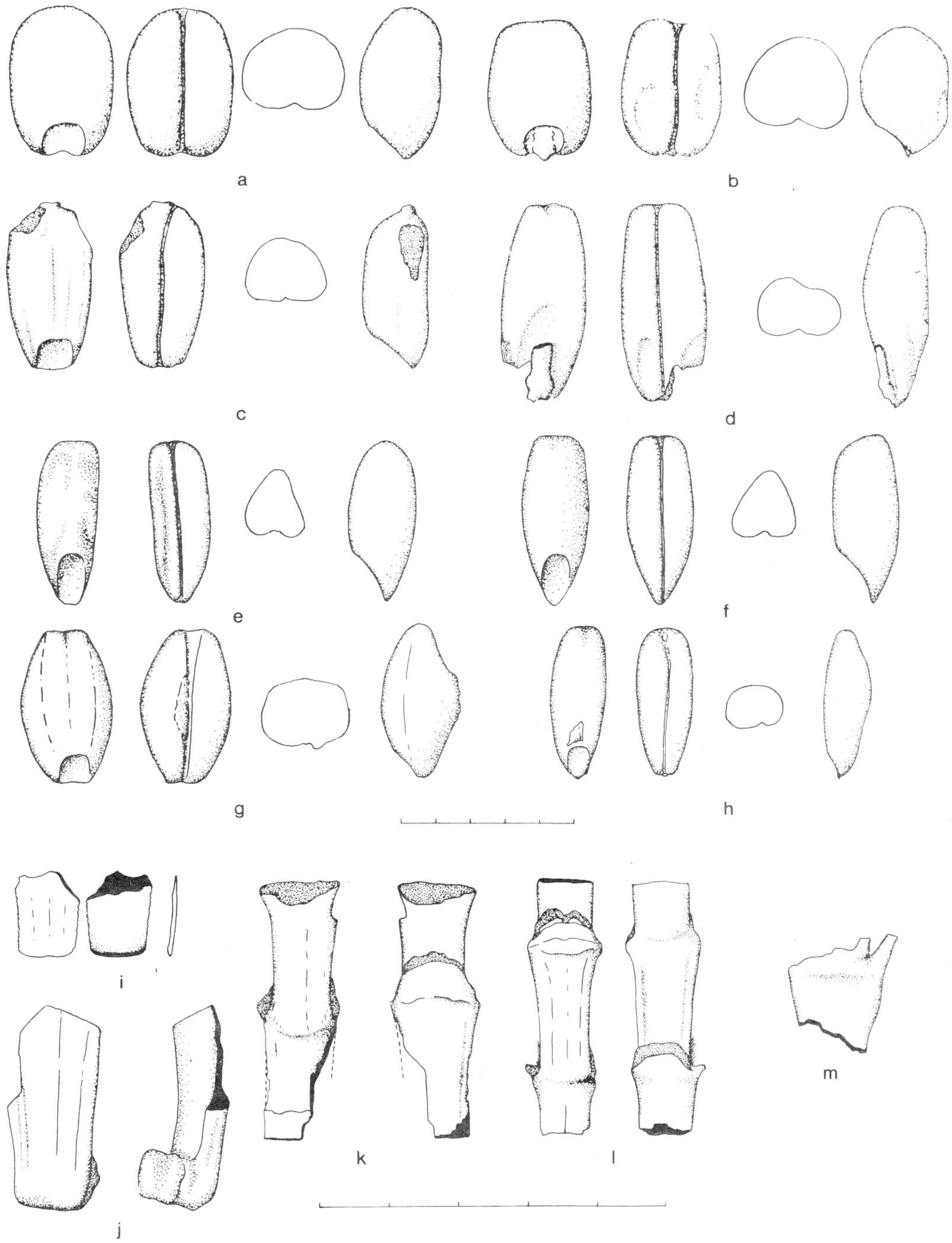

Fig. 9:4 Select charred cereal remains from West Stow. a,b Triticum aestivum s.l., caryopses. Hut 63, mid-5th century; c. Triticum c.f. spelta, caryopsis. Pit 472, Roman; d. Triticum c.f. spelta, Germinated caryopsis. Pit 472; e,f. Secale cereale, caryopses. Hut 63; g. Hordeum sp., caryopsis. Hut 63; h. Avena sp., caryopsis. Hut 63; i. Triticum cf. spelta. Rachis internode. Hut 63; j. Triticum spelta Glume base. Hut 63; k. Secale cereale. Rachis fragment. Pit 472; l. Secale cereale. Rachis fragment. Pit 59. Anglo-Saxon (unphased); m. Hordeum sp. Rachis internode fragment. Pit 472; Scales graduated in mm.

addition rare remains of free-threshing wheat (*Triticum aestivocompactum* Schiem.) and rye (*Secale cereale* L.) have been identified in Roman samples. Samples from Middle and Late Saxon sites contained mainly rye and barley with lesser amounts of free-threshing wheat and oats. A mid-5th century sunken hut at West Stow (63) produced remains of most of these cereal crops, and the sample from this context is thought to represent a transitional phase in the early Saxon period when spelt contained to be cultivated, prior to its later replacement by free-threshing wheats and rye. The status of oats as a crop or weed is uncertain at all periods, since this genus is represented only by naked grains or awn fragments. Since the oat plant requires more water than other temperate cereals (Renfrew 1973: 98) large-scale cultivation in the Breckland seems at first sight improbable, though the crop was certainly grown in the 13th century (Clarke, W.G. 1925: 23).

Crops other than cereals are, as usual at rural sites, poorly represented. The only pulse seeds are horse-beans (*Vicia faba* var *minor*) from Middle Saxon samples at Brandon and a vetch or small pea from West Stow. Charred seeds of flax (*Linum usitatissimum*) were also recovered at Brandon and a Late Saxon sample from St. Barnabas' Hospital, Thetford produced a fragmentary hemp fruit (*Cannabis sativa*). The production of these fibre crops in the Breckland must have been confined to the damper, more organic soils of the valley floors and terrace margins since flax is very shallow-rooted, requiring ample moisture during early stages of development (Renfrew 1973: 124) and satisfactory hemp crops can only be obtained on deep rich humic loams (Percival 1918: 349). Bradshaw *et al.* (1981) report results from pollen and macrofossil studies indicating hemp cultivation and retting probably of medieval date at two basin sites in East Anglia, one at Bugg's Hole, a small fen in the valley of the Little Ouse. A mineralised cherry fruitstone (*Prunus avium*) and apple seed (*Malus* sp.) came from Roman contexts at Icklingham and a wet Middle Saxon pit fill at Brandon produced a plum fruitstone (*Prunus domestica*).

The charred seeds and other remains of wild plants associated with the cereals from these terrace sites have been examined with a view to producing some information on soil conditions in the arable fields though few of the samples contain sufficient weed seeds for quantitative analysis. The majority of Middle Saxon samples from Brandon, for example, consist of fully-processed prime grain with few impurities. Other samples probably contain only a selected assemblage of larger weed seeds produced during crop processing (e.g. a rye sample from West Stow, Pit 59, in which seeds of *Agrostemma githago* and grass caryopses are almost the only contaminants present). However, in Table 9.4 the composition of weed seed assemblages from six of the larger samples is summarised. The species identified are divided into seven groups (Clapham *et al.* 1962; Hind 1889; Petch and Swann 1968):

1. Seeds of weeds with affinities for dry sandy soils. Some of these are calcifuges, others occur on both acid and basic dry soils

2. Heathland species

3. Catholic weed species

4. Weed plants with preference for heavy clay soils (*Anthemis cotula*)

5. Grassland species

6. Plants characteristic of damp or moist soils

7. Unidentified, or otherwise uninformative, plant remains.

The weed samples come from three separate terrace sites and range in date from the Iron Age to the early Saxon period, but nevertheless are broadly similar in composition. In each case, as would be expected, catholic weed plants, which occur in a wide range of disturbed habitats, are very common (68.6-88.4%). Weed species frequent on dry, sandy soils account for up to 16.9%, but are not present in all samples. The presence of *Anthemis cotula* in hut 63 at West Stow is unexpected and unexplained, since this weed is nowadays more frequent on heavy chalky till soils in East Anglia. Possibly it was a contaminant of seed corn imported from an area of heavier soils. The remaining plants are not characteristically arable weeds, but the seeds of these species are probably derived from plants growing at the field margins and sporadically in the crop, spreading from adjacent unploughed land. They include heath plants (up to 4.1%), grassland species (up to 3.3%) and plants characteristic of damp or wet soils (up to 2.0%). In summary, the relative frequencies of these groups are thought to indicate that the area ploughed at these different periods was largely on the dry sand soils of the terraces, but that cultivation extended onto the edges of river flood-plains, where some colonisation of field margins by damp-ground plants was possible. Charred capsules, leaves and shoots of *Calluna vulgaris* and pinnules of *Pteridium aquilinum* are thought to reflect the proximity to the arable land of heath at Brandon and West Stow.

At the present day one of the principle factors limiting cereal yields on sandy soils in the Breckland is moisture availability and this undoubtedly applied equally during past climatic phases as dry as or drier than that of today. In dry years nowadays barley yields can be reduced to less than half normal yields (Corbett 1973: 29). Trace element deficiencies may further reduce yields and wind erosion of surface soil in dry spring months occurs in most years, occasionally making re-drilling necessary. Moisture losses may be reduced by minimum cultivation, and, as Legge (1981) has pointed out, the relative inefficiency of early ploughs may in these circumstances have proved to be an advantage. Losses due to wind erosion and to drought at the seedling stage could have been reduced by cultivating winter crops, so as to ensure that the young plants would be relatively well-developed before spring and summer droughts (Limbrey 1978: 25) and that the soil surface would be stabilised by root systems during the spring. Spelt and rye, successively the most abundant cereals in the samples examined are both normally autumn-sown crops though fruits of *Galium aparine*, a weed which Reynolds (1981: 112) suggests is characteristic of autumn sown cereals grown under 'primitive' conditions are extremely rare in the samples examined. Rye is particularly well adapted to growth on dry soils.

Table 9:4: Species composition and ecological groups of selected charred 'weed' seed assemblages associated with cereals from three Breckland terrace sites.

	Iron Age West Stow		Iron Age West Stow		Iron Age Brandon		Iron Age Brandon		Roman West Stow		Early Saxon West Stow	
Context number	034		100		1362		1372		472(2)		63	
1. Papaver argemone	-		-		-		-		2		-	
Raphanus raphanistrum	-		-		-		-		6		2	
Arenaria serpyllifolia	-	(0%)	-	(11.1%)	-	(2.4%)	-	(0.9%)	34	(16.9%)	1	(12.8%)
Spergula arvensis	-		-		-		-		3		3	
Scleranthus cf. annuus	-		-		3		1		19		9	
Aphanes arvensis/microcarpa	-		-		-		-		1		3	
Rumex acetosella	-		17		13		3		356		214	
2. Calluna vulgaris (capsules)	-		-		27		10		2		6	
(Pteridium aquilinum) (pinnules)	-	(0%)	-	(0%)	(+)	(4.1%)	-	(2.2%)	-	(0.1%)	-	(0.4%)
Stellaria graminea	-		-		-		-		-		2	
3. Silene alba	-		-		1		1		16		8	
Agrostemma githago	-		-		-		-		-		3	
Stellaria media-type	-		1		-		-		30		9	
Chenopodium hybridum	-		-		-		-		-		3	
Chenopodium album	18		9		125		4		87		278	
Atriplex patula/hastata	-		1		7		2		502		215	
Chenopodiaceae indet.	102		63		65		183		157		269	
Malva sp.	-	(87%)	2	(68.6%)	7	(85.5%)	1	(88.4%)	2	(71.8%)	2	(71.3%)
Vicia sp.	-		-		2		-		-		3	
Medicago/Trifolium-type	-		8		286		63		415		238	
Polygonum aviculare	11		7		31		30		258		102	
Polygonum convolvulus	2		6		9		63		188		66	
Polygonum cf. persicaria	7		2		4		-		-		7	

	Iron Age West Stow		Iron Age West Stow		Iron Age Brandon		Iron Age Brandon		Roman West Stow		Early Saxon West Stow	
Context number	034		100		1362		1372		472(2)		63	
Polygonum lapathifolium	1		-		-		-		-		-	
Rumex sp.	1		1		9		24		56		33	
Polygonaceae indet.	9		3		7		20		60		44	
Anagallis sp.	-		-		-		-		1		-	
Myosotis sp.	-		-		-		-		-		1	
Galium aparine	1		-		-		-		1		-	
Tripleurospermum maritimum	-		-		-		-		3		-	
Bromus mollis/secalinus	2		2		12		4		9		9	
4. *Anthemis cotula*	-	(0%)	-	(0%)	-	(0%)	-	(0%)	-	(0%)	16	(0.9%)
5. *Ranunculus acris/repens*-type	-		-		-		-		25		5	
Prunella vulgaris	-	(1.7%)	-	(3.3%)	-	(0.3%)	-	(0.4%)	8	(1.7%)	4	(1.1%)
Plantago lanceolata	3		5		2		2		10		10	
6. *Ranunculus* cf. *flammula*	-		-		-		-		3		-	
Montia fontana subsp. *chondrosperma*	-	(0%)	-	(0.6%)	-	(0.3%)	-	(0%)	3	(2.0%)	6	(2.0%)
Eleocharis sp.	-		-		1		-		16		7	
Carex sp.	-		1		1		-		27		23	
7. Unidentified/incompletely identified	20	(11.3%)	25	(16.3%)	49	(7.4%)	36	(8.1%)	185	(7.4%)	207	(11.4%)
Total 'weed' seeds/fruits	177		153		661		447		2485		1808	

It has a more extensive root system than other cereals, extending to depths of 1.6-1.8m and, probably as a consequence of this, is less inhibited by drought (Renfrew 1973: 85). It is probable that the apparently increased cultivation of rye in the post-Roman period was one factor permitting an extension of the arable area: there is evidence from Hockham Mere for markedly increased pollen frequencies of cereals and arable weeds in the period 1300-1140 bp. (Sims 1978: 58).

The moisture-holding capacity and base status of these terrace soils could have been improved by marling. Chalk is not available in surface outcrops in the immediate vicinity of the terraces, but chalk lumps were present in pits at West Stow (WSW 030), having been imported to the site from slope areas above the river terraces. Undoubtedly, however, the principle adaptation ensuring adequate and consistent cereal yields must have been a close relationship between arable and pastoral farming, so as to ensure sufficient supplies of manure. Manuring would have increased the organic content and hence the moisture capacity of the soil, helping to reduce soil moisture deficits and stabilising the surface, as well as replacing minerals lost by cropping and leaching. There is, nevertheless, some evidence for soil nitrogen deficiencies. Although charred seeds of *Vicia* and *Lathyrus* never account for more than a few percent of the weed seeds associated with the cereals, some Iron Age deposits from Brandon and from the early Saxon deposit from hut 63 at West Stow contained large numbers of badly deformed small leguminous seeds of *Medicago* or *Trifolium*-type: as much as 48% of total weed seeds in one Brandon sample (1354).

The only substantial collection of faunal remains from a rural Breckland terrace site is that from West Stow. A few identifications of cattle, sheep, pig, red deer, beaver, dog and crane from the Iron Age site at West Harling are given by Clark and Fell (1953), but other recent excavations have either been too small to produce useful bone assemblages or else are not yet completed. The West Stow assemblage comprises some 200,000 bone fragments mainly from Anglo-Saxon features. Crabtree (1981) has discussed the results of analysis of material from this phase. Amongst the large domestic herbivores sheep are numerically predominant, followed by cattle, with some pig and goat. Estimates of ages at death, based on dental eruption and wear, indicate a high mortality for cattle in the first two years of life, but about 44% of the mandibles examined had a fully erupted dentition. Some cattle survived to well over five years of age. Measurements of metacarpals show that the majority of mature cattle were cows. These data suggest dairying in addition to meat production. For sheep the distribution of ages at death is interpreted as indicating a combination of meat production and dairying, with no sign of large-scale wool production. The highest mortality for pigs is seen in the early third year, corresponding to the stage when the animals reached bodily maturity. Crabtree notes that the open landscapes of the vicinity provide little or no pannage for pigs and suggests that sty husbandry, feeding on cereals and food wastes, may have been practised. Size estimates from bone measurements indicate that the West Stow animals were larger than Iron Age stock.

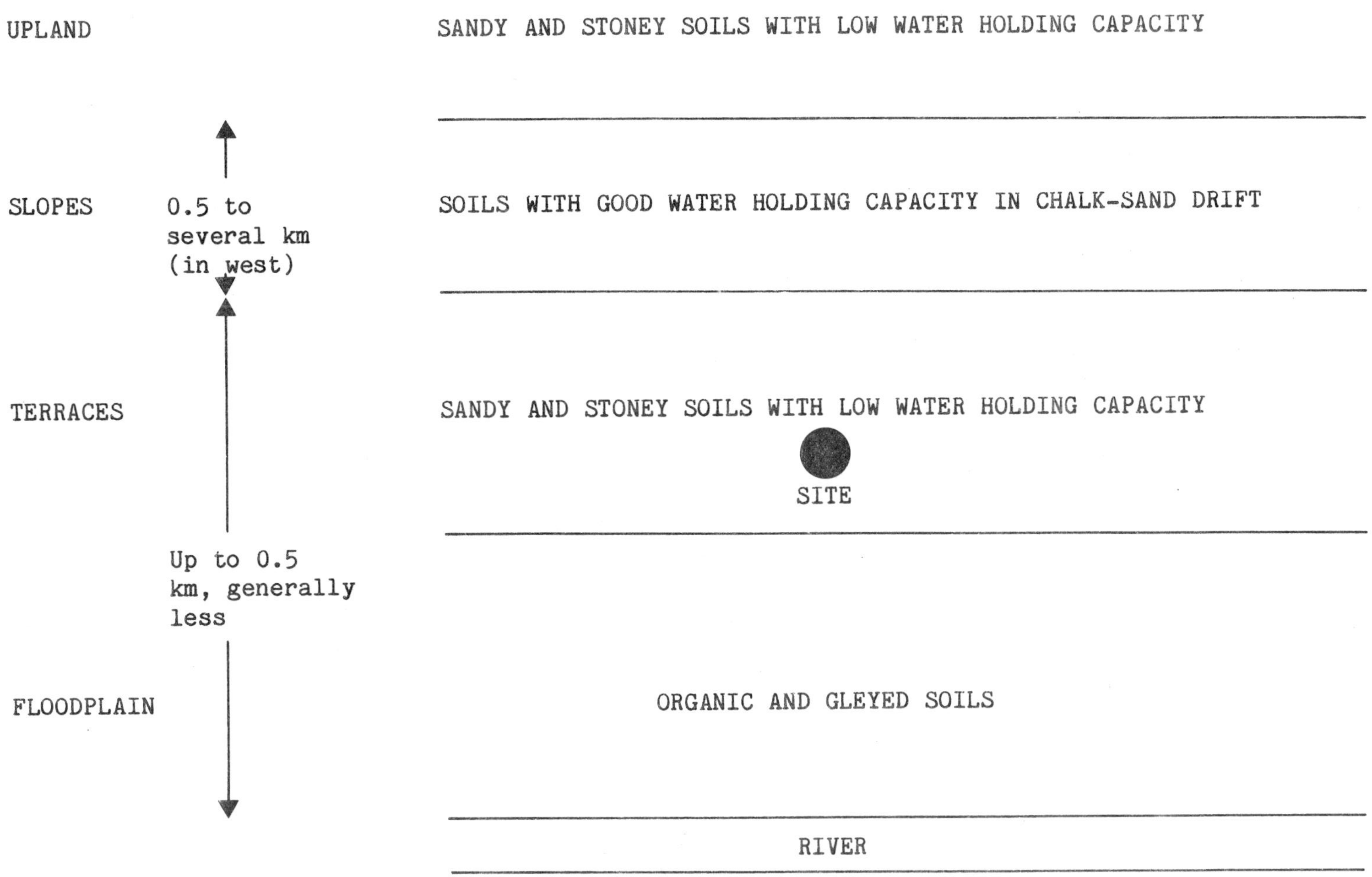

Fig. 9:5 Idealised location of a settlement site on a river terrace.

As is noted above the vegetation of the terraces today consists mainly of heathland with some grass-heath, and the presence of charred *Calluna* and *Pteridium* in archaeological features and in the peat at Brandon suggests the proximity of such communities between the Iron Age and Middle Saxon periods. Grazing in the immediate vicinity was therefore probably not of high quality and may locally have been dominated by bracken, which is poisonous to stock, but as is shown in Fig. 9:5 terrace sites were centrally situated to exploit both valley floor pastures and also the calcareous grassland on the slopes above the terraces. Evidence for grazing in the valley floors has already been discussed and land use on the slopes will now be considered.

SLOPE AREAS

Between the river terraces and the upland areas of the interfluves are gentle slopes overlying the chalk-sand drift, a deposit which has been interpreted as a Till (West and Donner 1956, Perrin 1956) or alternatively as a mixture of cryoturbated chalk with wind-blown cover sand (Watt, Perrin and West 1966). The drift surface shows undulations which have a surface expression in patterned ground, typically stripes on slopes and polygons on more level sites. There is marked small-scale lateral variation in soil types. Depending on the depth below the surface of the chalk-sand drift a rendzina, calcareous brownearth or sandy brownearth is found. The more calcareous soils support *Agrostis-Festuca* grassland with calcicoles, whereas *Calluna* may develop on the more acid soils (Corbett 1973; Duffey 1976). These plants often form a complex mosaic of communities.

Calcarous slope soils today provide much of the agricultural land in the Breckland. Under present-day climatic conditions and with modern crops these soils are more suitable for agriculture than are the terrace and upland soils, largely because of the superior moisture-holding capacity of their parent material, the chalk-sand drift, but also because of their higher base status (Corbett 1973: 61). It is therefore regrettable that there is very little direct evidence at present for arable farming on these soils during the period under consideration, largely because of lack of excavation of suitable sites. At Barnham, Suffolk excavation was concentrated on the defences of an Iron Age enclosure: extensive refuse deposits were not available for sampling. Only a few poorly preserved cereal grains were recovered. Rare wheat and barley grains were also present in the upper fill of a Bronze Age barrow ditch at Little Cressingham, Norfolk in association with Iron Age and Roman pottery. Further excavation and sampling will no doubt produce more evidence.

The slope soils also support some of the most productive grassland. Watt (1940) distinguished five main categories of grassland in the Breckland, related primarily to soil type. Vegetation on the deeper acid sands includes a very high proportion of the lichen *Cladonia*, and the immature *Festuca* grassland on the thinnest chalk soils includes up to about 50% bare ground, but on the deeper calcareous soils there is a continuous grass sward rich in species and dominated by *Agrostis* and *Festuca*. This latter grassland is the local equivalent of the Downland pasture of southern England.

Table 9:5: SELECTED IRON AGE SNAIL ASSEMBLAGES

	Barnham Inner ditch 160-170cm	Barnham Outer ditch 155-165cm	Fison Way Thetford 3309
Carychium tridentatum (Risso)	1	-	-
Cochlicopa sp.	6	36	(1)
Truncatellina cylindrica (Férussac)	6	11	-
Vertigo pygmaea (Draparnaud)	-	2	-
Pupilla muscorum (Linné)	35	165	2
Vallonia costata (Müller)	-	1	-
Vallonia excentrica Sterki	1	19	12
Vallonia sp.	28	64	17
Ena obscura (Müller)	6	-	-
Punctum pygmaeum (Draparnaud)	1	11	-
Vitrina pellucida (Müller)	2	4	44
Oxychilus sp.	5	-	11
Indeterminate Zonitidae	5	-	-
Limacidae	-	-	1
Cecilioides acicula (Müller)	1	1	13
Helicella itala (Linné)	22	46	4
Trichia hispida (Linné)	-	-	40
Cepaea nemoralis (Linné)	-	-	4
Cepaea sp.	-	1	4
Cepaea/Arianta	-	-	12
Indeterminate	8	7	-
Sample wt (kg)	1	1	5

1. Barnham. Inner ditch 160-170cm. Lower part of buried soil.

2. Barnham. Outer ditch 155-156cm. Buried soil.

3. Fison Way. Thetford. 3309. Post-hole of hut containing chalky material in-washed from erosion of associated dwarf wall.

Extensive areas are still used for sheep grazing today, and the Domesday Survey records exceptionally large numbers of sheep in the Breckland hundred of Lackford, with large flocks at Mildenhall, Icklingham, Downham and Eriswell (Darby 1935).

Evidence from sub-fossil molluscan faunas indicates open shade-free vegetation in the vicinity of several slope-soil sites. Of direct relevance to the period considered in this paper are mollusca from the upper fills of Bronze Age barrow ditches at Little Cressingham, Norfolk and Risby, Suffolk and from the ditches of an Iron Age enclosure at Barnham, Suffolk (BNH 009) (Table 9:5). The faunas from these sites are very poor in species and consist almost entirely of open-country xerophiles. Pupilla muscorum is invariably the predominant species, and this is thought to indicate that soil surfaces in the immediate vicinity of these sites were unstable. Vertigo pygmaea, which has a preference for stable soil surfaces and a complete grass cover (Evans: 1972 143) is very rare. These unstable conditions result, in part, from the unconsolidated nature of the chalksand drift into which the ditches were cut, but it is also possible that some of the sandy ditch fills represent wind-blown sand transported over short distances by saltation. High frequencies of Pupilla may also have been encouraged by the development on the earthworks adjacent to the ditches of immature calcareous grassland resembling the modern Breck Grassland A, which consists mainly of Festuca ovina with about 50% bare ground (Watt 1940). Faunas dominated by open-country snails occur throughout the fill of the Outer ditch at Barnham. There is no sign of scrub development on the buried soil in this ditch and the only faunal change detectable is the appearance at the base of the modern soil formed over the infilled ditch of the recently introduced Candidula intersecta. Similar faunas occur in most of the deposits of the Inner Ditch, but shade-loving taxa are present at low frequencies in the buried soil in this ditch, apparently reflecting some slight shading over of the habitat. This is thought to indicate that the Inner Ditch, partly protected by the Outer Ditch on one side and an internal bank on the other, was less accessible to grazing animals and some patchy development of taller vegetation could therefore occur. However there is no evidence for extensive scrub development at the sites investigated, and the maintenance of generally open conditions is attributed to continued grazing.

UPLAND AREAS

Flat upland areas at elevations of over 30m OD include some of the most impoverished soils of the Breckland. These are formed on deep glacial sands and gravels, and include acid brownearths and humus podzols (Corbett 1973: 49). These upland areas, correspond to the heathland nuclei discussed by Susan Limbrey (1978: 25); there is good reason to believe that upland areas remote from the river valleys were too dry to be worth the effort of reclamation for arable farming by manuring or marling.

Upland plateaux sites have been excavated at Gallows Hill, north of Thetford and at the nearby new industrial estate of Fison Way (5853 THD), but the upland area just to the north of Thetford is atypical in two respects. Firstly there is a discontinuous impervious sandy clay loam underlying the modern soil, which has led to the development of a perched water table, and secondly the sites are only about 1km from the River Little Ouse, with a very steep and narrow belt of slope soils. Consequently water supply was much less of a problem than in many upland areas, and this no doubt influenced the location of settlements.

Charred plant remains have been recovered from over eighty samples at Fison Way. Four phases of Iron Age-Roman activity are distinguished (T. Gregory, pers. comm.) though excavation is not yet complete, and interpretation is therefore provisional:

I Late Iron Age settlement, dating to between the mid 1st century B.C. and the early 1st century A.D.

II A large ditched enclosure with three round-houses, of the mid 1st century A.D.

III A complex defended ditched enclosure dating to the 50s and 60s A.D.

IV Upper filling and levelling of major ditches during the 4th century A.D.

Table 9:6 presents the results of botanical studies in the form of a 'presence analysis' (Hubbard 1975). Charcoal of the Ericaceae was present at all phases. Most of this consists of very small immature stem fragments which have not been identified to genus, but capsules of _Calluna vulgaris_ are certainly present in one phase II sample, and four very large dense deposits of _Calluna_ charcoal with mature stems, capsules, leaves and young shoots were isolated from phase III features. In addition there are some charred nutlets closely matching _Carex arenaria_, though possible deformation during charring means that this identification must remain tentative. It does seem clear, however, that there was heathland vegetation in the vicinity from at least the 1st century B.C. Charred seeds and nutshell fragments of scrub plants (_Crataegus monogyna_, hawthorn; _Corylus avellana_, hazel and _Sambucus nigra_, elder) were identified at low frequencies, but these are edible fruits or nuts which might have been brought to the site from other localities.

More detailed information on local conditions during the Late Roman period comes from the nearby site on Gallows Hill. This mound, constructed of stacked turves was originally thought to be a Bronze Age barrow, but a date of 1600 ± 70 b.p. or 350 a.d. was obtained for charcoal from its base. Sealed by the mound was an immature soil formed on blown sand influenced by a perched water table caused by an impervious sandy clay loam horizon at 1m depth (R.1. MacPhail, forthcoming). The layer of charcoal and other charred plant remains near the base of the turf-stack consisted predominantly of charred remains of heather, _Calluna vulgaris_. This was almost certainly growing locally: in comparable modern situations, for example at Lakenheath Warren (Watt 1937), _Calluna_ is the dominant plant on areas of thin blown sand and on sandy soils truncated by wind erosion. Charred remains of _Stellaria graminea_, _Juncus_ sp. and _Carex_ sp. (not _C.arenaria_) were also present, and it is thought that these were

Table 9:7: Presence analysis of charred plant remains from Fison Way

Phase	I	I/II	II	II/III	III	IV
Cereal indet. caryopses	11	16	10	3	8	3
Cereal indet. caryopsis fragments	15	20	15	3	7	2
Cereal indet culm nodes	1	3	4	2	2	-
Hordeum sp. caryopses	8	15	3	2	3	2
Hordeum sp. rachis internodes	-	2	-	-	2	-
Hordeum vulgare caryopses	-	-	2	1	-	-
Triticum sp. caryopses	3	13	6	2	6	3
Triticum sp. brittle-rachis internodes	1	1	-	-	-	1
Triticum sp. spikelet bases	2	4	1	-	1	-
Triticum sp. glume bases	1	4	2	-	3	-
Triticum spelta spikelet forks	1	3	-	-	-	-
Triticum spelta glume bases	3	7	1	1	-	1
Triticum dicoccum spikelet forks	-	2	-	-	-	-
Triticum dicoccum glume bases	-	2	-	-	2	-
Avena sp. caryopses	-	1	2	1	-	-
Avena sp. awn fragments	3	5	1	-	-	-
Ranunculus acris/repens - type	1	1	-	-	1	-
Brassica/Sinapsis sp.	-	-	1	-	-	-
Raphanus raphanistrum	-	2	2	-	1	-
Cruciferae indet.	-	1	-	-	-	-
Silence cf *alba*	-	1	-	-	-	-
Stellaria media-type	-	3	-	-	3	-
Stellaria graminea	-	-	1	-	-	-
Spergula arvensis	-	2	-	-	-	-
Atriplex patula/hastata	1	1	2	-	-	-
Chenopodium album	6	10	3	1	1	-
Chenopodium sp.	2	1	-	-	-	-
Chenopodiaceae indet.	1	1	2	-	-	-
Malva sp.	-	1	-	1	-	-
Medicago/Trifolium - type	-	1	-	-	-	-
Vicia sp. seeds	1	-	-	-	-	-
Vicia sp. cotyledons	1	1	1	-	1	-

Phase	I	I/II	II	II/III	III	IV
Vicia cf. *tetrasperma*	-	1	-	-	-	-
Leguminosae indet.	-	1	-	-	-	1
Crataegus monogyna	-	-	1	-	-	-
Polygonum aviculare agg	1	2	3	1	-	-
Polygonum lapathifolium	-	-	2	-	-	-
Polygonum lapathifolium/ persicaria	3	6	-	-	1	-
Polygonum convolvulus	5	12	3	-	3	-
Rumex acetosella	2	3	1	-	1	-
Rumex sp.	1	9	5	2	2	-
Polygonaceae indet	-	5	2	1	3	-
Urtica dioica	-	1	1	-	-	-
Corylus avellana nutshell frags.	1	5	1	-	-	1
Calluna vulgaris capsules	-	1	1	-	4	-
leaves and shoots	-	-	-	-	4	-
Ericaceae charcoal	6	15	10	3	12	3
Prunella vulgaris	-	-	-	1	-	-
Plantago lanceolata	2	4	5	-	2	-
Galium aparine	-	1	1	2	-	1
Sambucus nigra	-	1	2	1	1	-
Centaurea sp. frags.	-	1	-	-	-	-
Carex arenaria-type	-	3	-	-	-	-
Carex sp.	-	1	3	-	-	-
Eleocharis palustris agg	-	1	-	-	1	-
Bromus mollis/secalinus	8	10	4	1	2	-
Gramineae indet.	1	6	3	2	3	1
Indet. stem fragments	1	-	-	-	-	-
Indeterminate, seeds etc.	11	12	7	3	8	-
No. of samples per phase	19	23	20	4	13	4

derived from locally damp habitats, for example in blow-outs and hollows in the vicinity. Charcoal of hazel, oak and ash was identified from this deposit, but, given the evidently unstable soil conditions of the vicinity it is questionable whether these trees were growing locally.

At first sight these conditions seem unsuitable for arable farming. However, part of the site is under barley today, there is a mature pine shelter belt adjacent to the site and the presence of chalk fragments in the modern topsoil indicates marling at some date. So long as the topsoil was stabilised, for example by shelter-belts, and supplies of manure and chalk were available to improve soil fertility, the perched water-table would have provided a moisture reservoir for crop growth. Moreover, just downslope is a belt of more fertile slope soils over the chalk-sand drift. Cultivation would therefore have been possible in the vicinity, though the scale and complexity of the enclosures indicates that something more than simple farming settlements is represented in the later site phases.

The crop plant remains recovered (Table 9:6) are six-row hulled barley and spelt with rare remains of emmer and wild or cultivated oats. Ratios of wheat to barley grains fluctuate haphazardly from phase to phase: neither crop is clearly predominant. The associated charred weed seeds include a comparable range of species to those from Iron Age and Roman terrace sites. At Fison Way, the few remains of damp-ground plants (e.g. Carex sp., Eleocharis palustris) are thought to reflect the presence of damp hollows in the arable area. Seeds of Leguminosae are rare at all phases, and this appears to indicate that soil nitrogen levels were maintained in the arable fields although it is possible that a system of long fallows may have been in use, whereby cultivation ceased before leguminous plants became a dominant component of the weed flora on the nitrogen-depleted soils.

Unfortunately nothing can be said about animal husbandry since acid soil conditions had resulted in almost total destruction of bone, apart from some powdery remnants, but in view of the proximity of the Little Ouse valley mixed farming would clearly have been possible. With increasing distance from water supplies on the permeable deep sandy soils of upland areas remote from the river valleys land use potential must have been more restricted.

CLIMATIC CHANGE

Since stress has been placed on annual rainfall variations as a factor influencing crop yields and also the productivity of grassland, some assessment of the likely effects of long term climatic changes is necessary. To interpret climatological data in agricultural terms inevitably involves an element of conjecture: the evidence is frequently inadequate for detailed interpretation. For example, it is often the distribution of rainfall through the year which is of importance for crop growth, rather than the absolute rainfall. However, climate is potentially a major variable affecting agriculture, and its possible effects cannot therefore be ignored. During the period considered in this paper there is evidence for two wet climatic phases: one in the first millenium B.C., and a second in about the late 6th. to 7th. centuries A.D., separated by a period of relative warmth and dryness during the Roman period (Lamb 1981). Considered in isolation, increased rainfall may have been beneficial to arable farming in the area. Higher mean rainfall may be expected to have resulted in consistently higher yields on the better soils and to have permitted extension of the arable area onto some excessively-drained soils which in dryer phases were not cultivated. As has been noted, pollen analysis at Hockham Mere indicated phases of arable expansion, broadly corresponding to these wetter periods from about 800 b.c. and again from about 650 a.d., with apparently increased pastoralism in an intervening period, from about 340-650 a.d. (Sims 1978). Given the complex range of environmental and cultural factors involved one hesitates to suggest an exclusively climatic explanation for these changes: indeed post-Roman arable expansion may well owe something to increased reliance on rye growing as was suggested above. However, the broad correlation between land use changes and climatic change is worth noting. Wetter conditions might also be expected to have improved grassland productivity on the well drained soils and, by raising valley-floor groundwater levels, to have extended the area of wet pasture and meadow. There is no unequivocal direct evidence for this, though the Brandon peat section could be interpreted in this light.

The effects of other climatic factors are still less certain. The wetter phases are known to have been stormy and in coastal areas major episodes of sand dune formation occurred between about 700 and 900 A.D. Obviously this is potentially relevant to the Breckland, but blown sand deposits in the area are rarely precisely dated. The 12th-13th century ridge and furrow buried by up to 3 feet of blown sand at West Stow is exceptional (West 1969: 3). Whether the lower temperatures and consequent shortening of the growing season associated with these wetter phases would have had major effects on agriculture in the Breckland is questionable: this factor is obviously not limiting for agriculture today, and there seems no reason to suppose that it has been in the past.

CONCLUSIONS

It is known that large areas of poor, relatively infertile sandy soils were under cultivation in the middle ages, and evidence presented above indicates that arable exploitation of such soils dates back at least as far as the Iron Age. Remunerative yields would not have been consistently obtainable without adequate manuring. Patterns of animal exploitation were constrained by the scarcity or lack of pannage, by the restriction of high grade pasture to the valley floors and to the calcareous slope soils and by water supply. Within limits these features of soil, water and vegetation may be regarded as environmental constants which defined the broad outlines of farming systems (Table 9:7). The general similarity of this proposed scheme of land use to that inferred for the Bronze Age (Legge 1981) and known from historical sources for the medieval period (Clarke 1925) is quite clear. Broadly similar constraints on land use applied during all these periods, though several factors appear to have permitted some degree of agricultural change.

Table 9:7: Summary of known and inferred land use.

<table>
<tr><th></th><th>Valley floors</th><th>Terraces</th><th>Slopes</th><th>Uplands</th></tr>
<tr><td>Arable land use</td><td>Hemp and flax cultivation extending from terraces from at least Middle Saxon period.</td><td>Good evidence for cereal cultivation at all periods.</td><td>Cereal cultivation inferred, some slight evidence.</td><td>Some cereal cultivation on areas with moisture-retaining subsoil. Less practical with increasing distance from chalk supply for marling and from water supply.</td></tr>
<tr><td>Pastoral land use</td><td>Wet valley pasture and ? meadow.</td><td>Rough grazing on heath and grass heath. Folding on arable fields inferred.</td><td>Grazing on calcareous grassland. Less practical with increasing distance from water supply.</td><td>Rough grazing on heath and grass-heath.</td></tr>
<tr><td>Vegetation-types</td><td>Alder carr, reed-bed, grassland.</td><td>Calluna and Pteridium heath, grass-heath.</td><td>Grass heath and calcareous grassland.</td><td>Calluna and Pteridium heath, grass-heath.</td></tr>
<tr><td>Other forms of land use</td><td>Reed-cutting, wild-fowling, fishing.</td><td></td><td>Chalk extraction for marling.</td><td></td></tr>
<tr><td></td><td></td><td colspan="3">Hunting/pannage in areas of relict woodland</td></tr>
</table>

These variable factors were climate, the adoption of new types of crops and improved animal breeds, and general improvements in agricultural methods. Possible effects of climatic change have been discussed above, and are best summarised by suggesting that wetter climatic phases may have resulted in an extension of the arable area and in increases in grassland productivity whereas in dryer phases there may have been a shift to increased non-intensive exploitation of low-grade dry grassland. In short one might envisage successive phases of agrarian expansion and contraction from productive nuclei.

The effects of agricultural innovation are likewise difficult to assess in detail, but it is possible to point to some features which must have increased productivity. For example, increased reliance on rye, a drought-resistant crop, would have produced worthwhile yields on previously marginal dry, sandy land. The evidence of carbonised cereal remains suggests a shift to rye during the Early Saxon period. Some improvement of nitrogen-deficient soils could have been effected by legume cultivation, though there is no evidence for this before the Middle Saxon period in the Breckland. More intensive use of parts of the valley floors would have been possible by growing flax and hemp, though it seems questionable whether there would have been any incentive to convert extensive areas of scarce pasture to arable production, and indeed there is no evidence for these crops before the Middle Saxon and Late Saxon periods respectively. The influence in the Breckland of some other innovations in land management and agricultural technology may not have been so great as in other areas. Jones (1981:111) has suggested that a significant trend in agricultural improvements between the Iron Age and Late Saxon periods was towards increasing production on heavy and poorly drained soils. Such changes would have been largely irrelevant to Breckland farming, where indeed some positive advantages may have accrued from the retention of traditional techniques: as was noted above, and cultivation, by virtue of its relative inefficiency would have helped to retain soil moisture.

On balance the available evidence seems to indicate that later prehistoric and early historic farming in the Breckland was essentially conservative, limited by the environmental constraints discussed above to a pattern of the type outlined in Table 9:7. Crabtree (1981: 14) discussing the faunal remains from West Stow suggests that the Anglo-Saxon farmers of the area "appear to have been exploiting improved stock in traditional ways", and a comparable interpretation may be placed on the crop plant remains. To conclude by referring back to the opening paragraph of this paper, the study of ancient agricultural systems is indeed potentially easier in areas of marginal land than in more ecologically diverse landscapes. But inevitably farming in such areas has to be sensitively adapted to local conditions, and consequently may show an adherence to traditional forms of land use.

Acknowledgements

I am most grateful to the directors of the sites discussed, Bob Carr, Tony Gregory, Andrew Lawson, Edward Martin, Jude Plouviez, Andrew Rogerson and Dr. Stanley West for their practical help and for information on their excavations. A paper of this type inevitably leads to reliance on the work of others and in this respect I would like to record my particular debt to Pam Crabtree, Andrew Jones, Roger Jones and Dr. Richard MacPhail. Any interpretation placed on their results is, naturally, my own responsibility. My thanks are also due to Richard Darrah for information on Breckland vegetation and farming. Finally I would like to thank Mavis Bithray for her typing. The research for this paper was supported by grants from the Department of the Environment.

Bibliography

Anon. (1973) West Stow. Current Archaeology 4, No. 5 151-8

Bradshaw, R.H.W., Coxon, P. Greig, J.R.A. and Hall, A.R. (1981) New fossil evidence for the past cultivation and processing of hemp (*Cannabis sativa* L.) in Eastern England. New Phytologist 89, 503-510

Churchill, D.M. (1980) Post Neolithic to Romano-British Sedimentation in the Southern Fenlands of Cambridgeshire and Norfolk in: The Fenland in Roman Times (ed. Philips, C.W.) Royal Geographical Society Research Series No. 5, London pp. 132-146

Clapham, A.R. Tutin, T.G. and Warburg, W.F. (1962) Flora of the British Isles 2nd ed. Cambridge: University Press

Clark, J.G.D. and Fell, C.I. (1953) The Early Iron Age Site at Micklemoor Hill, West Harling, Norfolk and its Pottery. Proceedings of the Prehistoric Society 19 1-40

Clarke, R.R. (1960) East Anglia London: Thames and Hudson

Clarke, W.G. (1925) In Breckland Wilds. London: Robert Scott

Corbett, W.M. (1973) Breckland Forest Soils Soil Survey. Special Survey No. 7 Harpenden

Crabtree, P.J. (1981) The Archaeozoology of the Anglo-Saxon site at West Stow, Suffolk. Paper presented at the Sixteenth International Congress on Medieval Studies, Kalamazoo, Michigan

Crompton, G. and Sheail, J. (1975) The Historical Ecology of Lakenheath Warren in Suffolk, England : a case study. Biological Conservation 8 299-313

Darby, H.C. (1935) The Domesday Geography of Norfolk and Suffolk Geographical Journal 85 432-52

Darrah, R. Musgrave, M. and Rudley, G. (1980) West Stow Country Park : Vascular Plant Check-List

Duffey, E. (1976) 'Breckland' In: Washbourne, R. *et al.* (1976) Nature in Norfolk: a Heritage in Trust 62-77 Norwich: Jarrold

Evans, J.G. (1972) Land Snails in Archaeology London: Seminar Press

Godwin, H. (1944) Age and Origin of the Breckland Heaths of East Anglia Nature 154 6-7

Hind, W.M. (1889) The Flora of Suffolk London

Hubbard, R.N.L. (1975) Assessing the botanical component of human palaeo-economies Bulletin of the Institute of Archaeology. London University 12 197-205

Jones, A.K.G. (forthcoming) Fish Remains In: Rogerson, A. Thetford Excavations East Anglian Archaeology

Jones, M.K. (1981) The development of crop husbandry In: The environment of man: the Iron Age to the Anglo Saxon period (eds Jones, M. and Dimbleby, G.) Oxford: BAR 87 pp. 95-127

Lamb, H.H. (1981) Climate from 1000 BC to 1000 AD In: The environment of man: the Iron Age to the Anglo Saxon period (eds Jones, M. and Dimbleby G.) Oxford: BAR 87 pp. 53-66

Lawson, A.J. Martin, E.A. and Priddy, D. (1981) The Barrows of East Anglia East Anglian Archaeology Report No. 12

Legge, A.J. (1981) The Agricultural Economy In: Mercer, R. J. Grimes Graves Norfolk Excavations 1971-2, Vol. 1 London: H.M.S.O.

Limbrey, S. (1978) Changes in the quality and distribution of the soils of Lowland Britain In: The effect of man on the landscape: the Lowland Zone (eds Limbrey, S. and Evans, J.G.) CBA Research Report No. 21 pp. 21-26

Percival, J. (1918) Agricultural Botany (5th edition) London: Duckworth

Perrin, R.M.S. (1956) The nature of 'Chalk Heath' soils. Nature 178 31-2

Perrin, R.M.S. Willis, E.H. and Hodge, C.A.H. (1964) Dating of humus podzols by residual radiocarbon activity. Nature 202 165-6

Petch, C.P. and Swann, E.L. (1968) Flora of Norfolk Norwich, Jarrold.

Peterson, R. Mountford, G. and Hollom, P.A.D. (1954) A Field Guide to the Birds of Britain and Europe London: Collins

Renfrew, J.M. (1973) Palaeoethnobotany. The prehistoric food plants of the Near East and Europe. London: Methuen

Reynolds, P. (1981) Deadstock and Livestock In: Mercer, R. J. Farming Practice in British Prehistory Edinburgh: University Press pp. 9-54

Sims, R.E. (1978) Man and vegetation in Norfolk In: The effect of man on the landscape : the Lowland zone (eds Limbrey, S. and Evans, J.G.) London: CBA Research Report No. 21 57-62

Tansley, A.G. (1939) The British Isles and their vegetation Cambridge: University Press

Walters, S.M. (1953) Montia fontana Watsonia 3 1-6

Watt, A.S. (1937) Studies in the ecology of Breckland. II. On the origin and development of blow-outs. Journal of Ecology 25 91-112

Watt, A.S. (1940) Studies in the ecology of the Breckland. IV. The grassheath. Journal of Ecology 28 42-70

Watt, A.S. (1955) Bracken versus heather, a study in plant sociology. Journal of Ecology 43 490-506

Watt, A.S. Perrin, R.M.S. and West R.G. (1966) Patterned ground in the Breckland: Structure and composition. Journal of Ecology 54 239-58

West, R.G. and Donner, J.J. (1956) The glaciations of East Anglia and the East Midlands Quarterly Journal of the Geological Society of London 112 69-91

West, S. (1969) The Anglo-Saxon Village at West Stow. An interim report of the excavations, 1965-8. Medieval Archaeology 13 1-20

10. SIZE AND SHAPE, TIME AND PLACE:

SKELETAL VARIATIONS IN CATTLE AND SHEEP

B. A. Noddle

The detailed study of domestic animal bones from archaeological sites began about the middle of the nineteenth century, following the discovery of the neolithic Swiss Lake Villages. The pioneering work was carried out by skilled zoologists and the work tended to concentrate on the animals providing the bones rather than their relationship with man. Much effort has expended on distinguishing different types, mainly from cranial characteristics; there was considerable confusion as to what constituted a race and what a species. Further work on bones from North Germany was carried out in the hope that an elucidation of these different types might give some clue to the origins of the human populations of different areas, assuming they had brought distinctive livestock with them. This early work is summarised by Herre (1969) and Epstein (1971).

Considerable changes in livestock type have been demonstrated by means of these paleontological methods, but mainly for the later prehistoric period, for which numbers of more or less complete skulls are availabled owing to the number of whole animal burials that have been discovered. A distinct bovine type, the Celtic ox, has been described for the Bronze and Iron Ages which seems to have been widely distributed over Western Europe (Jones 1939). There was also a less well known neolithic type described by Degerbol and Fredskild (1970) using Danish examples; the best British reports are on the type-site of Windmill Hill (Jope and Grigson 1965, Grigson 1983) and the Orcadian site of Skara Brae (Watson 1931).

Craniology of the larger domestic animals has not been studied much in recent years, except by Grigson in a series of papers (1974-1978) and by Bartosiewicz (1980). However, considerable attention has been given to the subject of bovine horn cores, by Grigson (1983) and by Clutton-Brock and her colleagues (Armitage and Clutton-Brock 1976, Armitage and Goodall 1977). Recently Armitage has carried this study forward into the 18th Century (Armitage 1982, 1983).

Much of the work of archaeozoologists is now concerned with the massive bone deposits of urban sites. In these complete skulls are almost entirely lacking and horn cores are rare, apart from large industrial deposits which usually date from the 18th century and later (Armitage 1983 and Knecht 1966). By this time written descriptions of animals are available, although these require supplementation from skeletal studies. Much of this work has been summarised by Trow-Smith (1959).

The purpose of this paper is to attempt the distinction of livestock types using aspects of the post-cranial skeleton supplemented by horn core dimensions. The range of differences in modern breeds can be described in some detail for modern sheep, whose skeletons are readily available. A similar account cannot be given for cattle, but the limited material available to the author is set out in the hope that it can be added to by other workers. Three aspects of study have been employed, these comprise absolute size, proportions indicating bone shape and various non-metrical characters. The uses and limitations of these three will be discussed in turn.

Absolute size

The size of a mature bone is determined by its genetic inheritance, its sex and its nutritive status during the growing period. The genetic contribution is related to growth rate; the inheritance of this factor is complex and outside the scope of this paper. Certainly this aspect of size has been overemphasised by many workers, for example Jarman and Jarman (1968). This is not to say that absolute size has no value, Grigson (1969) demonstrated this in her paper on the distinction between wild and domestic cattle. However the influence of nutrition has been underestimated; the magnitude of size differences in the domestic pig using littermates on different rations has been spectacularly demonstrated by McCance and his colleagues (McCance 1960). A group of Aberdeen Angus cows kept under range conditions and given winter rations inadequate in energy and or protein content averaged 8 cm less in withers height at the time of 2nd calving (about 4 years old) than their fully fed contemporaries (Wiltbank et al. 1965).

It seems unlikely that the majority of domestic animals reached their full genetic size prior to the adoption of adequate winter feeding in the 18th century. This is borne out by the presence of the odd animal of much greater stature on a number of Roman sites including "Castella North of the Alps" (Herre 1969), Gedabridge villa, Hants (Harcourt 1974) and quantities of large animals at Segontium Fort (Noddle unpublished) which were presumably fully fed. Also the majority of metapodial bones from archaeological sites have a midshaft width which is less in proportion to the bone ends than that found in present day stock, and a number of young cattle in the author's collection which are known to have been suffering from prolonged malnutrition at the time of their death also have this narrow configuration.

The influence of sex on bone size in domestic animals has not been systematically studied. Higham produced data on bovine distal metapodials (Higham and Message 1969) Walker (1964) noted that there was a greater size difference between male and female in primitive African cattle than can be discerned in modern domestic breeds and this is borne out by the limited data on the metatarsal presented in table 10:10. Figure 10:1 presents the lengths of metacarpals from sheep of various breeds; male and female mouflon (a form of wild Ovis aries) are clearly separated, but this is not the case with domesticated sheep, even the primitive Soay, and there are no separate groups in the modern Clun Forest. All three groups of

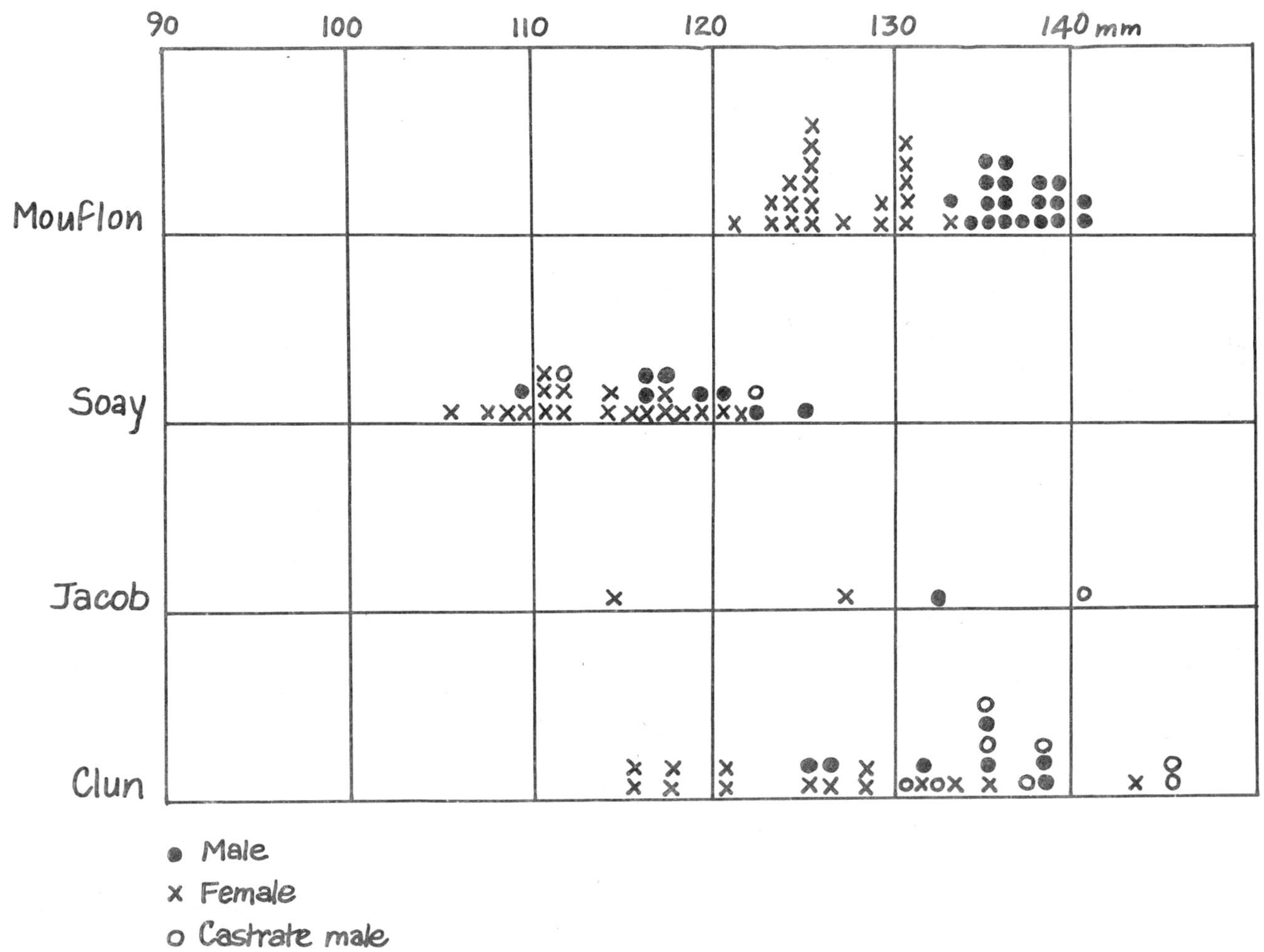

Fig. 10:1 The maximum length of the metacarpal bone in different breeds of sheep, arranged in assumed order of primitiveness.

animals come from single flocks; the Soays and Cluns came from the same establishment over the same period of time. As always with domestic animals, the situation is complicated by the presence of castrate males.

Bone Proportions

Though bone proportions can be influenced by nutrition as described above and by sex (in particular the horn core described below) there are some which appear to vary for genetic reasons only. Two of these are employed in this study, the neck of the scapula and the proximal metacarpal of the sheep.

The neck of the ovine scapula is relatively long and thin in the primitive short-tailed sheep (although less so in the wild mouflon than in the domesticated Soay) and becomes progressively shorter and thicker in the modern meat breeds. The actual measurements taken to establish a ratio, referred to hereafter as the scapula index, are the minimal neck width, which is divided into the distance between the glenoid cavity and the base of the scapula spine. Neither sex nor nutritional status seems to affect this ratio, and the primitive form is strongly inherited in first crosses between long and short-tailed types. Only when the short tailed genetic contribution is reduced to 25% is any change exhibited. This has been more fully discussed in Noddle (1978). This ratio is affected by the age of the animal, but only when it is very young, and any scapula with a fused coracoid process can be used.

O'Connor (1982) has studied the proximal metacarpus in the sheep, and has demonstrated that there is a genetic relationship between the proximal width of this bone and its proximal anterior posterior depth. This ratio has been designated the metacarpal index. The more primitive animal has a ratio tending towards unity, and neither sex nor nutritional status seem to affect it. Niether does age, given that the animal is old enough to have formed a proper cortex of compact bone in this region. Haak (1965) also examined proportions of the metapodials of different breeds of sheep, but his difference were affected by the sex of the animal.

The bone in which the most obvious genetic effects can be demonstrated is the horn core, though its shape is also affected by sex and age. Two numerical ratios are regularly in use and have been employed here. These are the length along the outer curvature divided by the basal circumference (termed length index here) and the greatest basal diameter divided by the least (termed basal index here). Other writers, in particular, Armitage and Clutton-Brock (1976) have employed basal index calculated the other way round.. The length index is affected by age, in that the young horn core increases faster in circumference than it does in the length, and it would appear that youn horn cores tend to be more circular than adult ones. However, these young stages can be readily distinguished using the criteria of Armitage (1983). Both ratios are affected by the sex of the animal, more so in some breeds than others.

The basal index for 6 breeds of sheep horned in both sexes (except the Mouflon) is set out in figure 10:2; no parallel exercise

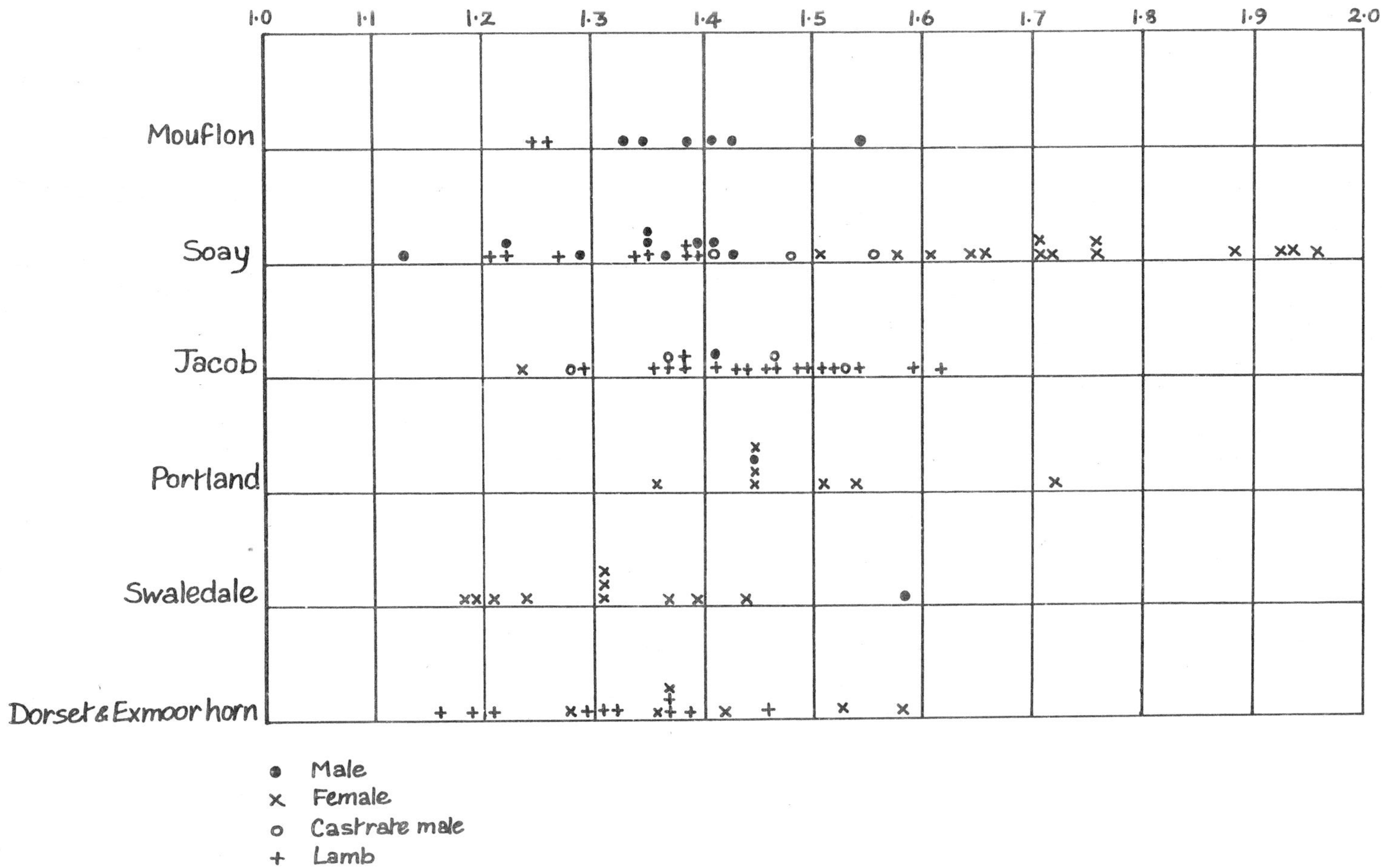

Fig. 10:2 The proportions of the base of the horn core in different breeds of sheep arranged in presumed order of primitiveness. The index is obtained by dividing the least basal diameter into the greatest. Details of breeds are given in appendix A.

on the length ratio has been carried out owing to the number of lambs in some of the groups. The details of these breeds are set out in appendix A. It can be readily seen that there is a wide range of horn forms in the primitive Soay, those of the female being strongly elliptical. By contrast, in the more modern breeds the ewe's horn core is more circular than that of the ram. It is unfortunate that the ewes in the flock of Mouflon available to the author are largely polled, carrying short scurs (rudimentary horns) at the most. It appears from preliminary studies (the test animals being at the moment fifteen months old) that the castrate male of this type will also only carry short scurs.

It is also unfortunate that bovine horn cores of the present day are so rare owing to the almost universal British practice of dehorning calves. Measurements of the author's own inadequate collection, supplemented by those of the Cambridge University Museum of Zoology and other museums, are set out in table 10:1, from this it can be seen that cows tend to have rounder horn cores than bulls, and that steers occupy an intermediate position; Armitage and Clutton-Brook (1976) came to the same conclusion. However, there are also breed differences, round horned breeds with a ratio below 1.2 and oval ones where it is higher. The most ellipical specimen here is a Friesian bull, and even this is less extreme than the Celtic ox; the average value for the seven specimens described in Armitage & Clutton-Brock (1976) is 1.57. The basal shape does not depend on size; the Spanish ox was a 19th century animal of enormous horn span, possibly related to the Texas longhorn, whereas the Channel Island breeds have the shortest horns in the collection. Long and short horn types are fairly well distinguished, as Armitage and Clutton-Brock pointed out. There may be some variation amongst animals of one sex and type; the auroxen all appear to be bulls with one possible exception. However, although undated these specimens probably come from widely separated points of time. Though most of them are described as being found in pre-Holocene Fen clay, the Lowe's farm specimen was Bronze Age in date (Shawcross and Higgs 1961).

It is clear from the above that a simple average obtained from all the horn-cores in an assemblage may need some qualification, and it should be ascertained whether there is a predominance of one sex, and very young specimens should be excluded. Armitage (1983) has defined the criteria upon which the age of a specimen should be decided in the case of cattle, and these may easily be used for sheep, although the chronological ages will be different from this animal. It is not easy to distinguish between female and male castrate horn cores, but the problem may be partially avoided in sheep. Hatting (1975) has shown the horn cores of wether sheep develop very thin walls, so that they are unlikely to be excavated intact. Thus it seems likely that the majority of complete horn cores will be female or intact male, easily distinguished, and that profiles of wether horn cores are most likely to be obtained from the broken stumps on the frontal bone.

Non-Metrical Differences. The Postion of the Nurient Foramen of the Femur

In early embryonic life the shaft of the developing long bone is supplied by a number of arteries, some of which penetrate the cortex to supply the marrow and endosteum (Pinard 1952; Brookes 1963). The number of these is reduced in later foetal life and after birth there is normally only one for each bone, the position of which is usually constant for the species. In the case of the artiodactyl femur there are usually three foramina at about the time of birth which are later reduced to one. It is possible for any one of the three to survive, and two do so occasionally, particularly in cross-bred animals. In sheep the three positions are the anterior proximal shaft, the posterior distal shaft and the posterior midshaft, in declining order of frequency, but all three positions are usually found in a large number of bones from any one breed, or any large archaeological assemblage. In short-tailed breeds other than the Shetland the position is usually proximal. With the exception of the Swaledale the distal position usually only dominates in small isolated or inbred groups. Evidence from a breeding trial which involved five sires from a number of breeds suggests the distal position is inherited, however. Data on various modern breeds and archaeological material is given in Noddle (1978).

In cattle the positions are usually either immediately above the supracondylar fossa on the distal posterior shaft or a few centimetres medial to this, but the foramen is occasionally found in the anterior proximal position as in the sheep. Double foramina are less common than in sheep and when they occur they involve the two distal positions. Table 10:2 presents the positions in a number of present day breeds and first crosses, supplemented by the Cambridge auroxen and a few other specimens present in museums or in private collections in Britain or on the Continent. The cross-bred animals are all from Friesian cows. The medial position predominates in modern breeds, with the exception of the Welsh Black and Charrolais crosses and possibly the ancient Chillingham breed. Five out of eight auroxen and medial are in the ubiquitous Friesian the medial position is just the most frequent. Wilson (1909) tried to derive a history of British breeds from their predominant colouring, black animals being the ancient British stock and red breeds being Saxon imports. Red breeds have a higher proportion of foramina in the medial position than the black breeds with the exception of Aberdeen Angus, which is a fairly recent breed deriving from a small number of foundation stock. However, almost without exception other than Orcadian animals, the lateral position predominates in archaeological material. It is true that a fragment of femur including the supracondylar shaft is easier to identify than the fragment containing the medial foramen and is probably more robust than the latter, but care has been taken to minimise this possibility.

Skeletal Profiles of Modern Breeds and Archaeological Assemblages

It is considered that by using a combination of measurements and other variables discussed above it might be possible to construct a profile of an assemblage which may distinguish it from those of other sites or periods, but it would not be possible to distinguish between two types found in the same assemblage. The extent of any differences found can be compared with similar profiles constructed from modern breeds, this is only possible in the case of sheep, owing to the large size and frequent lack of availability of complete adult cattle skeletons. Table 10:3 presents profiles of five modern sheep breeds, none of which possesses to any great degree a modern meat producing conformation (Hammond 1932). The breeds, in assumed order of primitiveness, are Mouflon, Soay (both short-tailed) Portland, Jacob and Swaledale. A more detailed description of these breeds is given in appendix A. There are two small breeds, the Soay and the Portland and the other three are of similar stature, though not body weight. The bones whose dimensions have been employed to give an estimate of stature have been selected partly on the grounds of their frequent occurrence in archaeological assemblages.

Considering the individual profiles, Soay and Portland are of roughly equal size, and both the metacarpal and average basal horn core index are also similar. However, the two can be differentiated on the scapula index and the femoral foramen position. The three larger breeds are all of similar size, but the Swaledale can be distinguished on the basis of its scapula index and the Jacob by means of the metacarpal and basal horn core indices. However, quite large differences in the living animal may show up as quite small differences in the skeleton. Thus only results from large samples are really credible.

Three groups of sites have been chosen to try out the method. The first group comprises two Orcadian sites, Skara Brae which is neolithic in date and Skaill which was inhabited by two succeeding populations, the Picts and the Vikings. The cattle profiles from these sites are presented in table 10:4 and those of the sheep in table 10:5. Secondly two Roman sites have been compared, separate 1st, 2nd and 5th century assemblages fro Wroxeter (Shropshire) and 4th century material from Segontium (North Wales). This material is set out in table 10:6 (cattle) and 10:7 (sheep). The third group comprises two thirteenth century sites, Hereford (urban) and Loughor Castle (South Wales) which are presented in tables 10:8 (cattle) and 10:9 (sheep). Further details of all these sites are set out in appendix B.

In the case of the Orcadian sites there is a gap of about 2000 years between Skara Brae and Skaill. Genetic change is to be expected and is readily apparent. For cattle apart from the considerable decrease in size there is a distinct change from a long horn to a short-horn type. It is not easy to decide whether there are genetic changes involved in the differences between Pictish and Viking Skaill. The size reduction indicated by the astragalus measurements could be nutritional in origin and the amount of Viking material is too small to decide whether the small differences in all other other profile constituents are significant. In the caes of sheep there is again a difference between neolithic and Pictish but it is less marked than that of the cattle. The evidence in the case of the Pictish and Viking assemblages is again unclear. The Pictish sheep are more like the neolithic in respect of the metacarpal index but more like the Viking in the case of the scapula. The horn core evidence is again inconclusive but the femoral foramen suggests a difference between all three periods. On balance it seems likely that the Vikings imported new stock but as might be expected there was interbreeding with the old.

Considering the Roman sites, it is apparent that there were different types of cattle at Wroxeter and Segontium during the 4th to 5th centuries. The large numbers of horn cores validate the small differences in the basal index and the femur differences are marked, although the size is similar. Changes that took place during the passage of time at Wroxeter are less clear. It would seem likely that the first settlement at Wroxeter, a military one, would aquire local native livestock. Horn core basal index and femur data suggest that changes had already taken place during the subsequent 100 years, along with a slight increase in size that might indicate improved husbandry. Changes during the following centuries are less marked, apart from a considerable increase in size. It is likely that new stock was imported, but the passage of time and selective breeding of existing animals might account for the differences. Armitage (1982) is of the opinion that the Romans did not make massive livestock importations, but the bulk of his evidence comes from South East England, a rather different area from the Welsh Borders.

The Roman sheep from Segontium differ from those at Wroxeter in that they were bigger. However, the slight differences in metacarpal and scapula indeces are based on very small samples and there were no horn cores. Again there seem to have been changes in stock during the first 100 years at Wroxeter. The metarcarpal index indicates a more advanced animal, and the scapula index range, though not its average, also suggests genetic change. The horn core basal index includes animals of a Soay type which are subsequently lost. In later centuries there is a change in size rather than type, which could be the result of improved husbandry. Ryder (1964) believes that sheep with long tails and white fleeces were imported by the Romans. Certainly they managed to import fallow deer, an animal less easy to manage than the fully domesticated sheep (Chapman and Chapman 1975).

The medieval stock at Hereford and Loughor are very similar, and the small number of speciments from Loughor does not warrant any firm conclusions. The sheep evidence is contradictory, with a more primitive metacarpus and a less primitive scapula at Loughor, but the two types probably were distinct. However, the possibility exists that stock bought in for the castle at Loughor was carefully selected, whereas that from Hereford town comprised a more complete cross section of local types.

It is concluded from the above examples that it is possible to suggest genetic differences between sheep and cattle populations from different locations or from different periods. The degree of certainty with which these suggestions can be made depends not only upon the extent of the differences in the various items comprising, but also upon the size of the sample available. This method should not be attempted with small assemblages unless they are of particular potential importance.

Skeletal Proportions in the Whole Animal

All the above calculations are based upon bone assemblages and measurements upon complete modern specimens have been amalgamated for comparison. However complete skeletons are occasionally found in an archaeological context, and other methods of comparison are possible using these. The data to be investigated here are the relative lengths of the long bones, a study suggested by Hammond's pioneering work on the conformation of the meat sheep (Hammond 1932). The idea has also been extended to cattle, but again lack of modern material is a problem. The somewhat meagre data available for cattle are set out in table 10:10, and are derived primarily from the literature, but I have been greatly assisted by the measurements made on the Chillingham cattle in the British Museum of Natural History, which were kindly supplied by Dr. Philip Armitage. Other sources are indicated on the table. It appears that there is little difference in bone proportions between any of the specimens other than aurox males and the early neolithic Danish bull which was likely to have been closely related to the wild type. The use of this method of approach in cattle may be limited to these early sites. The reasons why the domesticated bull assumed the proportions of the female aurox would be of great importance if they could ever be demonstrated, but since large modern animals tend to assume the proportions of the aurox, the cause may be nutritional.

The proportions of the long bones in sheep are rather more informative. Table 10:11 presents a number of breeds in assumed order of primitiveness, again including the short-tailed Mouflon and Soay and the relatively advanced Clun Forest, but no really modern types have been included such as the Down breeds or the very large lustrewools, neither of which are likely to have occurred before the agricultural revolution. There is little difference between relative lengths of the metapodials, but the bones of the middle segment, radius and tibia are relatively longer in the long-tailed breeds and this is even more marked in the bones of the proximal segment, the humerus and femur; this is in keeping with Hammond's (1932) theories on the selection of a good meat bearing conformation. The basal horn core and scapula indices are included in table 10:11 as a supplement to the relative bone lengths. Also included are a number of archaeological examples from the later prehistoric period and the Roman era. The New Wintles and Garton Slack specimens date from the Bronze and Early Iron Ages respectively, whilst the others are 4th century AD. The earliest specimens appear to be the most primitive and the New Wintles animal has a horn core shape similar to that of the Soay ewe. It would appear that the 4th century animals were of the short-tailed type, but it is thought that the Barnsley Park and the Etaples animal were crossbred specimens, the former because of the general robust nature of its bones, although it was female, and the latter because the tail vertebrae were actually excavated with the animal (Vadet 1978) and indicate that it had a tail of medium length. Ollercain et al (1974) and Shelton (1977) have shown that the crossing of short and long tailed breeds results in a tail of intermediate length, the exact proportion depending on the tail length of the mother. It has been demonstrated that the short-tailed skeletal characters predominate in first crosses with respect to the scapula (Noddle 1978), and calculations using 1st and 2nd crosses of

Finnish Landrace (a modern short-tailed breed) with large modern meet sheep indicate that the overall proportions of the relative bone lengths are inherited in the same way.

More material, both modern and archaeological, is necessary to demonstrate more clearly the differences in relative limb bone lengths between different breeds and types. However, it would appear that if modern proportions are demonstrated for a specimen, it is likely that long tailed characters are well established and the original introduction of this type may well have taken place some time previously.

Summary

Bones from different specimens of the same animal species may differ from each other in absolute size, in proportions which can be demonstrated by dividing one dimension into another to obtain an index, and in non-metrical ways. The proportional lengths of the different bones in a limb may also vary. These differences and the reasons for them are discussed. Profiles of different present day breeds of sheep and of archaeological assemblages of both sheep and cattle have been constructed using a combination of the above factors, in order to demonstrate differences comparable to those of recent breeds between different archaeological sites or different periods within the same site. This is possible with large assemblages but such differences may be influenced by changes in husbandry as well as genetic influences, and each case must be discussed individually. Whilst few conclusive distinctions can be made between animals from Pictish and Viking Skaill (Orkney) quite considerable changes seem to have taken place at Roman Wroxeter during the 2nd century.

The proportional lengths of the limb bones in the individual skeleton have also been assessed. In the sheep it would appear that there is no sex difference, but that modern long-tailed breeds have longer upper limbs than primitive short-tailed breeds and archaeological specimens from the later prehistoric and Roman periods. The position in cattle is less clear, but it would seem that males of <u>Bos primigenius</u> and recently domesticated stock have relatively longer upper limbs than primitive females or both sexes in later stock, but it is possible that absolute size may influence this.

Acknowledgements

The bones of the modern animals come from a large number of sources, but I am particularly grateful to the following institutions and individuals:- A.R.C. Inst. Animal Physisiology, Babraham (Mr. G. Embleton), ARC Meat Research Inst. Bristol (Mr. A. Brown, Mr. D. Brock), Meat and Livestock Commission (Mr. M. Owen, Mr. D. Treble), Zoological Society of London (Mr. R. Cinderey), Dr. J. R. Baker, University of Liverpool, Mrs. M. Arkle and Messrs. K. S. Briggs, P. Marsland and G. Purefoy, farmers. Also Dr. K. Joysey, director of the Cambridge University Museum of Zoology, Dr. G. M. King, Natural History Department of the Oxford University Museum, Dr. K. Morgan, Zoological Department of the Welsh National Museum, and the excavators of the sites mentioned in appendix B for permission to use data from unpublished reports.

Appendix A

Notes on the Breeds of Modern Sheep described and on the Archaeological Sites

The sheep breeds are described in assumed order of primitiveness.

Mouflon. This is one of three forms of the wild *Ovis aries* from which domesticated sheep take their origin. It is short-tailed and has a hair covered skin. The specimens described come from a small flock of long standing maintained by the Zoological Society of London. They are rather inbred. The original animals came from Corsica, where it is suggested the animals were never truly wild but became feral at a very early period (Ryder 1983).

Soay. This animal is generally believed to be the most primitive form of domestic sheep. It became feral at an unknown date and survived on the remote St. Kilda group of islands (Outer Hebridees). It is very similar to the Bronze and Iron Age sheep of S. W. England (Jackson 1971).

Portland. This small breed is believed to be a survivor of the medieval sheep breeds of Wessex (Ryder 1964). Very small numbers are still extant, but the material here described originates from two flocks, so it is not particularly inbred.

Jacob. This piebald sheep is well known in its polycerate form as a parkland sheep kept for its fleece and decorative character. However there is also a two horned commercial variety which is a profitable meat producer. All the horn material described is derived from the latter, but some of the limb bones come from polycerate animals. The origin of this breed remains something of a mystery despite a long essay on the subject by Werner (n.d.).

Swaledale. This mountain breed has recently become widespread. It is considered to be very old, but is not mentioned in 18th and 19th century literature possibly because it was then too localised. The characteristic grey muzzle of the animals is very similar to that of the primitive Orkney, suggesting considerable antiquity. Some of the animals here described came from pedigree flocks, but material has also been collected from moorland areas where the breed predominates.

Clun Forest. This is an upland breed of the Welsh borders. It is believed to have fairly early origins from other black faced breeds of Central Wales. At one time the rams were horned and a specimen preserved in the Clun museum has features similar to those found in a specimen from Saxon Hereford (Noddle in Shoesmith 1983). Nevertheless, the present day breed is quite modern in conformation, although none of the animals described in this paper have the extreme "meaty" conformation of the Downland sheep.

Dorset Horn (with Exmoor Horn)

The Dorset Horn came originally from the same type as the Portland, but may have an admixture of merino blood (Alderson 1978). However it has a more modern conformation than the Portland and is also larger in size. It and the closely related Exmoor are characterised by laterally projecting horns in the ewe. This and the very individual facial expression of this breed are portrayed in several medieval manuscripts.

Appendix B

Archaeological Sites

Barnsley Park. This is the site of a 4th century Roman villa in North Gloucestershire. It was excavated by Dr. Graham Webster. Part of the excavation report has already been published (Webster 1982) and the other part, to include the bone report, is due in 1984.

Garton Slack. This is a complex site dating mainly from the Iron Age in the Yorkshire Wolds. The site was later occupied by a Roman-British farm. The excavator was T. C. Brewster, whose report, including the author's bone report, is available (Brewster 1980).

Hayton. This is the site of a Roman fort in East Yorkshire, excavated by S. Johnson. The bovine skeleton was described in detail by Canby (1977).

New Wintles. This is a site in Oxfordshire excavated by M. Gray. The skeletons described here were found in association with some Bronze Age earthworks, but may be of a later date.

Loughor Castle was excavated by P. Lewis (Welsh National Museum). It guards a river estuary in South Wales, and dates from the 13th Century.

Hereford. A long series of excavations has been carried out by R. Shoesmith in Hereford. Part 1 of his excavation report has already appeared, Part 3, containing the bone report, will be published shortly (Shoesmith forthcoming). The material described here is medieval.

Skara Brae. This famous neolithic village was originally excavated by Childe in the 1930s. The bones here described result from a limited excavation by D. Clarke (Scottish National Museum of Antiquities) in 1974.

Skaill. This site has been excavated over a number of years by P. Gelling. It is on the opposite side of the Orkney mainland from Skara Brae, on the Deerness penninsula.

Seguntium. This Roman fort and associated settlement have been excavated over a long period by a number of workers. The bones here described were excavated from a relatively small area by P. J. Casey. The site is now part of Caernarvon, North Wales.

Wroxeter. The Roman town of Virconium is the subject of extensive excavations over a long period by Dr. G. Webster (earlier periods) and P. Barker (5th century). Barker (1982) has published a preliminary report.

Frocester Court. This is the site of a prosperous villa dating from the 2nd century, overlying an Iron Age site. The excavation is continuing, but an interim report including the author's bone report, was published in 1979 (Noddle in Gracie and Price, 1979).

Bibliography

Alderson, L. (1978) The Chance to Survive London: Cameron and Tayleur

Armitage, P. L. (1982) Developments in British cattle husbandry. The Ark 9 50-54

Armitage, P. L. (1983) A system for ageing and sexing the horn cores of cattle from British post-medieval sites (17th to eary 18th century) with special reference to unimproved British Longhorn cattle. In: Ageing and Sexing Animal Bones from Archaeological Sites (ed. Wilson, R., Grigson, G., Payne, S.) British Archaeological Reports British Series 109 37-54

Armitage, P. L. and Clutton-Brock, J. (1976) A system for the classification and description of the horn cores of cattle from archaeological sites Journal of Archaeological Science 3 329-348

Armitage, P. L. and Goodall, J. A. (1978) Medieval horned and polled sheep: the archaeological and iconographic evidence. Antiquaries' Journal 57 73-89

Barker, P. (1982) Wroxeter Roman City: Excavations 1966-1980 London: Department of the Environment

Bartosiewicz, L. (1980) Relationships between the cranial measurements of cattle Ossa 7 19-32

Brewster, T. C. L. (1980) The Excavation of Garton and Wetwang Slacks London National Monuments Record (available on microfiche)

Brooks, M. (1963) Cortical vascularisation and growth in foetal tubular bones. Journal of Anatomy (London) 97 597-609

Canby, J. I. (1977) The Animal Bone from Hayton Dissertation for B.A. degree, University College, Cardiff, Unpublished

Chapman, D. and Chapman, N. (1975) Fallow Deer Lavenham: Terrance Dalton Ltd.

Degerbøl, M. and Fredskild, B. (1970) The urus (Bos primigenius Bojanus) and neolithic domesticated cattle (Bos taurus domesticus Linn) in Denmark. Kongelige Danske Vidensk-aernes Selskab Skrifter 17 51-234

Epstein, H. (1971) The Origin of the Domestic Animals of Africa Volumes I and II New York: Africana Publishing Corporation

Grigson, C. (1969) Uses and limitations in absolute size in the distinction between the bones of aurochs and domestic cattle in Deomestication and Exploitation of Plants and Animals (eds Ucko P.J. and Dimbleby, G.W.) London: Duckworth

Grigson, C. (1974) The craniology and relationships of four species of Bos. 1 Basic craniology: Bos taurus Linn. and its absolute size. Journal Archaeological Science 1 353-379

Grigson, C. (1975) ibid 2. Basic craniology: Bos taurus Linn. Proportions and angles Journal Archaeological Science 2 109-128

Grigson, C. (1976) ibid 3. Basic craniology: Bos taurus Linn. Sagittal profiles and other non-measurable characters. Journal Archaeological Science 3 353-379

Grigson, C. (1978) ibid 4. The relationship between Bod primigenius Bojanus and Bos taurus Linn. and its implications for the phylogeny of the domestic breeds. Journal Archaeological Science 5 123-152

Grigson, C. (1983) Sexing neolithic domestic cattle skulls and horn cores. Ageing and Sexing Animal Bones from Archaeological Sites (ed. Wilson, R., Grigson, C. and Payne, S.) British Archaeological Reports British Series 109 25-35

Haak, D. (1965) Metrische untersuchungen an rohrenknochen bei Deusche Merionland-schafen und Heidenschnucken. Dissertation, University of Munich

Hammond, J. (1932) Growth and development of the mutton qualities of sheep Edinburgh: Boyd and Boyd

Harcourt, R. A. (1974) The excavation of the Roman villa in Gadebridge Park, Hemel Hempstead 1963-68: the animal bones Reports of the Research Committee of the Society of Antiquaries London 31 256-262

Hatting, T. (1975) The influence of castration on sheep horns in Archaeological Studies (ed. Clason, A. T.) North Holland Press

Herre, W. (1969) The science and history of domestic animals in Science in Archaeology 2nd edition (ed. Brothwell, D. and Higgs, E.) London: Thames and Hudson pp. 257-272

Higham, C. F. W. and Message, M.A. (1969) An assessment of a prehistoric technique of bovine husbandry in Science in Archaeology 2nd edition (ed. Brothwell, D. and Higgs, E.) London: Thames and Hudson pp. 315-330

Jackson, J. W. (1917) Domestic Animals in Glastonbury Lake Village Volume 1 (ed. Bullied, A. and Grey, H.Stg.) Glastonbury Antiquarian Society pp. 648-672

Jarman, M. R. and Jarman, H.N. (1968) The fauna and economy of early neolithic Knossos Annals of the British School in Athens 63 241-262

Jones, C. B. (1939) Notes on some skulls of cattle from Ballinderry Crannog No. 2 Co. Westmeath and Lagore Crannog Co. Meath. Irish Naturalists Journal 7 205-211

Jope, M. and Grigson, C. (1965) Faunal remains from Windmill Hill and Avebury in Excavations by Alexander Kieller 1925-1939 (ed. Smith I) Oxford: Clarendon Press pp. 142-167

Knecht, G. (1966) Mittelaltenlich-Fruhneuzietliche Tierknochenfunden aus Oberosterriech (Linz und Ens) Dissertation. University of Munich

McCance, R. A. (1960) Severe undernutrition in growing and adult animals British Journal Nutrition 14 59-73

Noddle, B. A. (1978) Some minor skeletal differences in sheep. Institute of Archaeology Occasional Publication 3 Research Problems in Archaeology (ed. Brothwell, D. R., Thomas, K. D. and Clutton-Brock, J.)

Noddle, B. A. 133-141 (1979). The animal bones from Frocester Court Roman villa In: 2nd Int. erim Report (ed. Gracie, H. S. and Price, E.G.)

O'Connor, T. P. (1982) The archaeological importance of morphometric variability in British sheep limb bones. Ph.D. thesis, University of London, unpublished.

Ollercain, E. A. B. and Boyle, W. J. (1974) Inheritance of tail-length in cross-bred Finn sheep Journal Heredity 65 331-334

Pinard, A. (1952) Structure et vaisseaux de la diaphyse des os longs chez le foetus humain Acta Anatomica 15 188-216

Pitt-Rivers, A. (1888) Excavations in Cranborne Chase Volumes 1-3. Privately published

Ryder, M. L. (1964) The history of sheep breeds in Britain Agricultural Review 12 65-82.

Ryder, M. L. (1983) Sheep and Men Lond, Duckworth

Shawcross, F. W. and Higgs, E. S. (1961) The excavation of a Bos primigenius at Lowe's Farm, Littleport. Proceedings Cambridge Antiquarian Society 54 3-16

Shelton, M. (1977) Studies in tail length in Rambouillet and Mouflon sheep Journal Heredity 68 128-130

Shoesmith, R. (forthcoming) Hereford City excavations 3. Council British Archaeology Research Report

Stampfli, H. R. (1981) Das "Torf" rindskellett in Museum Schwaab-eine Raritat von internationaler Bedeutung. Neus Bieler Jahrbuch 1980 Biel 45-54

Trow-Smith, R. (1959) A History of British Livestock Husbandry 1700-1900 London: Routledge and Keegan-Paul

Vadet, A. (1978) Ossements animaux et humains Gallo-Romaine an Nord et a L'Est D'Etaples Bulletin Commission Departmentale Monument Historique Pas-de-Calais 10 82-85

Walker, C. A. (1964) The growth and development of the beef qualities of the Angoin cattle (east African Zebu) Journal Agricultural Science (Cambridge) 63 135-172

Watson, D. M. S. (1931) The animal bones in Skara Brae, a Pictish Village Childe, V. G. London: Kegan Paul pp. 198-204

Webster, G. (1984 forthcoming) Excavations at Barnsley Park part 3. Transactions Bristol and Gloucester Archaeological Society 101

Werner, A. R. (no date) An enqiry into the origin of piebald or 'Jacob' sheep Rare Breed Survival Trust, Winkliegh, Devon

Wilson, J. (1909) The Evolution of British Cattle and the Fashioning of Breeds London: Vintner

Wiltbank, J. N., Bond, J., Warwick, E. J., Davis, R. E., Cook, A. C., Reynolds, W. L., Mazen, M. W., (1965) Influence of total feed and protein intake on reproductive performance of the beef female through second calving. United States Department of Agriculture Technical Bulletin 1314

Table 10:1: Dimensions of Bovine Horn Cores (mm.)

Breed	Sex, (Age in years) where known	Collection if other than authors	Length Outer Curvature (a)	Basal Circumferance (b)	Ratio $\frac{a}{b}$	Greatest Basal (c)	Least Basal Diameter (d)	Ratio $\frac{c}{d}$
B.primigenius		Cam	720	380	1.89	136	98	1.39
		Cam	640	340	1.89	115	96	1.20
		Cam	600	355	1.69	117	89	1.31
		Cam	600	325	1.85	107	89	1.20
		Cam	600	310	1.94	104	87	1.20
						106	84	1.26
						105	82	1.28
(Lowes Farm)			785	342	230			
		Wel	660	335	1.97	115	100	1.15
		Wel	660	335	1.97	115	100	1.05
		Cam*	575	295	1.95	100	91	1.05
						63	48	1.28
Modern								
Ayrshire	F		200	155	1.29	53	40	1.33
Cadzow	M	Cam				82	63	1.30
Chartley	M	Cam				80	65	1.15
Chillingham	M	Cam				58	50	1.15
Dexter	C(5)		305	165	1.84	60	45	1.33
	C(3 1/2)		220	160	1.38	55	41	1.34
Friesian	F		120	110	1.09	40	35	1.14
	M		210	180	1.17	64	44	1.45
	C(J)					42	39	1.09
Guernsey	F		95	100	0.95	31	30	1.03
Hereford	F(11)		240	155	1.54	52	43	1.21
	F	Wel	220	145	1.52	49	39	1.25
	M(1 1/2)		150	140	1.07	51	38	1.34
	M(3)	Wel	290	205	1.41	74	61	1.21
Hereford x Friesian	F		110	120	0.92	40	32	1.25
	C					55	45	1.22
	C(J)					35	32	1.09
x Welsh Black	F(3)	Wel	170	135	1.26	44	37	1.26
Jersey	F(9)		102	130	0.78	33	32	1.03
	C(2 1/2)					51	44	1.16
Longhorn	F	Cam				43	38	1.16
	M	Cam				66	50	1.22
North Devon	F		260	150	1.73	50	45	1.11
	C		280	165	1.70	55	47	1.17
Spain	C	Cam				102	98	1.04
Ankole (African)	F					115	90	1.28
Gaynah (Indian)	F(12)	Oxf	120	195	1.26	31	28	1.17

Key to symbols

J	Juvenile term defined Armitage 1983
Cam	Specimen from Cambridge University Museum Zoology
Oxf	Specimen from Oxford University Museum Natural History
Wel	Specimen from Welsh National Museum
*	Specimen labelled forgery
M	Male
F	Female
C	Castrate Male

Table 10:2: Position of Nutrient Foramen in Cattle

Breed	Number of Bones Examined	Foramen Medial Position	Foramen Supracondylar Position	Two Foramina	Foramen Proximal	% Medial
Purebred						
Aurox	8	5	3	-	-	63
Aberdeen Angus	37	36	-	-	1	97
Chillingham	4	2	2	-	-	(50)
Friesian	168	99	54	13	1	59
Hereford	52	39	12	-	1	75
Holstein	7	4	1	2	-	(55)
Jersey	10	7	3	-	-	70
Luing	1	-	1	-	-	(100)
Red Breeds	36	26	9	-	1	72
Welsh Black	38	23	10	5	-	60
Crossbred						
Charolais	25	11	13	-	1	44
Hereford or Simmental Cross	59	41	13	2	3	69
Red Breeds Cross	48	37	8	1	3	77
Welsh Black Cross	14	6	5	2	1	43

Red breeds comprise North and South Devon, Sussex and Limousin. The percentage is in brackets where the numbers are too small to be significant.

Table 10:3: Profiles of Modern Sheep Breeds

	Sample Number	Average measurement (mm.) Range in brackets
i. Mouflon		
Metacarpal length	32	131 (121 - 140)
Astragalus length	35	283 (26 - 31)
Metacarpal proximal index	56	1.30 (1.24 - 1.50)
Scapula Index	17	1.06 (0.90 - 1.35)
Horn core basal index (male only)	13	1.35 (1.00 - 1.56)
Femur. Position foramen	21	Proximal 86% Distal 14%
ii. Soay		
Metacarpal length	29	115 (105 - 125)
Astragalus length	42	23.06 (22 - 27)
Metacarpal proximal index	48	1.45 (1.27 - 1.54)
Scapula Index	22	1.17 (1.00 - 1.38)
Horn core basal index	33	1.51 (1.13 - 1.95)
Femur. Position foramen	79	Proximal 85% Distal 14% Midshaft 1%
iii. Portland		
Metacarpal length	14	111 (100 - 118)
Astragalus length	12	25.08 (24 - 27)
Metacarpal proximal index	13	1.42 (1.36 - 1.47)
Scapula Index	12	0.95 (0.87 - 1.00)
Horn core basal index	8	1.48 (1.35 - 1.71)
Femur. Position foramen	11	Proximal 17% Distal 73%
iv. Jacob		
Metacarpal length	9	128.08 (114 - 140)
Astragalus length	8	30.04 (29 - 32)
Metacarpal proximal index	9	1.45 (1.42 - 1.52)
Scapula Index	6	1.05 (0.96 - 1.22)
Horn core basal index	25	1.43 (1.24 - 1.61)
Femur. Position foramen	9	Proximal 66% Distal 33%
v. Swaledale		
Metacarpal length	14	122 (108 - 128)
Astragalus length	13	29.07 (28 - 31)
Metacarpal proximal index	25	1.35 (1.24 - 1.50)
Scapula Index	19	0.96 (0.90 - 1.01)
Horn core basal index	11	1.18 (1.18 - 1.58)
Femur. Position foramen	26	Proximal 50% Distal 46% Midshaft 4%

Table 10:4: Orkney Profile (Cattle)

1. Skara Brae — Neolithic
2. Skaill — Pictish
3. Skaill — Viking

		Sample Number	Average measurement (mm.) Range in brackets
Metacarpal length	(1)	4	195 (183 - 202)
	(2)	2	163 (160 - 165)
	(3)	1	182
Astragalus length	(1)	72	67.7 (60 - 75)
	(2)	25	59.8 (56 - 65)
	(3)	19	59.6 (56 - 63)
Horn Core Basal Index	(1)	-	
	(2)	12	1.42 (1.31 - 1.84)
	(3)	6	1.44 (1.14 - 1.58)
Horn Core Length Index	(1)	2	1.16 (1.13 - 1.18)
	(2)	1	0.74
	(3)	3	0.85 (0.74 - 100)
Femur Position Nutrient Foramen	(1)	41	Medial 44% Lateral 56%
	(2)	23	61% 39%
	(3)	12	68% 32%

Table 10:5: Orkney Profile (Sheep)

1. Skara Brae — Neolithic
2. Skaill — Pictish
3. Skaill — Viking

		Sample Number	Average measurement (mm.) Range in brackets
Metacarpal length	(1)	9	127 (120 - 202)
	(2)	2	120 (118 - 165)
	(3)	8	116 (100 - 121)
Astragalus length	(1)	76	27.6 (25 - 30)
	(2)	18	26.5 (23 - 28)
	(3)	10	26.6 (24 - 28)
Metacarpal proximal index	(1)	35	1.35 (1.19 - 1.50)
	(2)	31	1.36 (1.29 - 1.57)
	(3)	26	1.42 (1.25 - 1.50)
Scapula index Index	(1)	109	1.15 (0.92 - 1.41)
	(2)	26	1.07 (0.95 - 1.28)
	(3)	24	1.05 (0.95 - 1.29)
Horn core basal index	(1)	7	1.46 (1.19 - 1.70)
	(2)	11	1.39 (1.19 - 1.64)
	(3)	7	1.47 (1.29 - 1.91)
Femur Nutrient Foramen Position	(1)	112	Midshaft 45% Proximal 24% Distal 31%
	(2)	26	Proximal 50% Distal 50%
	(3)	15	Proximal 80% Distal 20%

Table 10:6: Roman Cattle Profile

1. Wroxeter Approx. AD 90
2. " 2nd Century
3. " 5th Century
4. Segontium 4th Century

		Sample Number	Average measurement (mm.) Range in brackets
Metacarpal length	(1)	22	174 (156 - 198)
	(2)	19	178 (162 - 197)
	(3)	25	183 (167 - 200)
	(4)	59	184 (168 - 201)
Astragalus length	(1)	14	59.5 (56 - 72)
	(2)	9	52.0 (57 - 67)
	(3)	92	61.2 (55 - 71)
	(4)	114	61.0 (54 - 72)
Horn Core Basal Index	(1)	26	1.43 (1.22 - 1.65)
	(2)	24	1.34 (1.17 - 1.61)
	(3)	35	1.32 (1.00 - 1.48)
	(4)	101	1.28 (1.14 - 1.62)
Horn Core Length Index	(1)	12	0.95 (0.66 - 1.07)
	(2)	18	0.95 (0.81 - 1.16)
	(3)	12	1.04 (0.82 - 1.25)
	(4)	8	0.98 (0.88 - 1.11)
Femur Position Nutrient Foramen	(1)	9	Medial 44%
	(2)	7	14%
	(3)	76	29%
	(4)	62	51%

Table 10:7: Roman Sheep Profile

1. Wroxeter — Approx. AD 90
2. " — 2nd Century
3. " — 5th Century
4. Segontium — 4th Century

		Sample Number	Average measurement (mm.) Range in brackets
Metacarpal length	(1)	11	111 (100 - 118)
	(2)	2	114 (112 - 116)
	(3)	3	123 (107 - 132)
	(4)	2	141 (134 - 148)
Astragalus length	(1)	11	25.8 (23 - 28)
	(2)	1	26
	(3)	23	24.4 (23 - 30)
	(4)	18	26.5 (24 - 31)
Metacarpal proximal index	(1)	17	1.45 (1.34 - 1.62)
	(2)	11	1.39 (1.36 - 1.50)
	(3)	43	138 (1.24 - 63)
	(4)	8	140 (1.29 - 154)
Scapula index	(1)	10	1.08 (1.06 - 1.14)
	(2)	10	1.08 (0.92 - 1.18)
	(3)	11	1.10 (0.82 - 1.27)
	(4)	6	1.07 (100 - 1.17)
Horn core basal index	(1)	7	1.52 (1.28 - 1.71)
	(2)	12	1.44 (1.00 - 1.74)
	(3)	7	1.43 (1.17 - 1.57)
Femur Foramen Position	(1)	19	Proximal 63% Distal 21% Midshaft 16%
	(2)	23	48% 52% 0%
	(3)	40	63% 29% 12%
	(4)	17	63% 35% 12%

Table 10:8 & 10:9: Profile Medieval Sites

1. Hereford C13
2. Loughor C13

		Sample Number	Average measurement (mm) Range in brackets	
Cattle				
Metacarpal Length	(1)	16	175	(159 - 189)
	(2)	13	173	(157 - 185)
Astragalus	(1)	16	59.7	(54 - 66)
	(2)	35	61.3	(55 - 65)
Horn Core Length	(1)	8	0.97	(83 - 118)
Index Core Basal	(2)	7	0.93	(74 - 113)
Horn Core Basal	(1)	39	1.21	(0.98 -1.39)
Index	(2)			
Femur. % medial	(1)	9	44	
Foramen	(2)	19	37	
Sheep				
Metacarpal Length	(1)	8	112	(104 - 120)
	(2)	1	117	
Astragalus	(1)	35	27.2	(24 - 31)
	(2)	9	26.4	(23 - 29)
Metacarpal	(1)	45	1.36	(1.28 -1.50)
Proximal Index	(2)	4	1.41	(1.32 -1.50)
Scapula Index	(1)	14	1.03	(0.09 -1.13)
	(2)	4	0.97	(0.95 -1.00)
Horn Core Basal	(1)	11	1.40	(1.32 -1.56)
Index	(2)			
Femur. Position	(1)	22	Proximal 45% Distal 50% Midshaft 5%	
Foramen	(2)	2	- 100% -	

Table 10:10: Cattle: Relative Proportions of Bone Lengths

Animal	Length Metatarsal	Humerus	% Metatarsal Length Radius	Metacarpal	Femur	Tibia	Source or Reference
Bos Primigeinus							
Danish male	287	150	126	87	-	167	Degerbøl 1970
Danish male	293	140	-	-	177	161	Degerbøl 1970
Lowe's Farm male	296	-	132	88	-	-	Shawcross & Higgs 1961
Danish Female	280	-	111	84	-	143	Degerbøl 1970
Neolithic							
Danish male	244	141	127	-	172	164	Degerbøl 1970
Danish female	236	129	116	87	157	148	Degerbøl 1970
Later Prehistoric and Romano-British							
New Wintles 1.	192	109	-	95	151	161	Noddle unpublished
New Wintles 2.	195	133	124	93	159	158	Noddle unpublished
Hayton	203	120	124	88	152	148	Canby 1977
Handley Down	188	129	126	88	-	143	Pitt-Rivers 1888
Schwaab Museum	160	131	125	-	165	161	Stampfli 1981
Modern							
Chillingham male	209	134	124	86	169	156	British Museum
Chillingham male	212	132	120	85	167	152	British Museum
Chillingham male	213	123	120	85	157	-	British Museum
Chillingham	210	143	124	85	175	152	Oxford Museum
Chillingham female	210	130	-	86	167	152	British Museum
Alderney female	227	141	132	91	168	158	Oxford Museum
Alderney female	228	131	126	86	167	149	Pitt-Rivers 1888
Friesian female	240	-	-	-	169	146	Noddle collection
Hereford female	232	134	134	87	172	155	Welsh National Museum
Hereford Cross Ayrshire castrate	255	135	133	84	177	166	Noddle Collection
Kerry female	203	129	124	90	165	147	Pitt-Rivers 1888
Non-European Modern							
Gaynah female (N.India)	185	116	116	81	146	141	Oxford Museum
Ankole (African Zebu) female	244	146	140	89	162	163	Noddle collection

Table 10:11: Sheep: Relative Proportions Limb Bone Length

Animal	Length Metatarsal	Humerus	Radius	% Metatarsal Length Metacarpal	Femur	Tibia	Astragalus	Scapula Index	Basal Horn Core Index
Mouflon (F)	137	104	109	91	129	151	20	1.14	-
Mouflon (M)	148	109	109	92	125	149	20	1.00	1.33
Soay (F)	120	102	105	88	123	145	18	1.06	1.07
Soay (M)	139	101	107	88	123	150	19	1.25	1.04
Soay (CM)	117	100	107	90	120	147	20	1.33	1.56
Portland (F)	115	111	115	90	133	155	21	100	1.53
Portland (M)	127	115	113	93	134	152	21	0.87	1.44
Jacob (F)	140	110	116	88	132	157	22	1.04	1.23
Jacob (M)	151	113	113	93	134	159	21	1.00	1.32
Swaledale (F)	137	112	114	96	134	157	22	1.05	1.43
Swaledale (M)	142	116	120	90	1.34	160	22	0.94	1.42
Clun Forest (F)	125	120	122	92	145	162	24	0.88	-
Clun Forest (M)	142	115	117	89	135	153	23	0.86	-
Clun Forest (CM)	140	119	126	98	135	164	22	0.91	-
<u>Archaeological Specimens</u>									
New Wintles	138	101	111	91	129	147	20	1.17	1.60
Garton Slack	130	103	108	90	126	145	20	1.15	-
Barnsley Park	126	109	111	90	-	145	21	1.00	1.47
Frocester Court	137	100	113	-	-	141	-	1.11	-
Etaples	144	104	118	90	124	141	21	1.04	-

11. MILK PRODUCTS

M. L. Ryder

INTRODUCTION

Once animals had been domesticated and it was no longer necessary to kill them to obtain their products, a new range of products became available (Fig. 11:1). One of these was wool, my usual topic, another was blood, and the third was milk, which is important for pastoral peoples who live on the products of their flocks. Milk production is a very efficient way of converting vegetable protein to animal protein, the efficiency being 27% compared with 6% meat, and milking is almost certainly much more ancient than the earliest depictions of about 4000 B.C. But since milk is rarely drunk fresh by primitive peoples archaeologists tend to look for stone or pot curd strainers as evidence for the use of milk in cheese making. It is not always realised, however, that cheese can be made entirely with perishable materials such as skin or wood containers and cloth or basket strainers, as vividly shown by Homer's story of the Cyclops. I have long thought that wool cloth must have preceded the more recent use of cotton cloth in curd straining, but only in the summer of 1982 did I see wool cheese cloths and wood strainers in Romania.

Environmental archaeology is describing more and more remains. We have had Andrew Jones at York demonstrating urine remains on fragments of pot, and Marilyn Plant in Cambridge is looking at pottery for evidence of milk protein. But what I am going to do is to describe briefly the wide range of milk products made by pastoral peoples today. In 1965 when I gave a paper on animal products to the British Association (Ryder 1966) I used Fig. 11:2 to illustrate the main processing of milk to give butter and cheese. I merely mentioned the baterial action that produces yoghurt, and the fermentation of mare's milk to make kumiss.

During the writing of Sheep and Man (Ryder 1983) I have come across variants of these main products in the literature as well as intermediate ones. I have also discovered others being made. The main descriptions are ethnographic sources; in these the observer usually records the local name and what he assumes the product to be. But the terminology is often so confused that it appears that some observers have mistaken the product, or that the same name is used for different products in different countries. For instance, often what literally translates as "sour milk" is described by the observer as yoghurt, whereas in some countries there is a distinct product known as sour milk which is different to yoghurt.

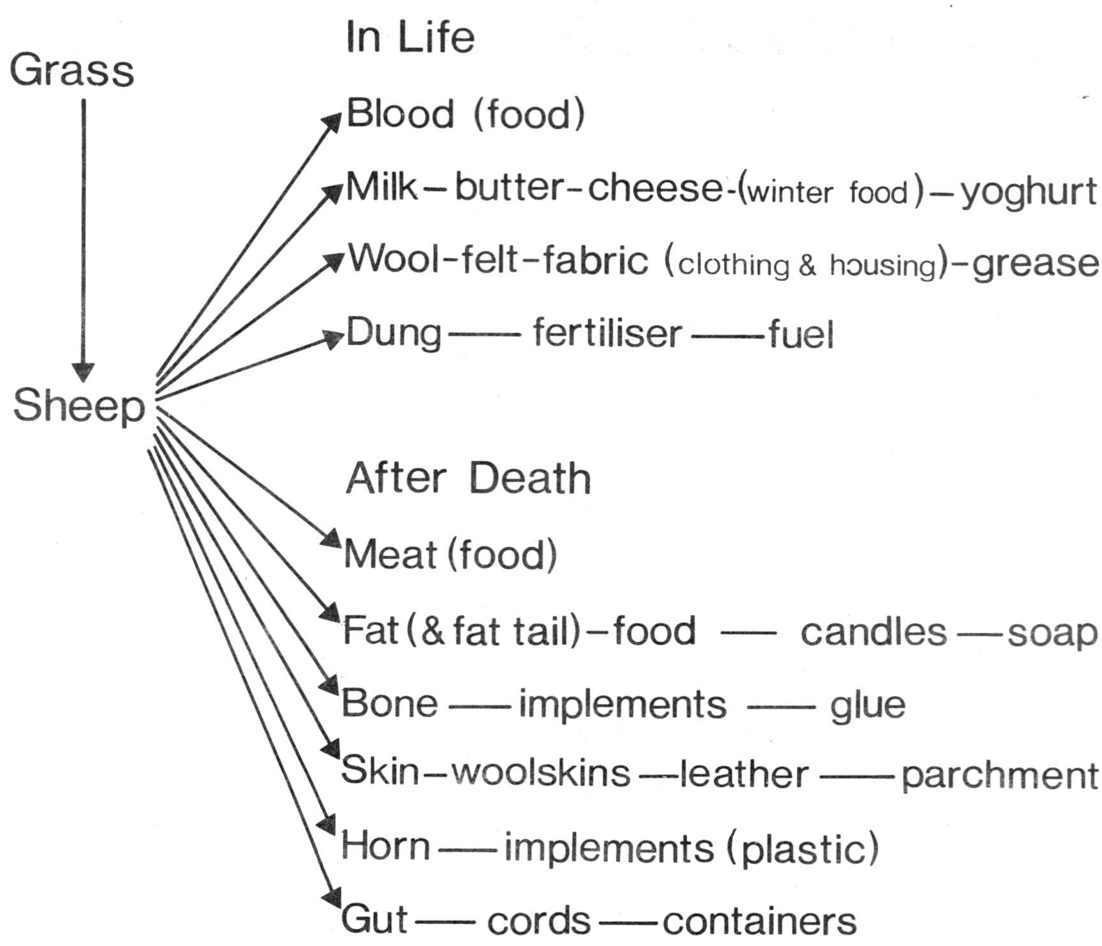

Fig. 11:1 The range of animal products (from Ryder, 1983).

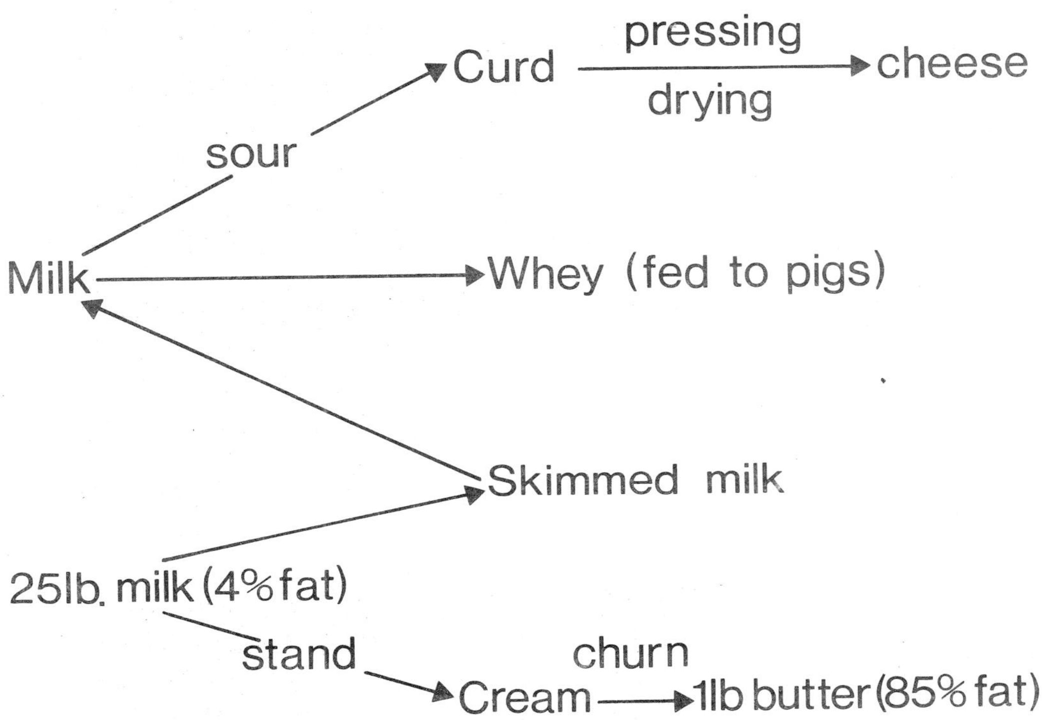

Fig. 11:2 Butter and cheese relationships (from Ryder, 1966).

Until recently modern dairy books covered only butter and cheese, but they are beginning to describe the manufacture of other products such as yoghurt. The trouble is that the modern technologist ignores history and traditional methods in his definition of modern laboratory-controlled and factory-made products. There is therefore considerable scope for biochemical definition of traditional products before they die out, which is apparently something that few ethnographers are able to do.

BUTTER

It will be useful to start with the pastoral nomads of Tibet since they make milk products in a logical sequence (Fig. 11:3). They use some milk to make yoghurt, which I will deal with later, and the rest to make butter. Cheese is then made from the remaining buttermilk. Whereas curdling, the first stage of cheese making, could have been discovered merely from milk left standing, the discovery that churning produces butter required something more positive such as carrying milk on a journey in a skin bag. Indeed, nomads still use skin bags as churns to make butter.

The first stage in modern butter manufacture is to let the cream come to the surface. The division here is into skimmed milk which can be used to make cheese, and cream which contains 25 to 50% fat. Modern technologists list several definitions from whipped cream with 35% fat, through double cream with 48%, to clotted cream with 55% fat. Churning of the cream produces solid butter with 85% fat and leaves buttermilk.

Butter used to be stored in the ground in places as far apart as Pakistan and Scotland where it occasionally crops up as bog butter. This incidentally provides a source of animal hairs for study. In the Middle East butter made from sheep's milk is distinguished by the "fuzz" of wool that can be seen along the edge when it is held against the light. Sheep are milked from behind, which means that wool is not the only foreign body that finds its way into the milk!

Because butter does not keep well in a hot climate it is made into a liquid known in India as ghee. This is clarified butterfat, and is made by heating butter or cream, and then filtering off the solids formed. Some versions might be regarded as a combination of butter and cheese because in Africa ghee is butter churned from whole milk that has been curdled with urine.

CHEESE

As I indicated the Tibetans make cheese from buttermilk, but skimmed milk is also used elsewhere as well as whole milk when butter is not being made, resulting in a richer product. In addition mixtures are often used. Cheese manufacture is basically a method of processing which allows milk to be stored for winter use. The solid product is highly nutritious containing 25% protein compared with 4 to 5% in milk, and 15 to 20% in meat. Cheese has 35% fat compared with 30% in meat.

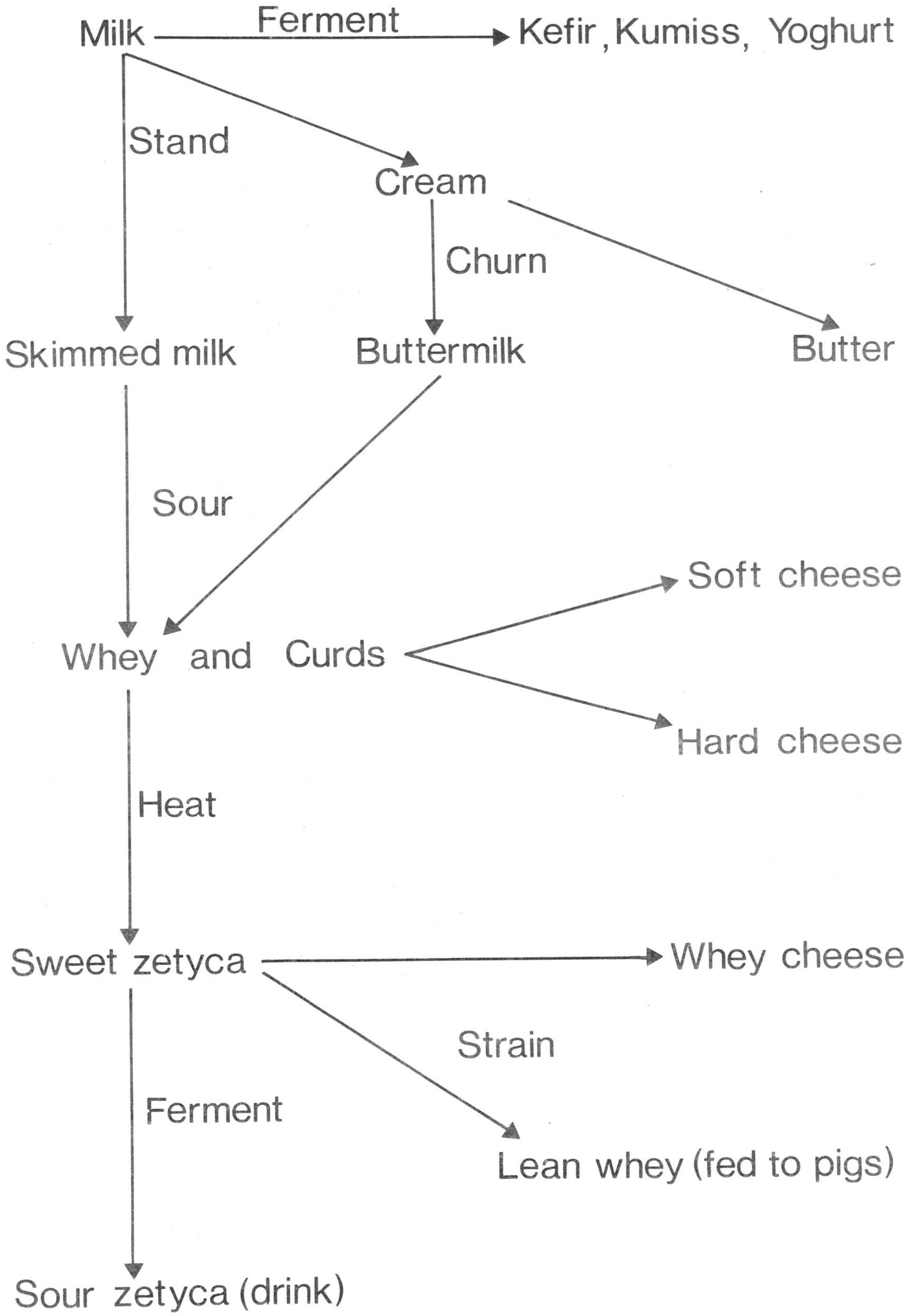

Fig. 11:3 Some relationships of milk products (modified from Ryder, 1983).

Ordinary souring produces an acid curd and usually soft cheese, but curdling can be speeded up with the enzyme rennin, and it is possible that this process was discovered by carrying milk in a bag made from an animal's stomach. Stomachs are commonly used as containers by nomads, and they are also used as a source of rennet, which is an impure form of rennin. Rennin acts to convert the soluble protein caseinogen into casein which forms an insoluble calcium casein compound.

Although there are many variations of the process the main stages of cheese manufacture are a straining to separate the whey from the curds followed by varying amounts of pressing of the curds to remove the last traces of the whey. This can be done with a large stone instead of an elaborate cheese press. In fact in Romania in 1982 one shepherd was using the large block of salt that would eventually be broken up for his sheep.

Soft cheese such as Greek fetta or Romanian telemea is pressed less, and is either eaten fresh or stored in brine. In Turkey they stuff such cheese into goatskins which pass in lorry loads, looking like large white balloons when the hair is inside, or like dead dogs when it is outside. With hard cheese the curds are repeatedly cut, and there is considerable pressing before a final maturation process.

Whey illustrates the inter-relationships of farming operations since cheese-makers have traditionally kept pigs to consume this by-product, as still done in Romania. The same system operates throughout the Carpathian region which extends through southern Russia and Czechoslovakia to southern Poland.

In Poland the cheese made is a hard one called oscypek which is then smoked in the roof of the mountain cheese-making shed. In Poland as well as Romania the whey is made into two drinks. What is known as sweet zetyca in Poland is made by heating whey. The precipitated solids rise to the surface and the drink contains these. It is the clear liquid remaining below known as "lean zetyca" that is fed to the pigs. The second drink, sour zetyca, is made by fermenting sweet zetyca for 2 to 3 days. The precipitated solids can be made into whey cheese which in Romania and Yugoslavia is called urda. Ricotta is a lacto-albumin coagulate, which is made in Sicily by heating the whey at 88°C often with 5 to 10% milk added. When acidified with vinegar the precipitated protein entangles with the less than 1% fat and floats to the surface. The Scandinavian version of whey cheese known as myoset contains brown caramelised lactose. Whey cheese was made in Scotland as late as the 18th century, probably with sheep's milk, and the widespread nature of these products suggests that they are very ancient.

A very hard, dry cheese is made from Mongolia throughout the Middle East as far as North Africa by drying precipitated casein on a tent roof. It is often carried as a powder and reconstituted with water to be added to various dishes. In Mongolia this dry cheese is known as besleg, and the Bedouin of the Middle East are said to make it from yoghurt by boiling, straining and drying, just as they are said to make ghee by churning yoghurt.

Cows' milk

↓

Fresh milk cool to 28°C

↓

Inoculate with starter

↓

Incubate at 28°C until

Acidity reaches 0·7–1·8% lactic acid

↓

Bottle

↓

Incubate for 2 hours

↓

Cool and store at 5°C

Fig. 11:4 Modern flow diagram of *kumiss* manufacture (from Hayes, 1981).

The involvement of the fermented product yoghurt adds a new dimension to the inter-relationships of the range of products, and there is at least one product that is intermediate between cheese and yoghurt. This is the Icelandic cream cheese known as skyr, which is produced by rennets containing lactic flora and so regarded as a yoghurt cheese.

FERMENTED PRODUCTS

I have so far found the fermented products less-well documented, and the inter-relationships less clear. Kumiss is made in Mongolia from mare's milk after churning to remove the butter. The skin bag in which kumiss is fermented is never cleaned out so the bacteria are always present. The buttermilk is stirred with a paddle and the kumiss is ready for drinking in 14 days when it is as strong as beer. Fig. 11:5 shows the modern technologists' imitation of the process, which incidentally is a biotechnological method of producing alcohol. This can be compared with the fermentation of fruit juice in the west for storage as wine. One European medieval commentator William of Rubruck, wrote of kumiss that "it bites the tongue like vinegar, but leave the taste of almonds, greatly delighting the inner man, and greatly provoking urine". The distillation of kumiss is described in the 19th century, the product being named arkhi or milk vodka.

With the other fermented products the fermentation goes further to produce acid (Fig. 11:5). Kefir is a liquid product of Russia and although I have drunk it I know nothing of the traditional method of manufacture. In modern technology the fermentation is carried out with kefir grains which are dried milk solids containing lactose-fermenting yeasts and bacteria. Fig. 11:5indicates a low alcohol content.

Mongolia has a sour milk named tareg which is said to be intermediate between liquid kefir and solid yoghurt. There is also a sour milk in Romania called covasit, and Russia has a distinct sour cream. Again I am not aware of any ethnographic definitions, but these are probably what the modern dairy technologist refers to as cultured buttermilk and cultured cream.

Until a few years ago yoghurt was not well known in the West. Yoghurt is the Balkan name and it is known by different names in India, and Arab countries including North Africa, and Turkey. It may have been discovered in the udder of a dead animal since the bacteria responsible for the formation of yoghurt grow best in the absence of oxygen. These are Streptococcus thermophilus and Lactobacillus bulgaricus with the help of Lactobacillus acidophilus (Fig. 11:6).

Although these organisms are found naturally in milk, in normal souring other bacteria grow first, so the essential first step in making yoghurt is to boil the milk to kill of unwanted bacteria. In Bulgaria (which gave its name to one of the organisms) the milk is simmered in a pot to reduce the volume by one third. The treated milk is then reinoculated with a mixture of the right bacteria, the source of which can be existing yoghurt. Yoghurt can therefore be made at home. S. thermophilus starts to grow first and grows best at 40°C, whereas L. bulgaricus, which grows later, prefers a temperature of 46°C.

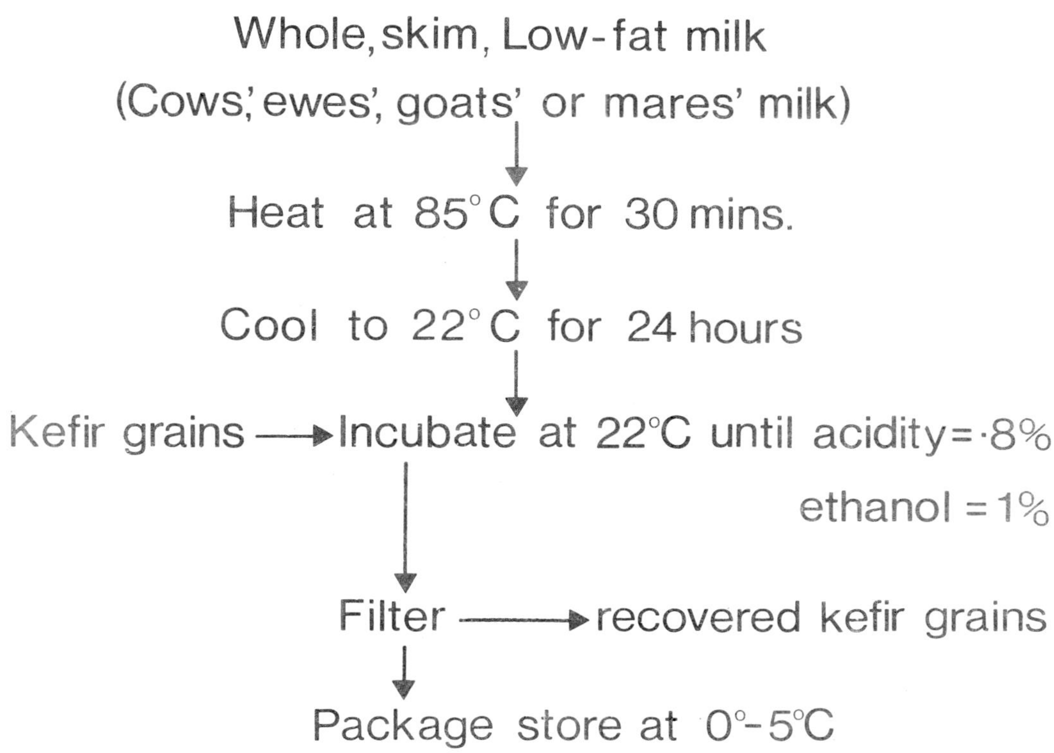

Fig. 11:5 Modern flow diagram of <u>kefir</u> manufacture (from Hayes, 1981).

Milk mix

↓

Heat treat

↓

Homogenize

↓

Cool to incubation temperature

↓

Inoculate with starter

↓

Incubate

↓

Cool

↓

Addition of fruit and or flavouring

↓

Package at 4·5°C

↓

Despatch at 2°C 4°C

Fig. 11:6 Modern flow diagram of yoghurt manufacture (from Hayes, 1981).

The best yoghurt I ever tasted was made from sheep's milk in Bulgaria. It was semi-solid and not very acid, and of course did not have the fruit flavouring which is a recent development in the West.

Since the yoghurt culture remains viable only 1 to 2 days, yoghurt making must have spread with the animals. Yoghurt also illustrates the importance of micro-organisms and their control in all these products. (I did not indicate that micro-organisms are involved not only in maturing some cheeses, but in "starting" others.) Milk products therefore provide an example of the biotechnological achievements of early pastoralists in which they were able to exert an amazing degree of control without recourse to a modern bacteriology laboratory.

LACTOSE INTOLERANCE

The milking of livestock is not universal. It is not carried out in parts of Africa nor in the South East Asia, and Europeans are so used to the drinking of milk that few of them realise that it is an unusual custom. Adult mammals in general have difficulty in digesting milk, and the aversion to milk shown by certain African and Asiatic races is due to a deficiency of the enzyme lactase. There are two schools of thought concerning the basis of this lactase deficiency which leads to lactose intolerance, an environmental one which I favour, and a genetic one which appears less likely. Many anthropologists (e.g. Simoons 1977) have given credence to the idea that lactose intolerance in non-pastoral peoples has a genetic basis, when there are several arguments favouring an environmental origin:

1) one does not have to be lactose-tolerant to consume milk products since cheese manufacture, fermentation and even boiling all destroy lactose, and pastoralists actually use such products as food rather than fresh milk.

2) The high proportion of lactose-tolerant individuals in modern dairying countries could result from the continued drinking of milk from infancy into adult life, and in countries such as China with few cows, there is little opportunity to acquire a taste for milk.

3) The lactose tolerance of peoples in northern Europe is not always associated with the milking of livestock, but helps to relieve vitamin D deficiency to which northern races are susceptible through lack of sunshine (Flatz and Rotthauwe 1973).

4) Finally, not only could these ideas be tested on nomadic pastoralists today, but a lactose-intolerant person has to drink an appreciable amount of fresh milk before an upset develops, and there is frequently very little milk available (see full discussion in Ryder 1983).

Bibliography

Cheke, V. (1959) The story of cheese-making in Britain London: Routledge and Kegan Paul

Davis, J.G. (1976) Cheese vol. III. Manufacturing methods Edinburgh: Churchill Livingstone

Flatz, G. and Rotthauwe, H.W. (1973) Lactose nutrition and natural selection Lancet 2 76-77

Hayes, S. (1981) Dairy microbiology London: National Dairy Council

Ryder, M.L. (1966) The exploitation of animals by man. Advancement of Science 23 9-18

Ryder, M.L. (1983) Sheep and Man London: Duckworth

Simoons, F.J. (1977) Geographic perspectives on man's food quest In: Food, Man and Society (ed. Walcher, D.N.) New York: Plenum Press

Veyret, P. (1951) Géographie de L'élevage. 6th edn. Paris: Gallemard

12. FAUNAL SAMPLE TO SUBSISTENCE ECONOMY: SOME PROBLEMS IN RECONSTRUCTION

James Rackham

INTRODUCTION

It is ten years since Hans Peter Uerpmann (1973) published his critique of the osteoarchaeological method and apart from a number of sophisticated developments of one of the techniques he considers, minimum number of individuals (Krantz 1968; Lie 1980; Fieller & Turner 1982), a study of recent reports indicates a minimum of methodological development. There has nevertheless been a growing realisation of the inadequacy of existing analytical techniques. Basic and very necessary work on the origins of the sample, methods of collection, fragmentation, differential preservation, degree of contamination and spatial similarities and dissimilarities are being carried out in an endeavour to understand and interpret the sample but the analysis is still dependant with little change upon those routines discussed by Uerpmann.

The new techniques for establishing the minimum number of individuals represented by a sample, although of importance in the analysis and study of hunter-gatherer communities have much less validity in agricultural and urban communities where the predepositional selection and taphonomic factors might be expected to be more complicated and the nature of the questions being asked of the sample are often different. Within agricultural societies one may well doubt that the target population, in this case the live herds and flocks of the farmers in their actual abundance, can ever be determined. It is a basic misconception that an 'archaeological' assemblage is the 'death assemblage', and that the death assemblage can be used to reconstruct the live populations. The only situations where this approaches the truth are the bison kill sites of North America (e.g. Frison 1974) and similar sites where the live population (target) is predictable because the whole herd was killed in a single event; but even these take no account of the other species important to man and their abundance.

In an endeavour to 'integrate' the results from faunal analysis into the subsistence economy of say, a 'prehistoric agricultural settlement' certain quantifications and details of the cull structure and herd management are essential. it is the object of the remainder of this article to assess the degree to which these details can be recovered from the analysis of the animal bones alone. Lyman (1982) has recently considered the path from archaeofaunas to subsistence studies and deals with many additional aspects to those considered below.

MODELS FOR THE FORMATION OF BONE ASSEMBLAGES

Meadows (1981) has published a model of 'the causation of animal bone data' (see Fig. 12:1B) that deals with the controllable and non-controllable factors governing the end product from a 'potential bone population'. His 'potential bone population' is defined as 'the sum total of all the bones of all the dead animals any of whose remains were deposited at the site', but not necessarily recovered. This corresponds to all the individuals at the level of 'butcher' (see Fig. 12:1A) one or more of whose bones contribute to the sample deposited at the site. This need not necessarily be the whole of the sample at the level of 'butcher', 'household' or even 'dustbin' since the bones of individuals represented at any of these levels may be removed before deposition. Meadows' model details the loss factors occurring between the generation of the bone population (i.e. after slaughter) and the recovery of the bone sample and its subsequent publication. The model does not consider the 'cultural practices' that might produce his 'potential bone population'. I have therefore constructed a simplistic model for the steps from flock or herd to primary disposal in an agriculturally based society (Fig. 12:1A) and placed it before a version of Meadows' model. Gilbert and Singer (1982) give a more detailed table of the biases and assemblage formation factors that is of more general application. The most significant difference between the model of Meadows and that presented here is that his 'potential bone population' corresponds to only a portion of the sample at 'butcher' and illustrates that the target population of the analysis itself is merely a sample of a target population at a higher cultural level. The number of stages in the model varies with the degree of stratification in society and in some hunter/gatherers and very basic subsistence farmers may be expected to reduce to only two or three where they may well correspond to activity areas within the site. This is very important in considering the degree to which different analytical techniques can be used for reconstruction at different levels of the model. At, for example, a mesolithic site the bone sample comes from animals taken directly from the 'wild population'. These animals may be 'butchered' on or off the site and distributed to further activity areas for consumption, bone working or with the skin. In each case the sample from the activity area is only two steps from the live animal which is itself a sample of the 'wild population'. On an urban site the animals and later their bones may pass through even more stages than those illustrated by Fig. 12:1A, making the bone sample many stages removed from the live flock. Each of these steps will have been subjected to specific and intentional biases with respect to the sample from which they come. It therefore follows that the fewer stages the greater the chance or ease of reconstruction and the greater the reliability/validity of the subsequent reconstruction if any. That is not to say that the analysis of a site whose model has many stages, such as an urban site, is of no use, but that its analysis can most usefully be targeted at the stages of the model immediately preceeding the 'primary depositional point' rather than the reconstruction of a far removed flock or husbandry pattern.

RECONSTRUCTION

When analysing that part of the subsistence economy based upon livestock it is potentially possible to reconstruct it at the 'consumer or household' level from a study of the 'bones recovered'. This sample is only two predepositional stages removed from the food itself. To extrapolate back beyond this stage even to the 'potential bone population' and certainly those earlier in the model becomes complicated by the inadequacies of the analytical techniques, by an increase in the number of imponderables and the need to make an assumption as to the number of stages between flock and household.

There are four major categories of observation and construction of relevance when considering an economic reconstruction from the bone evidence and these are:

1. Stratigraphy and spatial distribution.

2. Identification and quantification of species.

3. Sex of the animals from which individual bones derive, and quantification of the ratio.

4. Age of animals from which the bones derive and the quantification of the slaughter pattern.

Apart from minor specific anomalies such as pathology, and ignoring for the moment the major problems such as contamination and derived material, these categories encompass the maximum recordable and deducible information of relevance in an economic reconstruction. In certain situations, particularly urban, breed types and size may be relevant to the reconstruction, but for simplicity's sake we will exclude this category of information. Within these four headings are three levels:

(a) data - i.e. stratigraphy, distribution, identification
(b) level one of analysis - inferred age and sex from morphological or metrical data (in some situations sex may be determined with certainty rather than simply inferred)
(c) level two of analysis - quantification of species, sex ratio and age or slaughter pattern from data and level one information.

Ignoring for a moment the practical degree of accuracy or otherwise that these successive levels of recording and analysis may produce let us presume a 100% recovery of information under (a) and (b) above, i.e. we have all the information on stratigraphy, distribution, species, sex and age, to what degree, theoretically, can we reconstruct the animal economy.

We have determined precisely the material in the deposits of a particular locale within the site and in the site in general. We must now consider the degree to which the traditional analytical tools of osteoarchaeology can fashion our desired result.

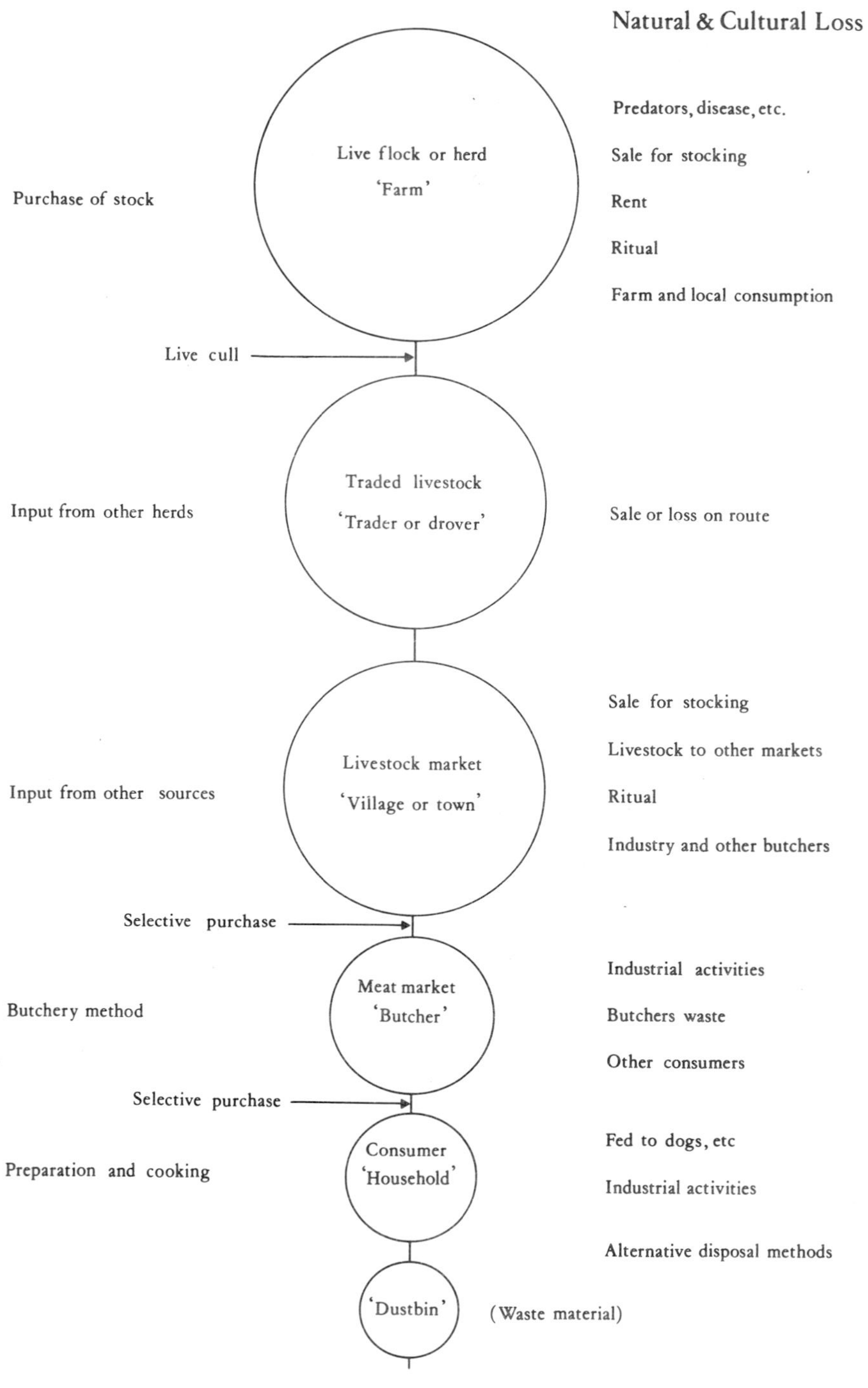

Fig. 12:1a A simplistic model of the stages from a flock of domestic animals to the waste from consumption or commercial activities.

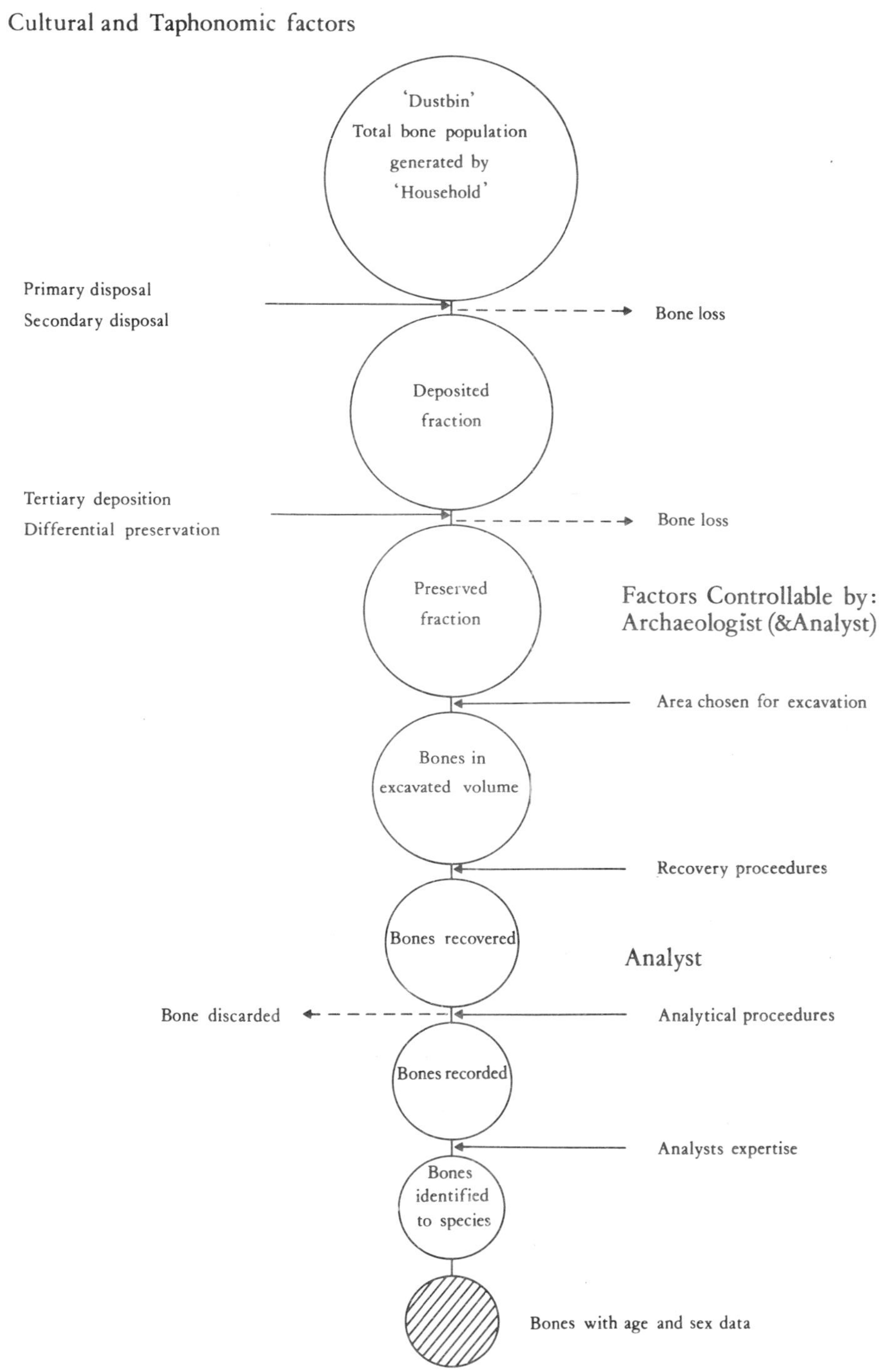

Fig. 12:1b A model of the cultural and taphonomic factors and recovery and analysis procedures that act on a sample between its generation as waste and its final recording (developed from meadows 1980).

ABUNDANCE

We will deal initially with the abundance of species, as indicated by their bones, on archaeological sites. This is the methodological aspect which has received most discussion over the years (White 1953; Chaplin 1972; Clason 1972; Grayson 1973; 1979; Kubasiewicz 1956; and others). It is discussed in many individual analyses and most recently it has been summarised by Gilbert and Singer (1982). The variety of techniques, although ostensibly with the same object in view, frequently calculate 'different' things and therefore not unexpectedly give differing results.

I don't propose to deal with the inadequacies of the different methods, since readers can find these discussed elsewhere (Casteel 1978; Grayson 1979; Gilbert and Singer 1982) but I have summarised them in Table 12:1. It is however of interest here to consider what each method is actually calculating since this is important to the discussion. I will use the model in Fig. 12:1A & B to illustrate at what level a technique acts.

1. Fragment counts - are the actual number of identified fragments of each species at the level of 'bone recorded' (see Fig. 12:1B) which in most instances should be the same as 'bone recovered'. They can be used to indicate species proportions in the sample only with a number of assumptions. Perkins (1973) suggests that all fragments derive from different individuals and therefore the count of fragments represents the number of animals (at least proportions) of each species (i.e. Meadows' 'potential bone population'). This is blatantly an unreasonable assumption for many sites and makes no allowance for pre and postdepositional factors that may act differentially upon species.

2. Joint counts - by an analysis of the fragments it is possible to reconstruct the relative and actual abundance of different parts of the skeleton in the sample. Being a reconstruction it belongs to a stage in the model prior to fragment counts and intends to construct the actual 'deposited fraction' and under the best circumstances may indicate the 'dustbin' or even the 'household'. This is something fragment counts could not do except where no immediately pre- and post-depositional fragmentation had occurred. As such it must be viewed as the primary information level for economic reconstruction in contrast to fragments which are the primary level for determining the factors of survival, damage, loss and recovery (i.e. taphonomic) that have acted upon the 'deposited fraction' to produce the recovered sample. They should be used together in order that the intervening stages may be reconstructable, a necessary prerequisite to validate the construction of the 'deposited fraction'.

3. Minimum number of individuals, MNI - most frequent skeletal part (see Clason 1972: 141 a). Although in certain circumstances this figure may approximate to the real or actual value of the minimum number of individuals present in the sample, in most circumstances it will illustrate a culturally selective or post

Table 12:1: Some of the problems associated with the different techniques for estimating species abundance and relative abundance from a sample of faunal remains.

Problems	Fragment counts	Epiphysical counts	Joint counts+	Bone weight	MNI (GMT) Chaplain 1971 +	MNI (using all bones)+	Estimation + Fieller & Turner 1982	RF Perkins 1973	Relative abundance (c)
Controllable factors									
Fragmentation during recovery	X							X	
Recovery	X	X	X	X				X	
Sample Size					X	X	X		X
Identification expertise	X	X	X	X	X	X	X	X	X
Thoroughness of application					X	X	X		
Archaeological unit definition (ie sample aggregation)					X	X	X		
Adhering soil matrix				X					
Non-controllable factors									
Preservation	X	X	X	X	X	X	X	X	
Fragmentation	X	X	X		X	X	X	(X)	
Butchery	X	X	X		X	X	X	(X)	
Pre-depositional selection	X	X	X		X		X		
Presence of skeletons	X	X	X				X		
Pre-depositional processing (food prep.)				X					
Variability bone/carcass weight relationship				X					
Pre- & post-depositional scavenging	X	X	X	X	X	X	X	X	
Mineral replacement & salt deposition				X					
Bones/individual in each species	X	X							
Independence of specimens	X	X					X		

X - indicates that the problem is applicable
\+ - constructs, other columns are primary information

All the problems listed above can be differential both within species (ie between bone elements or age groups) and between species. Thos problems whose biases are adjustable or irrelevant for a particular technique are left blank.

depositional bias for the bone and as such is informative but not as a measure of abundance.

4. <u>MNI - number of mandibles</u> (see Clason 1972: 141 c). This method falls into the same category as 3 above although since the same bone is always selected the bias may be positive or negative.

5. <u>MNI - Grand minimum number</u>, GMT (see Chaplin 1971: 74). Chaplin's method gives a figure for the smallest number of animals that could have produced the recovered sample. A more accurate method for determining this figure is to include all the bone elements in case animals of different age or sex are unrepresented by the most numerous element, this may add individuals to Chaplin's GMT. The result generally has no place in the model (Fig. 12:1A & B), neither is it equivalent to Meadows' 'potential bone population' although this is what it pretends to represent. It is not the actual number of animals represented in the sample since it will vary with the efficiency of the worker and the ease of recognition of paired elements

The problems with this technique are that:

(a) it presumes that for every animal slaughtered to produce meat for the 'household' one or more of its bones are represented in the 'recovered sample'. (Notice the difference in this definition to that of Meadows' 'potential bone sample' on page 252.)

(b) the most frequent bone upon which the method largely relies is subject to the same cultural and post-depositional biases alluded to in 3 above.

(c) when used as a basis for a ratio of species rather than actual numbers or rank ordering it relies on the assumption that selectivity or bias in the bones of one species is mirrored in all the others being analysed whereas the contrary might be expected. This is by no means a comprehensive criticism, Grayson (1979) discusses the validity of this measure in some depth.

The occasions when this technique is the most suitable for determining abundance are those when there is every reason to suppose that point a, above is true. On these occasions the relationship between MNI and actual abundance is one and the technique intends to reconstruct to the level of 'butcher' (or potential bone population). This level in agricultural and urban sites may still be a number of steps below the level of flock or herd. Fieller and Turner (1982) in the most recent consideration of MNI consider this index to be 'unambiguously the precise number of animals whose bones actually lie in front of the analyst', i.e. Meadows' 'potential bone population'. This statement is not true <u>in practice</u> and Turner's earlier article (1982) is much more to the point on this measure. It may conceptually be possible to arrive at such a figure, if Perkins (1973) is to be believed it is equal to the number of individual bones recovered, but in practice the MNI index is the smallest number of different individuals that the analyst has been able to identify in the sample and not the actual number of different individuals represented in the sample. This makes it an extremely doubtful measure for inter- or intra-site comparisons particularly since factors such as increased fragmentation and modern breakage (Wilson 1975) makes the exercise more difficult and tends to reduce the number of recognised individuals. One of its most awkward aspects from an interpretive point of view is that it has no known or theoretically constant relationship with the actual MNI of the 'deposited fraction', 'dustbin', 'household' or 'butcher'. Yet since it counts individuals and not bones or joints it must surely purport to represent abundance of carcasses and therefore 'butcher' level in the model.

6. Number estimation (Krantz 1968; Lie 1980; Fieller and Turner 1982). All techniques which endeavour to estimate the population from which the bone sample comes are faced with the same problems as methods 3 - 5 above. Whereas the methods above may only give good results where the assumption mentioned under 5a applies, estimation methods endeavour to reconstruct the level of 'butcher' in Fig. 12:1A without making the presumption that all animals are represented but rather the opposite. It can be seen yet again that although the objective is identical, quantification of species, the products of the analysis act at different points in the model. For Chaplin's (1971) GMT (or the more accurate method noted above in 5 to be a real estimate of the thanatocoenosis and not a 'totally artificial figure' it presumes that all the bones not being utilised in the analysis derive from the same animals that produced the bones that are and that the assumption 5(a) is true, a situation clearly unlikely to apply on any archaeological site. The estimation techniques of Krantz, Lie and Fieller and Turner on the other hand make no such claims, but on the contrary utilise the degree of loss as the basis for estimating the original populations. The technique as presented by Fieller and Turner presumes a random and independent loss of left and right bones of all elements between the stage of 'butcher' and 'bones recovered'. If there is any active selection or bias (non-random) at any of the various stages between butchery and sample (such as differential distribution of right and left carcass sides) whether positive or negative the concept falls down.

Theoretically the most interesting aspect of Fieller and Turner's technique is that it makes an estimate of an original population (death assemblage) that may include individuals none of whose bones were ever deposited upon the site because they were subject to preconsumption and pre-depositional losses to other locations. If a different set of random factors had acted on the original population then different individuals could have been represented in the sample but the statistic would still be calculating the original population. The technique presumes random loss not only after deposition but also after slaughter and butchery. It is therefore our only technique that endeavours to construct actual numbers of animals in terms of carcasses being distributed, but it acts from the moment the left and right sides of the carcass of each individual become independent and not before since the association of left and right bones would negate the statistical principle upon which the method is founded. Therefore the result, N, for rabbits or lamb will reconstruct to 'household' if whole carcasses were consumed and deposited on site whereas it will reconstruct to 'butcher' for cattle since left and right bones are probably separated before distribution to the consumer. This is obviously a complication for the assessment of relative abundance of species and shows that the target population might not be at the same level for all species even when using a single technique. The method is less reliable if there are no matched pairs of bones and in pratice a high level of accuracy in the assignment of paired elements is important, something extremely difficult to arrive at for all fragments of a bone. Various other problems which make the technique impracticable on large sites are presented by Fieller and Turner (1982: 57-58). The random nature of the loss and the independence of the left and right bones are the most problematic aspects. In contrast to MNI which is directed towards the 'potential bone population' of Meadows (1980) the estimation method pertains to <u>all</u> animals that <u>could</u> have contributed bones to the 'deposited fraction', rather than restricting itself to those that actually did.

7. <u>Weight method</u> (Kubasiewiez 1956; Casteel 1978). This method is specifically directed towards the calculation of the relative meat contribution of each species which as the example below illustrates need not be equivalent to the proportions of each species slaughtered. To take the simplest example, if one species arrives on the site as whole carcasses and a second as left halves but twice as many, then abundance from the point of view of the consumer is equal yet the MNI indicates that the latter species is twice as abundant in the sample. This is important at the level of diet and consumption since both species are equally represented. However at the level of 'butcher' or 'slaughter' the minimum numbers give a better estimate of abundance. The example indicates the degree to which the MNI can be changed by pre-depositional cultural selection in a way that misrepresents reality, and the theoretical premise that the weight of the bones of each species is a more appropriate measure for assessing the relative abundance in terms of the meat contribution of each species. The criticisms of the method have been detailed by Chaplin (1971) and more recently by Casteel (1978) and Grayson (1979) all of whom reject its use. Some of the problems associated with this method are summarised in Table 12:1. Leaving aside the practicalities of weighing 'clean' archaeological material and overcoming problems of differential preservation the most conspicuous problem with the method is that the ratio of bone to tissue or dressed carcass must be assumed to be constant for all age groups, breed types, sexes and sizes within a species. Otherwise an endless number of multiplication factors must be used for each species, each of which must be arrived at experimentally (see Smith 1975 for an application of this sort to minimum numbers rather than weight of bones). As a theoretical premise, if factors could be arrived at for each age

group, size and sex then the technique would allow the amount of meat that the excavated bone sample of each species represents to be predicted. However, the impracticality of the technique under most situations would seem to make it unusable. Apart from the joint method it is the only method that applies itself to the construction of the relative amounts of meat of each species eaten at the 'household' level.

8. Relative frequency, RF (see Perkins 1973). Perkins' approach was to discover the relative abundance of individual species not the 'actual' number of animals or any estimate of such. He compared the frequencies of different skeletal elements taking the average of these frequencies as the indication of the species abundance. This figure was then compared with those obtained for the other species. This method can therefore give an estimate of the general proportions in the sample and endeavours to bypass problems of predepositional element and recovery bias. Presuming no differential preservation it gives the average occurrence of elements of each species which may be taken to represent the average contribution to the diet of each species and if element selection is random, the proportions of the species at the level of 'butcher' The criticism of the method is that he does not use all the bones but only those elements least effected by cultural or preservational factors. Which elements these are is arrived at subjectively. Such a procedure is unnecessary (see below) and since these factors may vary between species the estimate for some species may be based upon fewer and different diagnostic elements from another.

9. Relative abundance. Payne (1972: 69) presents the ratios of particular bone elements between two species and apparently utilises these to make a subjective decision as to which ratio is the best measure of relative abundance. The subjectivity of this decision is unnecessary. A technique which on similar principles to that of Perkins (1973) and Payne (1972) is illustrated below. It is a measure of relative abundance and not actual numbers of animals and is illustrated below by reference to the Lerna sample that Payne uses and a sample from Beach View, Birsay, Orkney (Lamden, in preparation).

   In the Lerna sample (Fig. 12:2) the ratio sheep/goat to cow of each bone type in the sample is plotted on a bar diagram. The resulting distribution is broadly dispersed and uneven with a mode between 1 and 2 illustrating that many of the bones have a ratio in this range. The four entries to the left of the diagram which give ratios of sheep/goat to cow of between 1:4 and 1:16 are tarsals and phalanges. These low ratios contrasting with the main bulk of the bones may have a number of interpretations but certainly the most obvious is differential recovery, the smaller elements of sheep/goat being missed. This probable recovery problem makes the use of the whole distribution inappropriate for estimating the relative abundance because the median will be displaced to the left and not actually represent the proportions of the bones in the 'deposited fraction'. In this instance an actual measure of

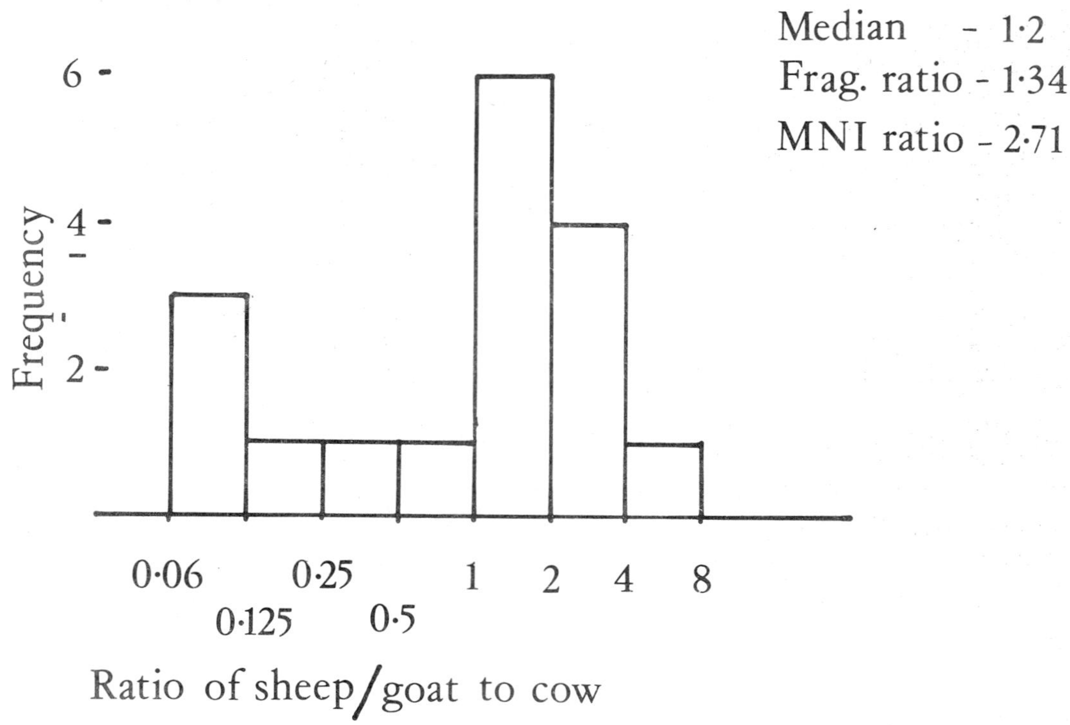

Fig. 12:2 A frequency plot of the ratios of the individual bone elements of sheep/goat and cow from Lerna. (data taken from Payne 1972.)

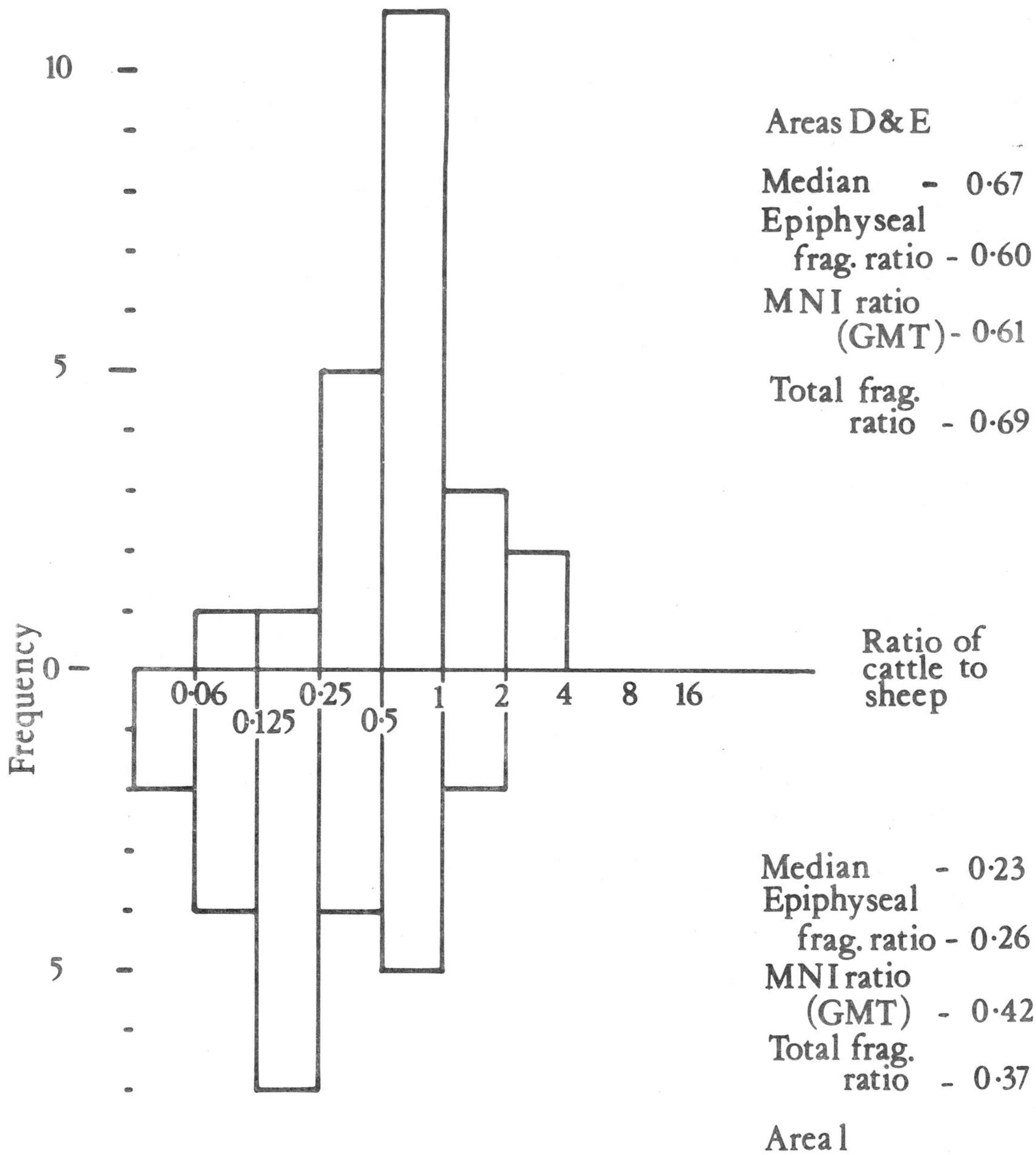

Fig. 12:3 A frequency plot of the ratios of the individual bone epiphyses of cattle and sheep from two areas at Beach View Birsay, Orkney (unpublished data supplied by B. Lamden from excavations directed by C. D. Morris for The Scottish Development Dept. Ancient Monuments Inspectorate).

abundance may be inappropriate but Payne's (1972) estimate of 4:1 sheep/goat to cow would at any rate appear to be an overestimate. In the absence of satisfactory explanations for the broad and irregular distribution a measure of abundance should be based upon non-parametric estimates.

In contrast to the Lerna sample the ratios from two excavation areas at Beach View, Birsay, Orkney (Fig. 12:3) are distributed more uniformly about a mode. They illustrate that a change in the relative frequency of species between the two areas is a characteristic of all the elements and not just the most common; the whole distribution has moved. The distribution indicates the most frequent range of ratios and therefore gives a measure of relative abundance. Distributions that show a dispersed and uneven range can be calculated upon the median but it must be remembered that this dispersion would suggest a sample heavily biassed by pre- and post-depositional factors and it may be unwise to even try and assess the relative abundance of species. The larger the number of elements the more credible the distribution and if the ratios are based on repeatable diagnostic zones such as those suggested by Watson (1979) a number of advantages accrue:

(a) 3 or more zones per bone allow a larger number of ratios and therefore a more reliable distribution.

(b) the smaller the fragment, the less chance of further fragmentation and the need to ensure that the same zone of a single bone is not counted twice.

(c) the great similarity (morphological) of the fragment between species will minimise post-depositional bias.

(d) the measure of relative abundance is supplied by the whole sample not just a portion of it. In Fig. 12:3 the distributions are different not just the modes.

(e) whether bones are left or right becomes irrelevant to the exercise, although the ratios can be calculated for left and right bones if desired.

This last method of assessing abundance (9) has a wide range of interpretable applicability far beyond its estimate of relative abundance.

(i) It allows an assessment of the degree to which pre- and post-depositional factors may have differentially affected the sample. Outliers on the bar diagram require some explanation.

(ii) It permits an assessment of whether species proportions would be a meaningful exercise on the sample or not.

(iii) It allows a real assessment of the validity of intra- and inter-site comparisons by allowing direct comparison at the data level rather than the construct level (i.e. Fig. 12:3). It can utilise any previous analysis if the primary data record is available.

(iv) It can operate at the level of 'bones recovered', 'potential bone population', 'household' and 'butcher'.

(v) It is a technique least affected by the biasses inherent in these types of studies (Table 12:1).

(vi) Where samples are large enough it can be used to work out the relative proportions of different age categories and sex groupings within a single species.

(vii) Recovery bias, such as lack of sieving, can be minimised by selection of appropriately sized diagnostic zones thereby permitting comparisons between sites with differential recovery.

(viii) Skeletons can be included within the counts with minimal bias since only one or two bones will appear in each ratio.

(ix) Left and right categories of bone and fused and unfused can be considered individually where sample sizes are large enough.

The major drawback of this method is that it technically gives a range for the proportion of species to species rather than a figure (unless the median is used), but this may in fact be a more realistic result than a specific ratio, and I would encourage people away from the spurious sense of accuracy implied by the ratio obtained from fragment counts or the MNI index. It is interesting to note that Gilbert and Singer (1982) found RF more accurately estimated species ratios than MNI in computer simulation experiments and this would tend to support the alternative method of determining relative abundance suggested here.

I have spent some time on the consideration of abundance since this aspect is central to all lines of analysis of faunal material. Despite the 100% data recovery from our original sample we have been unable to reconstruct the actual number of animals or the relative proportions of species with any great confidence.

It is all too often that having criticised the count of fragments as a measure of the abundance of species and used some other technique an author has gone on to use the number of fragments in particular age categories or sex groupings as a measure of the abundance of these groups. The following discussion on age and sex should therefore be considered in the light of the techniques available for quantification.

AGE

There have been many discussions and articles on the methods of ageing, their uses and interpretations. The quantification of age categories may be considered to be subject to all those problems and different techniques that have already been outlined above for species. Different age groupings when establishing relative or absolute frequencies can be viewed as analogous to species and therefore have the same quantification problems. However age is not a discrete variable like species but is a continuous one. Owing to the effects of the age of site, sex, breed, nutrition, etc. on maturation rate, fixed relative age categories that are equivalent for intersite or interspecies comparisons cannot be made. No matter at what level the category is determined, stages corresponding to weeks, months or years in modern animals, the absolute age of the animals for each stage will not be consistent through period, sex, breed etc.

The determination of age categories in a bone sample is of interpretative value at a number of levels. At the level of bone sample it is important as a basis for consideration of the differential survival of the bones of different age groups (see Binford and Bertram 1977) and for construction of the 'deposited fraction'. Differential occurrence of bone elements from different age categories (see Binford and Bertram 1977; and Maltby 1982) may be important at the 'household' level in assessing the relative contribution of different age classes to the diet and how they are supplied. At the level of 'butcher' the relative abundance of age categories is necessary for a reconstruction of the kill population. Depending on the number of stages prior to this level this may indicate the live cull or merely a later cultural selection process. At no time except for those sites already mentioned do the different age classes in the sample necessarily reflect the original live population.

It may now be worth considering the degree to which different ageing techniques and their assessment can be used within these interpretive brackets. There are four main methods of ageing animals from their bones and teeth. These are epiphyseal fusion and suture closure, tooth eruption, tooth wear and incremental layers. A fifth method based upon measurements will be discussed that has more limited applicability.

1. Epiphyseal fusion and suture closure. (see Silver 1969 for a description.) Watson (1978) has detailed many of the problems associated with the use of fusion data. Equally these apply to suture closure in which sex is also very important (Bullock and Rackham 1982). The main element is that the age at which fusion of individual epiphyses occurs varies between individuals in the same population and may differ in some cases by well over a year (see Bullock and Rackham 1982). It may be supposed that this variation is distributed normally about a mean ignoring for the moment variation between sexes, and that this mean be used as the average age at fusion. However as Watson (1978) has also indicated in a sample of animals at this mean age, up to 50% may already be fused and therefore cannot be identified as of this

age. The mean age is therefore only of use when the fusion is visible and seen to be in progress. The proportion of the sample showing this condition will depend on the time fusion takes from its inception to completion. As a result interpretation must be based upon 'older than' and 'younger than' data in which one bone can appear simultaneously in a number of the age classes drawn up to describe the data. However Maltby (1982) has shown inference can be usefully drawn merely from the relative frequency of fused and unfused epiphyses of each element, but not readily extended by quantification through the whole age range. To consider this point further, for a random slaughter (with respect to time of fusion) of animals at the mean age of fusion of a bone there will be an approximately equal number of fused and unfused bones. It is therefore essential to be aware that an 'age class' may be represented by both the adult and the juvenile condition. Epiphyseal fusion therefore poses considerable problems for quantification and at present it would appear that inference and interpretation must be based on the frequency of the fused and unfused condition in individual elements. It would be incorrect to group epiphyses of similar fusion ages and use the percentage of unfused as an estimate of the proportion slaughtered before and after the 'fusion stage'since as has been indicated above many of the fused elements may be at the same 'fusion stage' (i.e. age or age class) as those unfused and the figures would be heavily biassed towards the older age groups. This is likely to be one of the major contributors to incompatibility with ages obtained from tooth eruption data. It may be possible to arrive at a method of quantification by using a succession of epiphyses whose age range for fusion does not overlap at all, this would ensure that younger classes were not mistakenly counted into older groups but should only be applied where there was no apparent skeletal selection since this may exaggerate an older or younger group within the succession.

2. Tooth eruption (see Silver 1969 for a description). Chaplin (1971) discusses this method of ageing as do a number of authors in a recent volume (Wilson, Grigson and Payne 1982). It relies on an assessment of the number of animals at each developmental stage of the jaw (Ewbank et al. 1964) and presumes that those at the same stage are of the same age. As Chaplin (1971) has explained there is some variation and an age attributed in modern stock to the eruption of a tooth is a mean age for an event whose distribution is likely to be normal. Therefore those animals in an archaeological sample that have reached a specified stage are potentially only a proportion from that age class, many of which may not yet have reached the stage, or have already passed it. Those animals actually at the stage may also contain individuals from younger or older age classes if these classes are close together in absolute age. One may argue that these effects will be evened out throughout the age classes but this can only be true where slaughter is constant for all age classes; but where selective slaughter occurs if non-overlapping developmental events are not used the pattern may go unnoticed. In a sample of twenty sheep of mean age of eruption of a particular tooth approximately half will have erupted and the

remaining ten will still have this tooth unerupted or in the process of eruption. Therefore when counting tooth developmental stages the slaughter figures will be biassed in the direction of juveniles (in contrast to epiphyses) thereby exaggerating the discrepancy between epiphyseal fusion and tooth eruption. It is therefore important to realise in any quantification of developmental stages that it is not age classes that are being considered and if this is remembered then inference and interpretation can be made.

3. Tooth wear (see Payne 1973; Grant 1975; 1982). This method of age determination is inevitably subject to the same problems and inadequacies already alluded to for tooth eruption. However it has additional problems many of which are discussed by Grant 1978. Grant (1982: Fig. 4) in her recent review of the technique illustrates the result of the analysis in the form of a bar diagram of the frequency of mandible wear stages (MWS) based upon the wear stages of the individual teeth in the jaws. The problem with these as with freqency diagrams of eruption stages is that the intervals are wear stages and not age classes and are not mutually exclusive in terms of age, a fact well illustrated by Payne (1982, Figs. 13-15) where as many as seven different wear stages on a single tooth were found in female goats of the same age, and one particular wear pattern on the M_2, 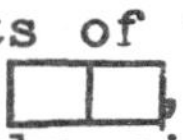, was found in animals from two and a half years to nearly eight years of age, all within a single population. These factors although unlikely to effect the interpretation of a pronounced pattern visible in the wear and eruption stages may well disguise a pattern and give the appearance of a 'constant rate' of slaughter such as that suggested for Barley (Ewbank, et al, 1964). The problem is still one of getting groups of data that fall into mutually exclusive 'age classes'.

4. Incremental Layers. Stallibrass (1982) has recently reviewed the literature on this aspect of ageing and its problems. The principle is that the incremental layers are added at uniform seasonal intervals and the technique should be the osteoarchaeologists lifebelt since it endeavours to estimate the absolute age of an animal from which a tooth or jaw survives and not an age relative to the other remains in the sample. This would immediately answer, at least with regard to the teeth and jaws, all those questions relevant to the age of the animals at death and would give some of the data necessary to reconstruct the cull and husbandry practices for a species, and permit detailed intersite comparisons. Although many authors have used the technique on modern wild species there is some doubt as to its suitability for archaeological domestic material (Coy, Jones and Turner 1982) where sex, nutrition and overwintering conditions which are thought to influence the formation of cementum incremental layers, are unknown.

It is surprising that osteo-archaeologists have not followed up this area of study (Saxon and Higham 1969). Certainly in temperate areas where incremental lines might be expected to form, it is potentially of greater interpretive value in

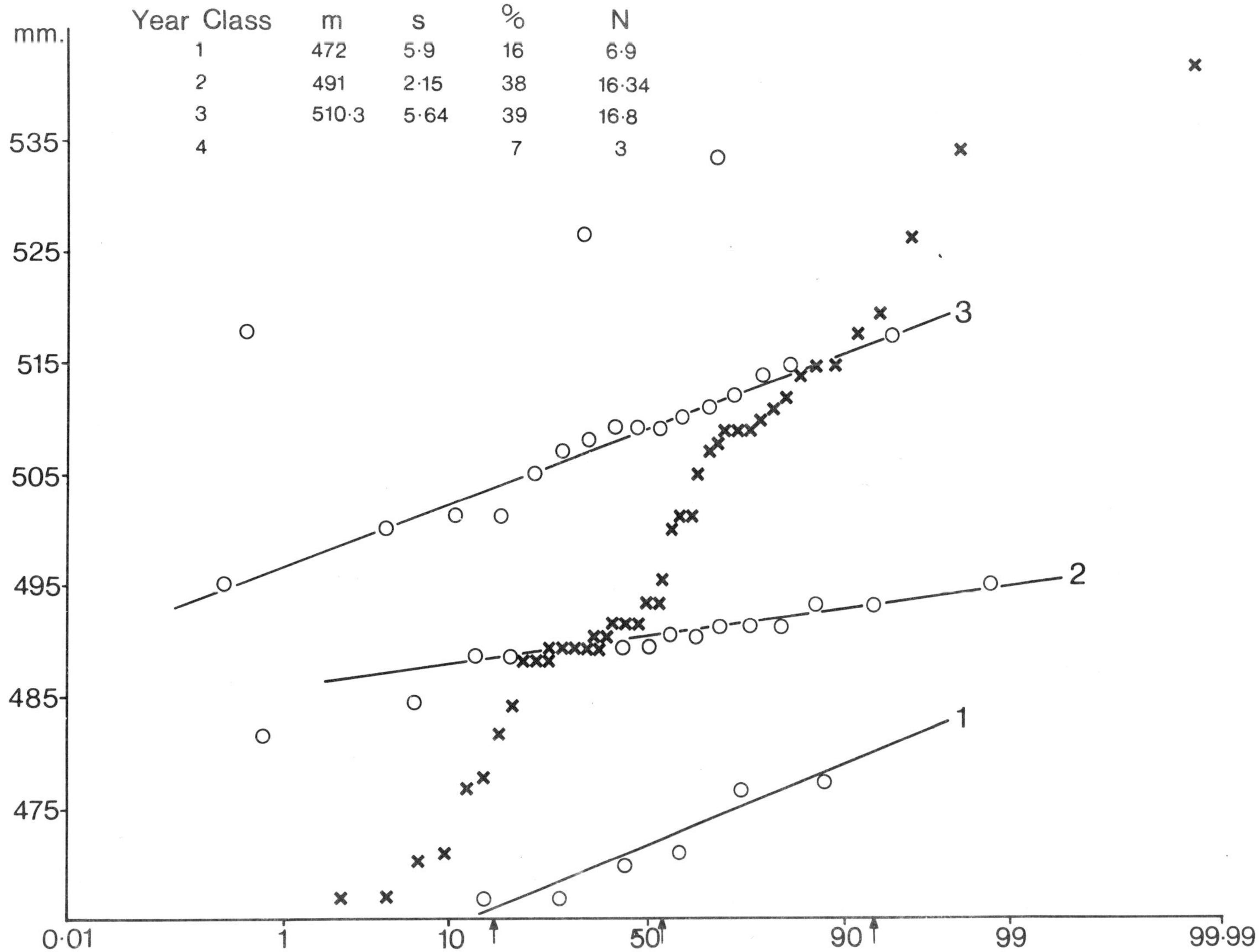

Fig. 12:4 Cumulative frequency plot on probability paper of the minimum shaft widths of male Bison metacarpals from Isleworth and Tattershall Castle (taken from Rackham, 1982).

economic and agricultural reconstruction than the other ageing techniques.

5. Other methods. Although there are other indicators of age in animal bones, some qualitative features, incremental lines in the long bones and jaws, antler development, etc. I will consider only one more of very restricted use and that is growth. Increase in size is only a useful tool for the estimation of age where seasonal occupation or slaughter might be expected. In these circumstances the normal size range of one year group is sufficiently distinct from others for it to be recognised and is useful up to the age at which increase in size diminishes or ceases. Where sexual dimorphism occurs it is essential that the sexes are separated. Midshaft width of metapodials (Wilson 1975; Rackham 1982) has proved useful for separating size classes in bison after the manner of Cassie (1954) (see Fig. 12:4). These classes are interpreted as age groupings whose abundance in the sample can be calculated.

There are three points worth reiteration that must effect any reconstruction and subsequent interpretation of ageing data. The first is that the techniques are not readily comparable. It is not even sufficient to use only the jaws as these need not characterise the whole sample (Maltby 1982). Secondly, the archaeological data is information on the developmental stage of the animal, not its age. Thirdly, a fundamental point, all these techniques make use of fragment counts within each category in their analysis which we have already seen in the discussion on relative abundance to have inherent problems. Chaplin (1971) recommended the ageing of each of the individuals in the minimum number. Techniques relying on the jaws should take account of possible matched pairs, and Fieller and Turner (1982) would recommend the use of the estimation technique on each wear stage before the distribution is plotted, holding that the number of left, right and matched pairs (of jaws) is central to the numerical representativeness of the sample to the age category of which it is part.

Despite the already depressing collection of problems associated with the ageing of archaeological material another problem arises at the interpretive stage. The so called 'survivorship curves' (i.e. Lawrence 1980, Fig. 4.2) imply the survivorship of the contemporary population of a species. What they actually represent is the age structure of a sample of bones deriving from the slaughtered portion of the live population. As is indicated by Fig. 12:5 this death assemblage need have no consistent relationship with the live population.

Despite 100% recovery of the information related to the age at death of animals represented in our sample only the incremental layers in the teeth have actually given an age structure for this part of the sample and even had we determined the age structure for the death assemblage the structure of the contemporary live population is unattainable. It may nevertheless be possible to determine what man selected from this live population for slaughter and the reasons for this

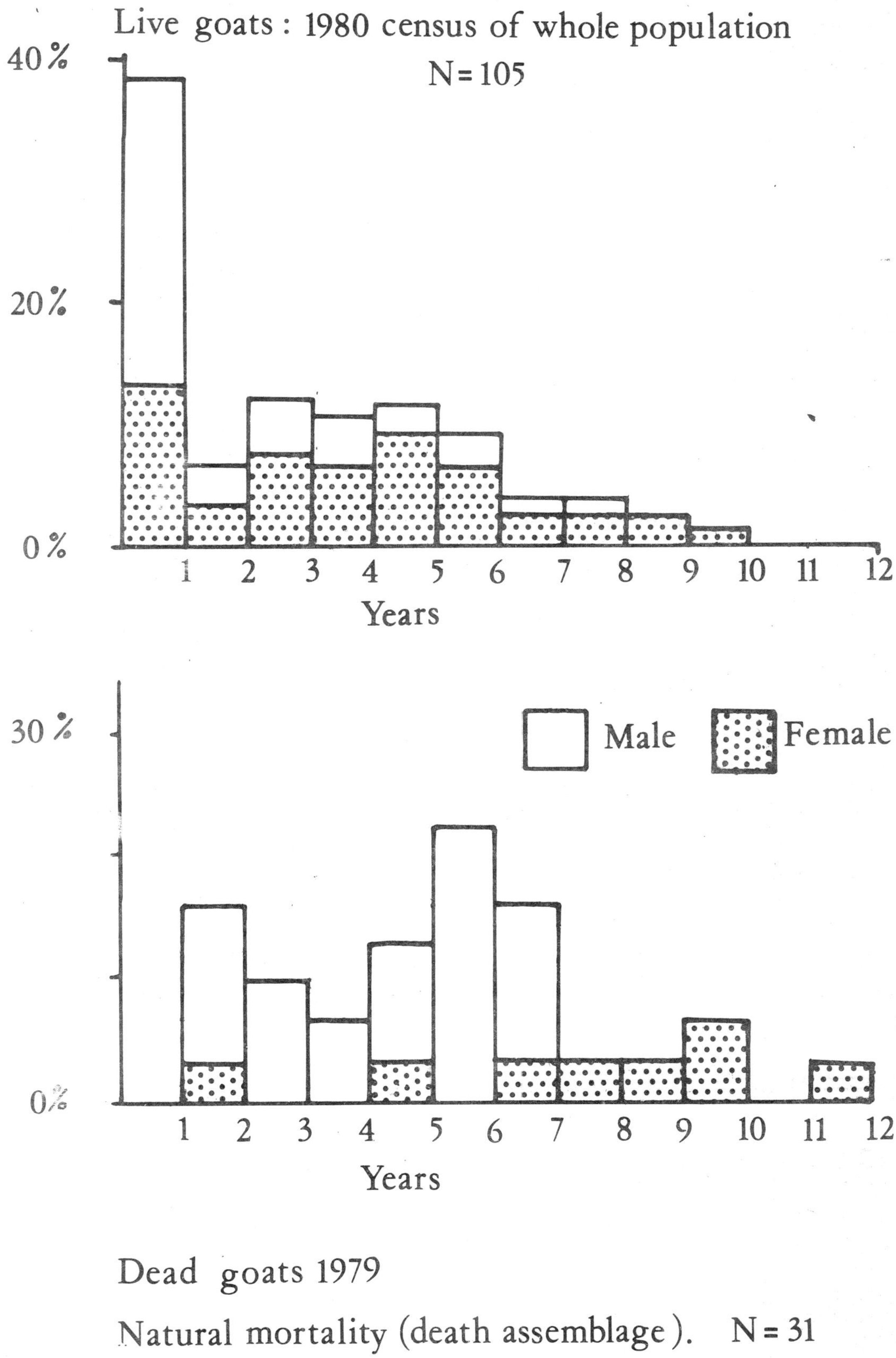

Fig. 12:5 Age class frequencies for live and dead feral goats from Moffatdale, Scotland. (data from Bullock and Rackham, 1982; and S. Pickering, unpublished.)

selection. It is no doubt possible that a computer simulation flock could be constructed that could continually respond to this slaughter pattern, and still survive, but so much would need to be programmed in, lambing time, lamb yield, lamb survival, climate, stocking input, livestock removal, disease, predation etc. that one is inclined to think that we must have known what the flock was before ever analysing the bones.

SEX

The determination of the sex of individuals that have contributed to the bone sample whilst useful in considering meat and dietary contributions by different species at the 'household' level is more particularly an essential prerequisite of any reconstruction of the structure of the meat supply or cull. Only with this information can any assessment and interpretation of prehistoric mans decisions with regard to his livestock be made.

The details of the methods for the determination of sex are described by Chaplin (1971) and for specific species in other papers (Boessneck 1969; Grigson 1982). Rackham (1982) found dimorphism attributable to sex in many of the bones of Bison cf priscus (Boj). In domestic animals castration severely complicates the picture.

However it is not so much the technical problem of sex determination but those of the subsequent analysis of the data that concern us here. In the analysis the quantification of the data on sex mirrors that for species, in that the individual elements, like species, are mutually exclusive, in this female, male and castrate. However in the distribution of the metrical data for a bone type there may well be overlap between the sexes which if large will prohibit the assignment of individual bones to a sex class. Where the metrical data for different sexes does not overlap severely, cumulative frequency distribution plots (e.g. Fig. 12:4) can be used to calculate the proportion of each sex in the sample (Cassie 1954). Quantification of this level one information throughout the sample also has its problems (See above). Maltby (1982) has illustrated how a consideration of the age structure of different skeletal elements in a sample produced contradictory results that could be attributed to selection bias due to different behaviour patterns with respect to carcasses of different ages. He also illustrated how the age structure can vary spatially on a site, explaining his example in terms of differential preservation. It may well be the case that such discrepancies also occur with sex, particularly cull behaviour will be selective for both age and sex. It is therefore important to establish the sex ratio for individual elements rather than merely total the fragments assigned to each. This inevitably means that the sex ratio must be estimated for a number of elements particularly those likely to be selectively distributed, and the ratios compared in the manner already suggested for species relative abundance.

Assuming that the technical difficulties of determining sex can be overcome the sex structure of our sample with 100% recovery of information can be calculated using any or all those methods described under 'Abundance', although method 9 is most suitable for establishing the proportions of the sexes in the sample. Subsequent interpretation may be complicated by a lack of ageing information.

CONCLUSIONS

It is apparent from the discussions above that many of the analytical techniques currently being applied in osteoarchaeological studies are inadequate as a basis for the interpretive consideration of man's husbandry practices, the establishment of the structure of the death assemblage, and the behaviour which generated it and no doubt many of the other interpretive ends to which faunal material is put. If these goals are to be approached in any degree a greater level of detail will be required in site bone recording and analysis such that age and sex can be established more reliably for a higher proportion of the sample than is current. Of course much of this will only be possible with a considerable amount of research into the ageing and sexing characteristics for known age and sex modern animals, of those species and breeds likely to be most relevant to archaeological material.

In the meantime, much more work can be usefully done on the later portions of the model (Fig. 12:1B). Those taphonomic processes now receiving so much attention in America (Binford 1978; 1981; Binford and Bertram 1977; Behrensmeyer and Hill 1980; Shipman 1981), the spatial distribution of element and age classes illustrated by Maltby (1982) and Wilson (1978), fragmentation analysis (Wilson 1975) are all important in this respect. We need to know why the distribution of sheep/goat and cattle elements at Lerna is so unpatterned in contrast to that from Beach View, Orkney before we extrapolate to cull and husbandry practices. If we cannot understand our sample at the level of household we should certainly not proceed to the live flock. Certain situations may be exceptional, such as some hunter gatherer or palaeolithic sites, but the approach should be something similar to that advocated by Grayson (1978) who considers that the faunal sample must first fulfill certain requirements before a technique, in this case minimum numbers, is used. Notwithstanding their validity it is revealing to note that all the methods of estimating abundance ar complimentary rather than alternative since they relate to different target populations.

In conclusion I concur with Brothwell (1978) in his article on animal growth and demography that a diversion, in this case to solving the problems outlined above, would perhaps 'stop some people writing science fiction when it comes to demographic interpretation'.

Bibliography

Behrensmeyer, A. K. and Hill, A. P. (eds.) (1980). Fossils in the Making: Vertebrate Taphonomy and Paleoecology. University of Chicago Press

Binford, L. R. (1978) Nunamiut Ethnoarchaeology. New York. Academic Press

Binford, L. R. (1981) Bones, Myths and Men

Binford, L. R. and Bertram, J. B. (1977) Bone Frequencies - and attritional processes. In For Theory Building in Archaeology: Essays on Faunal remains, Aquatic Resources, Spatial Analysis and Systematic modeling (ed. L. R. Binford) New York: Academic Press, pp. 75-153

Boessneck, J. (1969) Osteological differences between sheep (Ovis Aries Linné) and Goats (Capra Hircus Linné). In Science in Archaeology (eds. J. Brothwell and E. Higgs) London: Thames and Hudson, pp. 331-358

Brothwell, D. (1978) Growth and Demography; neglected aspects of mammal studies in relation to Archaeology. In Research Problems in Zooarchaeology (eds. D. R. Brothwell, K. D. Thomas and J. Clutton-Brock) London: Institute of Archaeology Occasional Publication No. 3 pp. 91-96

Bullock, D. and Rackham, J. (1982) Epiphysial Fusion and tooth eruption of feral goats from Moffatdale, Dumfriess and Galloway, Scotland. In Ageing and Sexing animal bones from Archaeological Sites (eds. B. Wilson, C. Grigson and S. Payne) Oxford BAR. BS 109 pp. 73-80

Cassie, R. M. (1954) Some uses of probability paper in the analysis of size frequency distributions. Australian Journal of Marine & Freshwater Research, 5: 513-522

Casteel, R. W. (1978) Faunal assemblages and the 'Wiegemethod' or weight method. Journal of Field Archaeology 5 71-7

Chaplin, R. E. (1971) The study of animal bones from archaeological sites. New York and London. Seminar Press

Clason, A. T. (1972) Some remarks on the use and presentation of archaeological data. Helenium XII 139-153

Coy, J. P., Jones, R. T. and Turner, K. A. (1982) Absolute ageing of cattle from tooth sections and its relevance to archaeology. In Wilson, Grigson and Payne (eds.) pp. 127-140

Deniz, E. and Payne, S. (1982) Eruption and wear in the mandibular dentition as a guide to ageing Turkish Angora goats. In Wilson, Grigson & Payne (eds.) pp. 155-206

Ewbank, J. M., Phillipson, D. W. and Whitehouse, R. D., with Higgs, E. S. (1964) Sheep in the Iron Age: a method of study. Proceedings of the Prehistoric Society 30 423-426

Fieller, N. R. J. and Turner, A. (1982) Number estimation in Vertebrate Samples Journal of Archaeological Science 9 49-62

Frison, G. C. (ed.) (1974) The Casper Site: A Hell Gap bison kill on the High Plains New York, Academic Press

Gilbert, A. S. and Singer, B. H. (1982) Reassessing zooarchaeological quantification. World Archaeology 14 21-40

Grant, A. (1975) The animal bones. In Excavations at Portchester Castle, Vol. 1 Roman, (ed B. W. Cunliffe) London. Society of Antiquaries pp. 378-408

Grant, A. (1978) Variation in dental attrition in mammals and its relevance to age estimation. In Research Problems in Zooarchaeology (D. R. Brothwell, K. D. Thomas and J. Clutton Brock, eds.). London. Institute of Archaeology Occasional publication No. 3 pp. 103-106

Grant, A. (1982) The use of tooth wear as a guide to the age of domestic ungulates. In Wilson, Grigson and Payne (eds.) pp. 91-108

Grayson, D. K. (1973) On the methodology of faunal analysis. American Antiquity 39 423-9

Grayson, D. K. (1978) Minimum numbers and sample size in vertebrate faunal analysis American Antiquity 43 53-65

Grayson, D. K. (1979) On the quantification of vertebrate archaeofaunas. In Advances in Archaeological Method and Theory Vol. 2 (ed. M. B. Schiffer). New York. Academic Press, pp. 199-237

Grigson, C. (1982) Sex and age determination of some bones and teeth of domestic cattle: a review of the literature. In Wilson, Grigson and Payne (eds.) pp. 7-24

Krantz, G. S. (1968) A new method of counting mammal bones. American Journal of Archaeology 72 286-288

Kubasiewicz, M. (1956) O metodyce badan wykipaliskowych szczatkow kostnych sqierzecych. Materialy Zachodnio-Pomorskie, pp. 235-244

Lawrence, B (1980) Chapter IV Principal food animals at Cayonu. In Prehistoric Village Archaeology in South-Eastern Turkey (L. S. Braidwood and R. J. Braidwood, eds.) BAR S 138 pp 175-199

Lie, R. W. (1980) Minimum number of individuals from osteological samples. Norwegian Archaeological Review 13 24-30

Lyman, R. L. (1982) Archaeofaunas and Subsistence Studies. In Advances in Archaeological Method and Theory Vol. 5 (ed. M. B. Schiffer) Academic Press, pp. 339-393

Maltby, J. M. (1982) The variability of faunal samples and their effects on ageing data. In Wilson, Grigson and Payne (eds.) pp. 81-90

Meadows, R. H. (1980) Animal bones: Problems for the archaeologist together with some possible solutions. Paleorient 6 65-77

Payne, S. (1972) On the Interpretation of bone samples from archaeological sites. In Papers on Economic Prehistory (ed. E. S. Higgs). Cambridge, CUP, pp. 65-81

Payne, S. (1973) Kill off patterns in sheep and goats: the mandibles from A van Kale. Anatolian Studies 23 281-303

Perkins, D. Jr. (1973) A critique of the methods of quantifying faunal remains from archaeological sites. In Domestikations forschung and Geschiehte der Haustiere (ed. J. Matolcei) Budapest: Akadémiai Kiadó, pp. 367-9

Rackham, D. J. (1982) Mid-Devensian Mammals in Britain. Unpublished M.Sc. thesis, University of Birmingham

Saxon, A. and Higham, C. F. W. (1969) A new research method for economic prehistorians American Antiquity 34 303-311

Shipman, P. (1981) Life History of a fossil: An Introduction to Taphonomy and Paleoecology. Harvard University Press

Silver, I. A. (1969) The ageing of domestic animals. In Science in Archaeology (D. R. Brothwell & E. S. Higgs eds.) 2nd Ed. London. Thames and Hudson, pp. 250-268

Smith, B. D. (1975) Toward a more accurate estimation of the meat yield of animal species at archaeological sites. In Archaeozoological Studies (A. T. Clason ed.) Amsterdam. Elsevior pp. 99-108

Stallibrass, S. (1982) The use of cement layers for absolute ageing of mammalian teeth: a selective review of the literature, with suggestions for further studies and alternative applications. In Wilson, Grigson and Payne (eds.) pp. 109-126

Uerpmann, H.-P. (1973) Animal bone finds and, economic archaeology: a critical study of the 'osteo-archaeological' method. World of Archaeology 4 307-22

Watson, J. P. N. (1978) The interpretation of epiphyseal fusion data. In Research Problems in Zooarchaeology. D. R. Brothwell, K. D. Thomas and J. Clutton-Brock (eds.). London, Institute of Archaeology, Occasional publication No. 3 pp. 97-102

Watson, J. P. N. (1979) The estimation of the relative frequencies of mammalion species: Khirokitia 1972. Journal of Archaeological Science 6 127-37

White, T. E. (1953) A method for calculating the dietary percentages of various food animals utilised by aboriginal peoples. American Antiquity 18 393-399

Wilson, B. (1978) Methods and results of bone analysis. In The excavation of an Iron Age Settlement, Bronze Age Ring-ditches and Roman features at Ashville Trading Estate, Abingdon (Oxfordshire) 1974-76 (ed. M. Parrington). Oxfordshire Archaeological Unit, Report I. CBA. Research Report 28, pp. 110-126

Wilson, B., Grigson, C. and Payne, S. (eds.) (1982) Ageing and Sexing Animal Bones from Archaeological sites. BAR BS 109

Wilson, M (1975) The Casper local fauna and its bison. In The Casper Site: A Hell Gap Bison Kill on the High Plains (G. C. Frison ed.) Academic Press.

13. AGRICULTURAL AND SUBSISTENCE SYSTEMS OF THE THIRD MILLENNIUM B.C. IN NORTH-WEST PAKISTAN: A SPECULATIVE OUTLINE

K. D. Thomas

PROLOGUE AND OUTLINE OF PAPER

This paper contains both information and speculation on settlement, agricultural and subsistence systems of the third millenium B.C. in Bannu District, north-west Pakistan. The integration of such systems is a major theme which is discussed in two stages: the identification of the individual elements of the system and then the construction of models which place these elements into some dynamic relationship with one another. This raises such questions as how the boundaries of the various systems should be defined and what are the hierarchical and spatial relationships between groups of elements (or sub-systems).

Much of the discussion is focused upon individual nuclear settlements which are related to economic systems based upon dry-land farming activities. The integration of these systems is analysed in terms of movement of energy and materials between settlements and their territories and between the identified, and inferred, elements of these systems. The role of the secondary products of both agricultural and pastoral activities is considered to be fundamental to integration.

Perturbations in these systems may result from a wide-range of environmental, social and political factors. Inter-annual and spatial variation in rainfall is a particularly significant factor in the area today (and this was probably true in the past), demanding flexibility of response by farmers. Such flexibility rests upon decision-making, particularly regarding the annual scheduling (and therefore the potential for inter-annual rescheduling) of arable activities. The annual scheduling of activities, with associated age- and sex-related divisions of labour, may be an important factor in integrating agricultural activities into the subsistence system, and subsistence activities into the economic system (see section VI for a discussion of these terms).

The main sections of this paper are:

I Introduction. Why is South Asian agriculture, and particularly that in the north-west of the sub-continent, of interest to the archaeologist?

II The Fieldwork Area. An outline of archaeological work in Bannu District

III Environment and Subsistence in the Bannu Basin today

IV Past Environments in the Basin

V Some theoretical considerations

1. Definitions and approaches
2. Levels of analysis
3. Focusing on the site
4. Quantification
5. Simple models

VI Agriculture and Subsistence in the Bannu Basin in the third millennium B.C.

1. Introduction
2. Crops
3. Products
4. Arable activities
5. Pastoral activities
6. Secondary products
   (i) from arable activities
   (ii) from pastoral activities

VII Approaches to the Integration of agricultural, subsistence and economic systems

VIII Postscript

The presentation includes the use of palaeoenvironmental and palaeoeconomic data, analogy with present-day systems in the region, and hypothetical models. This paper does not represent the final synthesis of a long-term and wide-ranging programme of fieldwork; it is more of an heuristic exercise. The field-work upon which it is based is as yet in its infancy.

I INTRODUCTION

Agricultural and economic systems of the prehistoric and protohistoric periods in the Indian Sub-continent are of interest for a number of reasons, and an understanding of the origins of present-day agriculture in the region is one of them. As Joseph Hutchinson (1976) pointed out, modern Indian agriculture is notable for its diverse range of crops and domesticated animals. This diversity applies not only to the range of species exploited but also to the supposed areas of origin of the various domesticates. While the potential ancestors of many (e.g. rice (*Oryza* sp.), some legumes, mustard (*Brassica* spp.), cotton (*Gossypium* sp.), sheep (*Ovis*), goat (*Capra*), cattle (*Bos*), water buffalo (*Bubalus*) may occur in the Sub-continent, this need not imply indigenous domestication of all or any of them. Rice and water buffalo may have come from China or South-east Asia. Sheep, goat and cattle could have come from western Asia, along with lentils, peas, wheat, barley and oats. *Sorghum* and

Eleusine probably came from Africa. The New World cultigens, such as maize, potatoes and chili peppers, pose few problems of interpretation. The history of Indian agriculture, although rather obscure (Hutchinson 1976), is of great interest, and archaeology is in a unique position to cast light on this.

The Indian Sub-continent is extremely varied in terms of climate, topography and ecology. Sopher (1980) has noted that there is a marked east-west gradient in humidity which is associated with a cline in agricultural activities, from rice-dominated agriculture in the east to mixed-grain farming with an increased pastoral component in the west. This paper is concerned only with this latter area.

Recent archaeological research in Baluchistan (south-west Pakistan) has produced compelling evidence for the indigenous domestication of certain plants and animals (Jarrige and Meadow 1980). Excavations in the deep neolithic deposits at the site of Mehrgarh (the location of this site, and of others mentioned in this paper, is shown in Fig 13:1.b), on the Kacchi Plain, have produced abundant remains of animals and plants. Analysis of the animal bones suggests early husbandry and perhaps local domestication of sheep, goat, cattle and water buffalo before the sixth millennium B.C. - and probably much earlier (Jarrige and Meadow 1980; Meadow 1981). Early crops include 2-row and 6-row barley (both hulled and naked varieties) and einkorn (*Triticum monococcum* L.), emmer (*T. dicoccum* Schubl) and bread wheats (*T. aestium* L.). It would appear therefore that not only is mixed farming of great antiquity in the north-western Indian Sub-continent, it may also have evolved *in situ*.

Agriculture in north-west India is remarkable for its diversity, conservatism and high degree of integration at the village level. When rainfall patterns are favourable and irrigation possible, farming is particularly intensive with both a winter crop (*rabi*) and a summer crop (*kharif*) being grown each year. Winter crops include wheat, barley and legumes while rice and millets are the main summer crops. Rice generally requires irrigation but *Sorghum* and the millets may be grown on the dry-lands. The antiquity of double-cropping systems is largely unknown. Plant remains from Harappan (Indus Valley Civilization) sites of the mid-late third millenium B.C. (see Allchin and Allchin 1982 for a recent discussion of the Harappan Civilization) are dominated by wheat, barley and legumes (Vishnu-Mittre and Savithri 1982) but rice has been reported from Lothal and Rangpur in Gujarat. The rice remains are in the form of imprints of husks and spikelets in baked clay and may be of wild rice, which is common in the local marshes (Vishnu-Mittre and Savithri 1982: 207). Other possible summer crops reported from Harappan sites include c.f. *Setaria* and c.f. *Eleusine coracana* from Surkotada and *Sorghum* from Late Harappan Daimabad (Vishnu-Mittre and Savithri 1982). These identifications are tentative and refer to only a few specimens; it appears that summer crops were not an important element in Harappan agriculture. The earliest evidence for summer and winter crops in Pakistan comes from sites of the second millennium B.C. Rice has been reported from Loebanr III and Aligrama in Swat (Costantini 1979), dating from the late second millennium B.C., and rice, *Sorghum* and grape (*Vitis*) have been reported from Pirak in Baluchistan (Costantini 1981). It has not been demonstrated

Fig. 13:1a (above) Map of Pakistan, showing the location of Bannu District. Fig. 13:1b (opposite) Map of Pakistan and north-west India showing the location of regions and sites mentioned in the text. 1: the Bannu Basin; 2: Mehrgarh; 3: Lothal; 4: Rangpur; 5: Surkotada; 6: Daimabad; 7: Loebanr; 8: Aligrāma; 9: Pirak; 10: Kalibangan; 11: Bahawalpur District; 12: Rehman Dheri and 13: Sarai Khola.

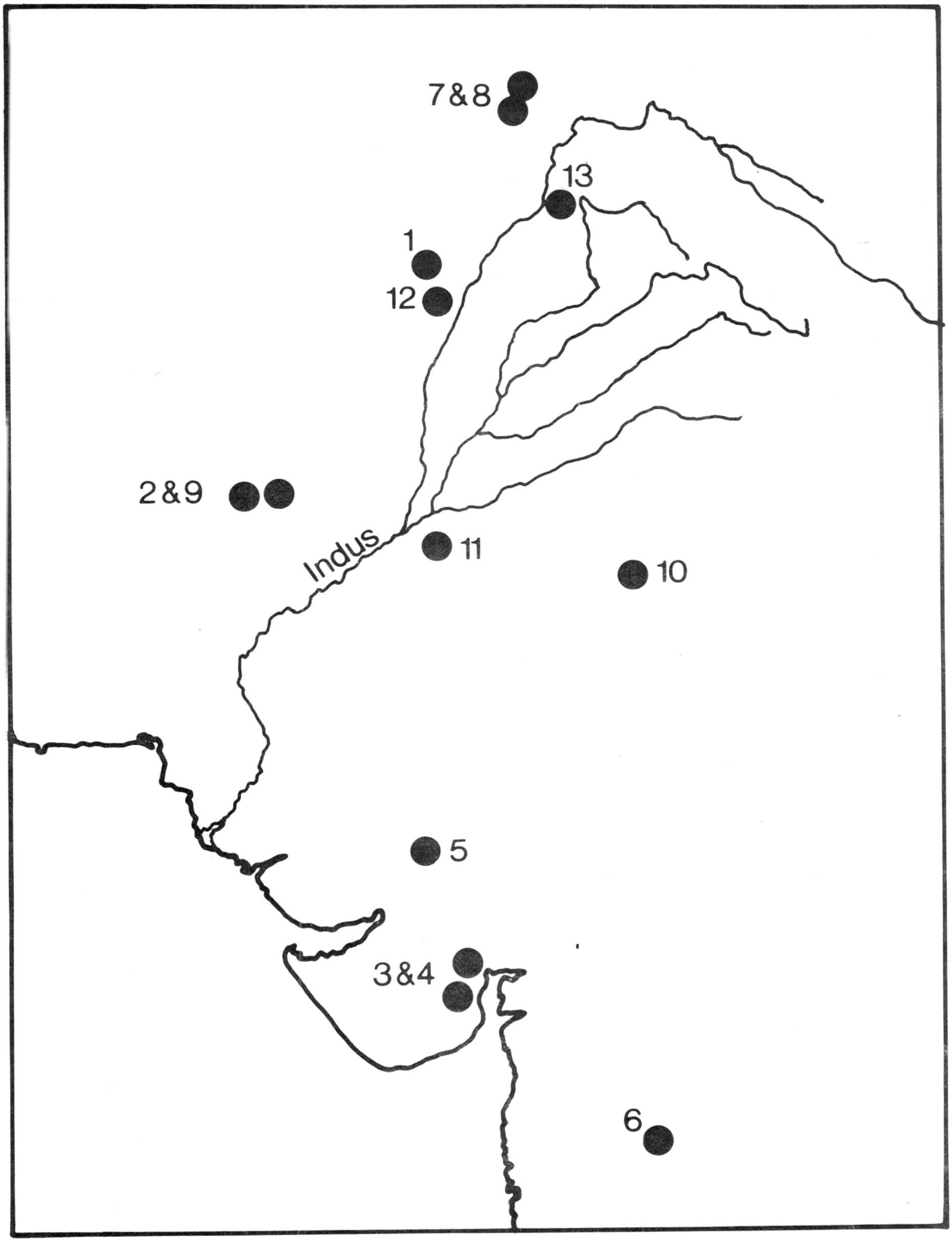
7&8
13
1
12
2&9
Indus
11
10
5
3&4
6

archaeologically that these summer crops were grown using irrigation technology. Although much more work needs to be done on the diversification of cultigens in north-west India and Pakistan, particularly in the period from the fourth to late second millennia B.C., it would appear that the major determinants of cropping patterns were environmental constraints and the availability of both suitable cultigens and irrigation technology.

Double or single cropping patterns are, of course, only one aspect of the diversification and integration of north-west Indian agriculture. Analyses of recent agricultural communities by Blaikie (1971a, b) have demonstrated the high level of organization between the diverse elements of the agricultural system - and how these interact with the environment and the overall socioeconomic system. Such agricultural systems, in all their variety and complexity, provide excellent examples of the integrated exploitation of crops, animals, wild biological resources and the landscape.

II THE FIELDWORK AREA: BANNU DISTRICT

Bannu District is in the North-West Frontier Province of Pakistan, occupying a topographical basin located to the south of the Salt Range and the Peshawar Valley and to the north of the Gomal Plain (Figures 13:1 and 13:2). The basin is drained by a major river, the Kurram, which flows east to join the Indus. The general characteristics of the basin have been outlined elsewhere (Rendell 1981; Thomas 1981 and in press) and further details are given below.

In the winters of 1977-8 and 1978-9 a team consisting of Pakistani and British archaeologists undertook site-surveys, and two limited excavations on sites, in the basin. Bannu District had been selected for investigation for two main reasons. First, apart from limited surveys by Raverty (1878), Stein (1928) and Dani (1971), it was an area relatively unknown from the viewpoint of early settlement. Second, the geographic location location of the basin, in the piedmont zone between the Indus plains and the hills and mountains of the Afghan borderland, was of considerable interest in that it resembles that of the Kacchi Plain in Baluchistan, where the site of Mehrgarh was discovered (see section I, above). The Kacchi Plain lies at the entrance to the Bolan Pass, which has been a route between India and Afghanistan for centuries, if not millennia. Comparable passes occur in the Bannu basin; the valleys of the Kurram and Tochi have long been used as a means of trade and contact between the Indian Sub-continent and Afghanistan.

We had hoped to locate early settlements in Bannu District which were comparable to Mehrgarh in terms of chronology and evidence for early domestication, but so far the earliest sites we have found date from the late fourth millennium B.C. (Fig 13:2 and 13:3). Of these, Lewan and Tarakai Qila (Fig 4, site numbers 5 and 13) have been subject to exploratory excavations (Allchin and Knox 1981a, b; Thomas 1981 and in press). (These sites are discussed briefly in section VI.1, below.) In this paper I will consider these early sites from the viewpoint of the agricultural and subsistence systems on which they depended. The data presented are provisional upon more extensive and intensive fieldwork and laboratory analysis.

III ENVIRONMENT AND SUBSISTENCE IN THE BANNU BASIN TODAY

Only a synopsis is presented here as a more detailed treatment will be published elsewhere (Thomas, in press).

The climate is arid or semi-arid and rainfall, at all times of the year, is highly unpredictable in quantity and spatial occurrence. The monsoon rains of July-September provide the heaviest rains of the year, although the monsoonal influence is very attenuated in this north-westerly part of the Indian land-mass. High losses of water by evapo-transpiration mean that agriculture without irrigation is not possible in the summer. Modern agricultural settlements are concentrated in the north-west of the basin, where permanent and ephemeral streams emerge from the hills and where the irrigation systems (mostly developed since the mid-nineteenth century) are concentrated. The north-west of the basin is also watered by rain in the winter as a result of the eastwards penetration of Mediterranean-type depressions. This winter rainfall, coupled with low evaporation rates, provides the basis for rain-fed agriculture (*barani*) in both the irrigated northern and western hills and the eastern plains of the basin.

Bannu City is the focus for intensive, market-based, land-use, with a wide range of summer, winter and vegetable crops being grown in the area. Elsewhere in the basin, away from the irrigated areas, agriculture depends either upon direct rainfall (dry-land farming) or on concentrated run-off at the bottom of slopes or in stream channels. In these areas only a single (winter) crop is grown each year and pastoralism is an important element in the economy. Further details of recent farming practices are discussed in relation to prehistoric farming (section VI, 4 and 5).

IV PAST ENVIRONMENTS IN THE BASIN

There is abundant evidence of changes in the pattern of water discharge and stream incision in the basin during Holocene times (Thomas, in press). It is not clear if these changes were due to secular variations of climates or to factors associated with the torrential nature of (especially summer) rainfall and the correlated instability of channelled run-off. Inter-annual variation in rainfall at the present day is enormous and probably exceeds in magnitude any likely longer-term shifts in climate. Therefore, short-term, year-to-year, shifts in the weather are more likely to be manifest in depositional and erosional sequences than are subtle longer-term trends. It has not, as yet, proved possible to date the various events of incision and deposition which are evident in the north-west of the basin.

Evidence for the climate of north-west Pakistan in the third millennium B.C. is lacking; the nearest evidence comes from Rajasthan in north-west India. Here, pollen analyses of the deposits in salt lakes (Singh *et al.* 1974) suggest that the climate was markedly wetter during the period 5000-3500 B.P. than it is today. It is not certain whether these changes in local floras in Rajasthan were due to an increase in the mean annual rainfall or to a change in the *distribution* of rainfall. Today, the summer monsoon provides the

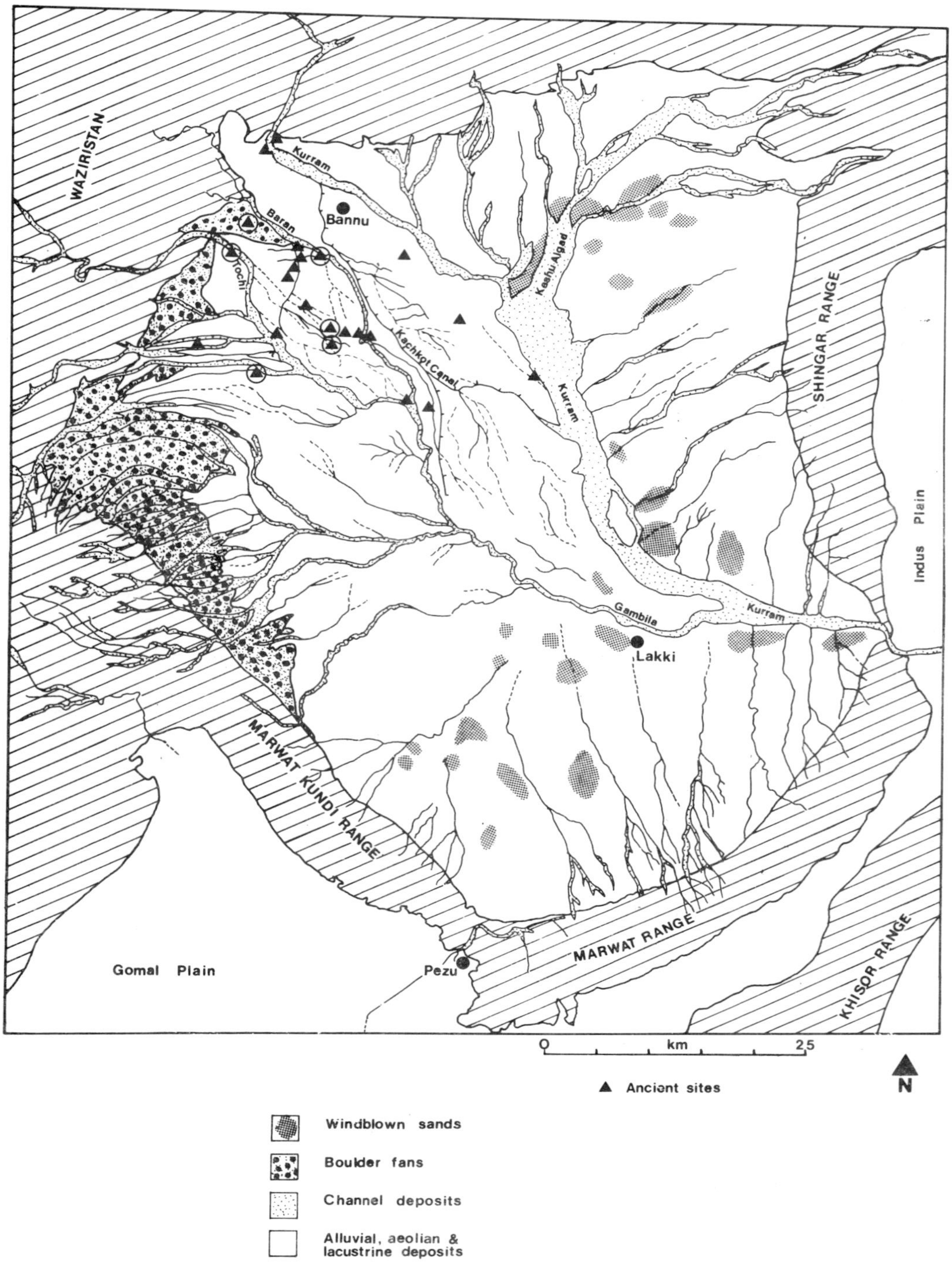

Fig. 13:2 Map of the Bannu Basin showing the major geomorphological featues, drainage channels, ephemeral torrents and the distribution of archaeological sites. Sites of the third millennium B.C. are circled.

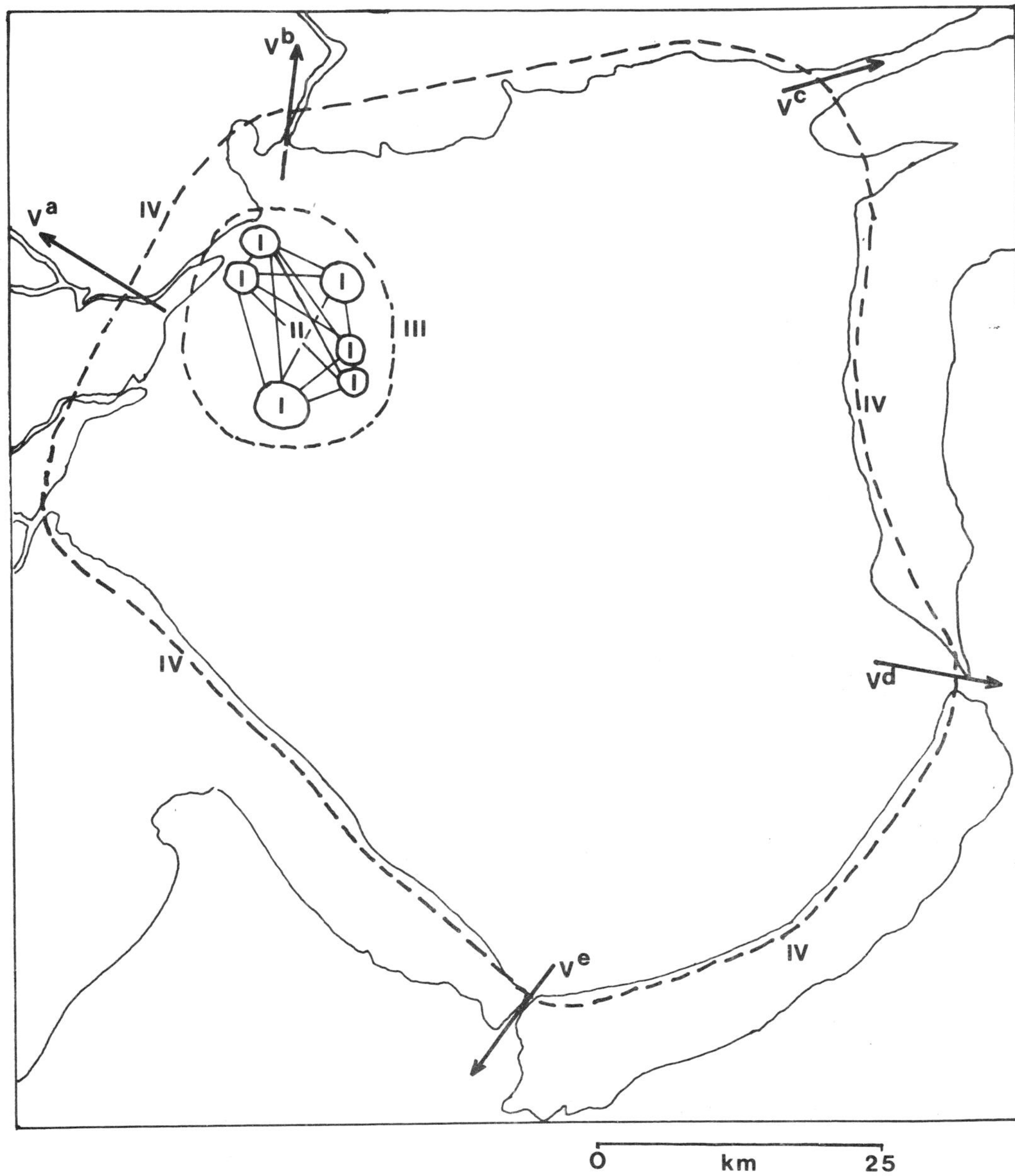

Fig. 13:3 Sites of the third millennium B.C. in the Bannu Basin: hypothetical boundaries which relate to various levels of analysis (see text). I: the local resources catchment; II: the inter-site nexus; III: the settlement pattern; IV: the regional unit, defined by the topographic limits of the basin; V: inter-regional contacts with Afghanistan, via the Tochi and Kurram valleys (Va and Vb, respectively), with Kohat and the Peshawar valley (Vc), with the Indus valley (Vd) and with the Gomal plain and Baluchistan (Ve).

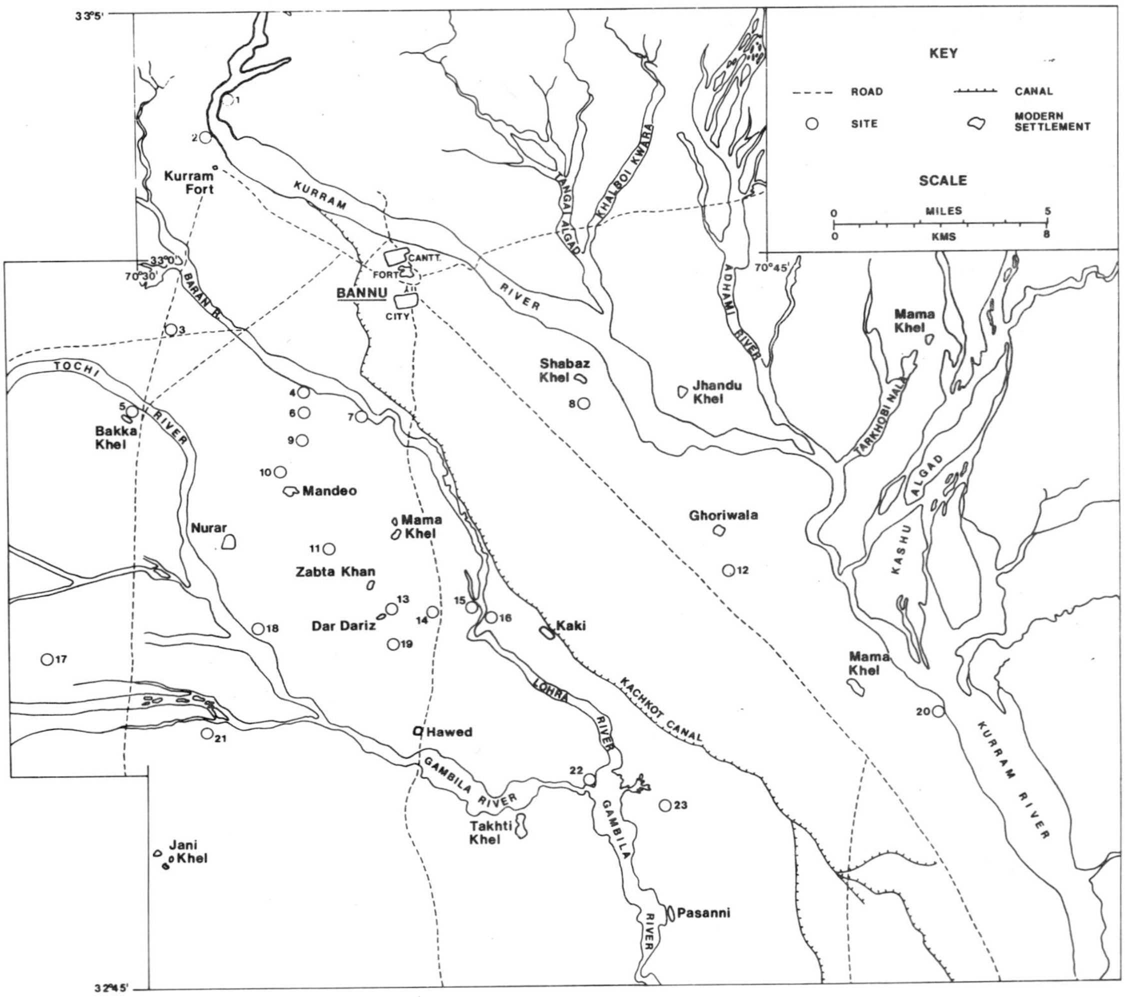

Fig. 13:4 Map of the north-west of the Bannu Basin, showing the distribution of archaeological sites and recent settlements. The sites discussed in this paper are: 3: Islam Chauki; 5: Tarakai Qila; 7: Dad Kala Kachkot; 13: Lewan; 19: Mirz Ali Khan Dheri; 21: Laklargai.

effective rains of the years, as far as plant growth and agriculture are concerned. Further, it is not clear if these trends reflect global changes, thus having had an impact in the plains and foothills of the North-West Frontier. Even if the monsoon had a more effective penetration into these north-western ares, it's impact upon non-irrigation arable agriculture may have been minimal, unless evapo-transpiration rates were also correspondingly lowered. Increased range-land production could have had a positive effect on pastoral output. If climatic changes were manifest in a more even distribution of total annual rainfall, or in a more effective eastwards penetration of winter depressions, winter crop productivity could have been greatly increased.

Thus, although climate in the third millennium B.C. may have differed to some extent from today, we have no reliable data for this from within the Bannu basin. Many of the observable environmental changes can be related to man's activities, particularly the degradation and removal of the vegetation. Most plant communities represent grazing climaxes and many species of trees and shrubs survive only because they are useful to man as a source of fruits, fuel or animal fodder. There is little reason to doubt that primary (non-agricultural) productivity and range productivity were higher in prehistoric times than they are today.

Lastly, it is worth noting that the observed distribution pattern of archaeological sites of all periods (Fig 13:2) is remarkably similar to the distribution of agricultural settlements in the basin today. This suggests that the general factors influencing settlement location in the recent past were broadly similar to those in the earlier historic and prehistoric periods. Access to reliable sources of water appears to be a major determinant.

V SOME THEORETICAL CONSIDERATIONS

1. Definitions and approaches

Although 'subsistence' is often equated with a 'minimal diet' (e.g. Clark and Haswell 1970: xi; Schneider 1974: 252), in the present context I consider that subsistence and production are synonymous. This is similar to the approach of Dennell (1979: 122) who believes that the analysis of subsistence should include all products as well as the technology needed to obtain them.

Subsistence systems, or economies, are taken to be those in which the units of production are the same as the units of consumption. As such they are susceptible to analysis by the use of systems models (Spedding 1979), involving inputs, through-puts and outputs. Throughout this paper I use various terms, such as agricultural systems (or sub-systems), subsistence systems and economic systems, as carefully as possible. My understanding of these terms is that, taken together, they form a nested hierarchy relating to human activities, or behaviour as follows:

agricultural systems: functional associations between man and domesticated crops (plants and animals) in which the latter are manipulated for their products and in which all the elements are united by the movement of energy and movement or cycling of materials.

subsistence systems: functional associations between man and the landscape in which the latter is exploited for products. Agricultural systems are nested within such systems, along with exploitation of natural materials.

subsistence economies: systems in which subsistence systems are embedded and which involve the manipulation and management of products, surpluses, loans, debts, and possibly some trade with other systemic units. The units of production are essentially the same as the units of consumption.

Practical applications of these concepts remain difficult. One major problem lies in defining the units of study on the ground. As Harris (1978) has pointed out, the definition of boundaries of systems or, in practical terms, of territories should be done from the known or inferred behaviour of a population. In the present context the individual behavioural (i.e. subsistence) unit is difficult to define. Is it the household, the village (or site) or a localised group of settlements (or sites)? If the latter, how 'localised' do they have to be in order to be incorporated into one particular subsistence system as opposed to another? Clearly, the question of boundaries and their definition is closely related to the scales or levels of analysis (c.f. Butzer 1982, 233-4, who discusses organization at various levels).

2. Levels of analysis

I intend to focus the rest of this theoretical discussion on the particular problems of the Bannu basin. Fig 13:3 shows the six sites which date from the third millennium B.C. in relation to theoretical boundaries which described potential levels of analysis.

Level I: the site in relation to its localised resource catchment. The site may be considered as an autonomous unit of production and consumption. A major problem is that while the boundary of the settlement, as it survives today, is reasonably well-defined, that of it's exploitation territory will not be. This level of analysis is considered in more detail in the following section (V.3).

Level II: comparisons between sites may reveal the variation in subsistence systems within the population of sites. There are three major problems here: we do not know if all the sites were occupied at the same time; only two sites have been excavated (and only in a preliminary fashion), so we do not know the individual subsistence base of each site; lastly there is the problem of assessing functional relationships between all or some of the sites.

Level III: the settlement pattern in relation to local environmental variables. Irrespective of assumed relationships between the sites, the gross distribution pattern may indicate those factors which were favourable, or limiting, to agriculture. Analysis at this level must assume that the survey data are adequate, i.e. that sites in other areas have not been overlooked or destroyed by natural processes. The availability of relevant palaeoenvironmental data is a further constraint.

Level IV: the total diversity of landscape units available for exploitation, as defined by the topographic units of the whole basin. A number of problems exist at this level, particularly the absence of archaeological data which would permit inferences about land-use and resource exploitation outside the apparent core-area of settlement.

Level V: systems of inter-regional contact. There are only a few easily negotiated routes into and out of the basin and we may reasonably assume that any contacts with settlements, or settlement systems, in other regions occurred via these. Contemporary settlements occur in the Punjab, the Indus Valley and the Gomal Plain but it is difficult to assess the level of economic or socio-political contact between these areas.

According to where we choose to draw arbitrary boundaries, each of the levels described above (except for level V) may be analysed as subsistence systems or economies.

3. Focusing on the site

The site is selected as the unit of analysis because it is the only simply-defined and least ambiguous spatial unit that we have. Tarakai Qila (Fig 13:4, site number 5) is the only permanently occupied nucleated settlement so far excavated in the basin, therefore comments apply especially to this site.

Some models illustrating approaches to the interpretation of non-nucleated urban settlements are shown in Fig 13:5 (Fig 13:5.4 may also apply to the urban case). In Fig 13:5.1, the nucleated settlement (S) consists of a number of fairly independent subsistence units (household or family units). Each of these contributes some fraction of their production to the 'village economy' (represented as a function of (B_1--B_n)) but are otherwise autonomous. This may be a realistic model for Tarakai Qila, but the size of the production units within the site, and the area and distribution of the farm plots within the catchment or territory, are difficult to define.

This problem may be overcome by analysing the site as a 'black box', representing the sum of the production units within the settlement (Fig 13:5.2). In such a model, heterogeneities in the sizes and functions of the individual production and consumption units are ignored. This model may also apply to communal settlements where there is joint productive endeavour and shared consumption of products. Since the excavations at Tarakai Qila were essentially vertical, rather than horizontal,

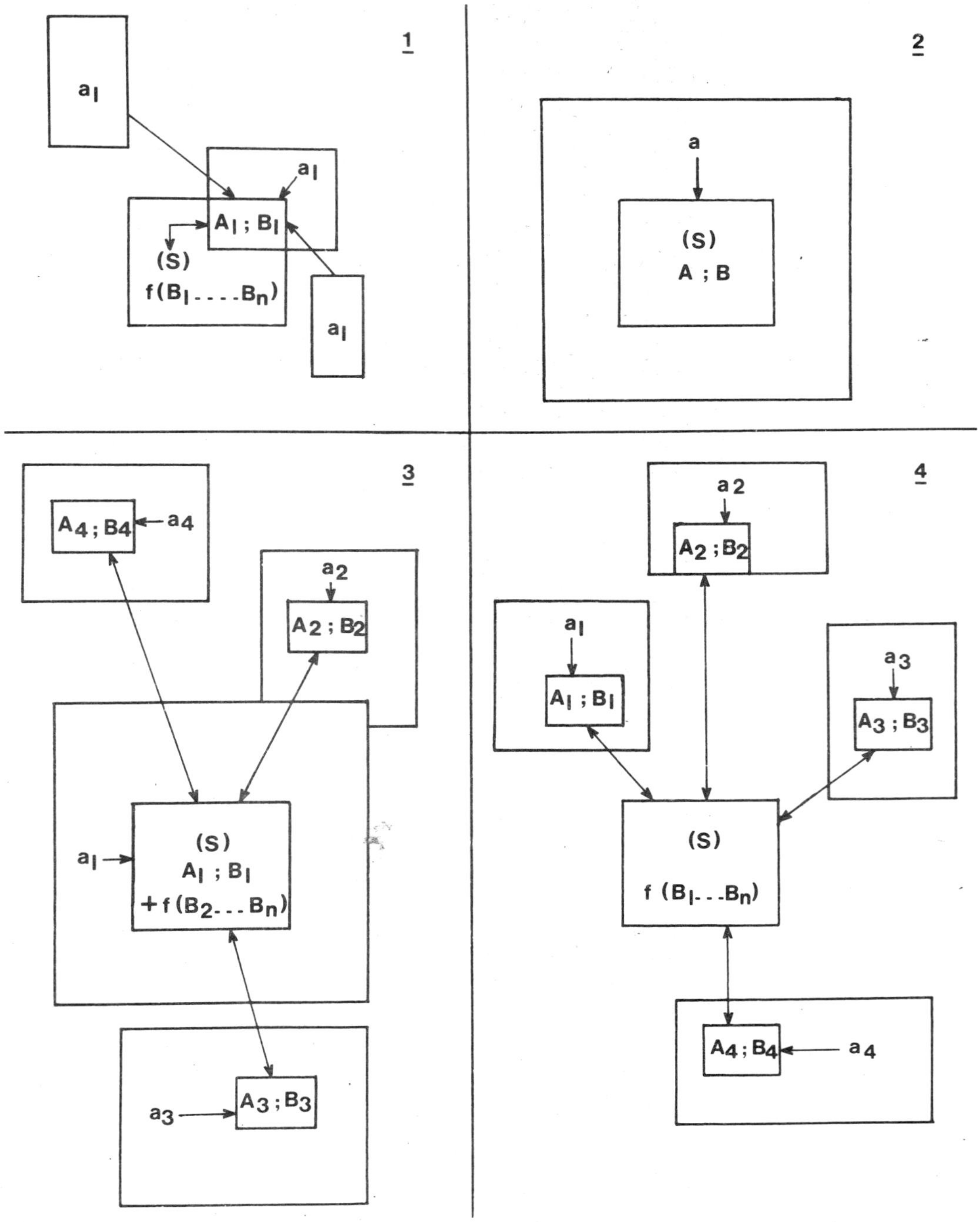

Fig. 13:5 Models of nucleated settlements in relation to the subsistence base. Nucleated settlements (S), individual production and consumption units (A/B) and the area of land exploited by such units (a) are enclosed in boxes; arrows indicate the net flow of energy and materials between such units, as well as the hierarchical relationships between them. Symbols: (S): the nucleated settlement; A_n: the units of production; B_n: the units of consumption; a_n: the area of land exploited by each individual production/consumption unit; $f(B_1 \; ..... \; B_n)$: the proportion of the total production of 'n' production/ consumption units which contributes to the maintenance of the nucleated settlement (represented as some function, f, of the sum of the individual units).

we know little about the intra-settlement social organization of this site.

Other, more complex, models may also apply to Tarakai Qila. Fig 13:5.3 shows a situation in which a nucleated settlement engaged in agricultural production interacts economically with other, satellite, subsistence units. These satellite units may represent farming settlements or temporary settlements occupied seasonally, or intermittently, by transhumant groups. The site of Lewan may be an example of this. The transhumant groups could be from the nuclear settlement, from the satellitic permanent settlements, or from elsewhere.

Fig 13:5.4 shows the most extreme development of the model in which the nucleated settlement does not represent a subsistence unit as it is not engaged in primary agricultural production. Instead, it acts as a market for the produce of the various satellite subsistence systems, or units, and as such represents a proto-urban trend, albeit on a small ('village') scale.

Which of these models are most applicable to Tarakai Qila? Analysis of the plant remains recovered from the site suggests that crop-processing (especially threshing) was carried out. This would suggest that the model outlined in Fig 13:5.4 does not apply to this site. Model 3 (Fig 13:5) may be relevant but substantive data are lacking because of the low archaeological visibility of 'farmstead' units. Models 1 and 2 (Fig 13:5) are closely related: model 1 is potentially very interesting but, as we have seen, the units are difficult to define. Even in the modern context, studies of 'households' or 'farms' as production units is difficult. Farm plots associated with dry-gain farming in semi-arid zones are usually non-contiguous and, as Hill (1982: 53) emphasizes, a "Eurocentric notion of 'farm' is generally inapplicable to our dry-grain mode". Therefore, the 'black box' approach is, at present, the only practical solution to our problem.

The difficulty of defining the 'catchment area' or 'subsistence territory' remains and here, again, a pragmatic approach is adopted. No attempt is made to 'define' the catchment of sites by arbitrary spatial or temporal parameters, such as 1 kilometre radii or one-hour walking distances. The use of natural boundaries to define territories is constrained by the geomorphological instability of the Bannu basin; great changes, particularly in the location of permanent and ephemeral stream channels, have occurred in the Holocene. Anyway, the use of detailed site-catchment models based upon modern ecological data (c.f. Higgs and Vita-Finzi 1972) may not be appropriate (even if the environments had not changed over time) because, as Hill has pointed out for villages in India, we do not know if nucleated villages are "embedded in a tract of arable land, with visible or well-known boundaries, which pertain to that particular village" (Hill 1982: 81).

In view of these problems, the catchment or territory in ths study is not viewed as a defined spatial unit but rather as consisting of a variable set of environmental opportunities and limitations (c.f. Butzer 1982 Table 13.3). Therefore, the characteristics of subsistence catchments are deduced from the environmental data associated both with general settlement patterns and with the various levels of analysis outlined above (section V, 2 and Fig 13:4).

4. Quantification

For a number of reasons, both theoretical and practical, the approach adopted is explicitly non-quantitative.

We have seen how difficult it is to define the boundaries of systems. Therefore, quantification in terms of output or production is very difficult. Added to this is the problem of how closed the systems were; movements of products to and from other subsistence units would make quantifation almost impossible, or of dubious validity.

We do not know if the technology for agricultural production, and the methods of its application, in the past sufficiently resembles that of today to permit the use of comparative (comparable?) modern data on crop yields, livestock densities, etc. Such modern data are anyway unreliable (Hill 1982: 18) and usually do not take into consideration the enormous interannual variation in output in these semi-arid regions.

Usually the archaeological, palaeoenvironmental and palaeoeconomic data from excavated sites are not adequate for detailed quantitative analysis, such as the determination of the relative importance of pastoral and arable production. Such questions may not even be meaningful in semi-arid zones with high levels of environmental uncertainty. The idea of average levels of production, say of grain and meat or milk, and the average contribution of these elements to the economy, does not validly apply to high-risk agriculture, where systems are based upon risk-spreading and decision-making strategies. In such systems, the levels of input, especially for arable production, often bear no close relationship to the achieved output (in terms of crop yields), so simple energy-budgetting and energy-flow models are not easily constructed. The scheduling of subsistence activities can change greatly from year-to-year and re-scheduling will occur if the winter rains fail.

5. Simple models

The subsistence system may include all those factors necessary for sustaining existence: food, fuel, building materials, stone and artefacts, clay for pottery, etc. For simplicity, in this paper I am concentrating on the biological elements of subsistence systems, comprising all biological products and including the soil. The main elements are shown in Fig 13:6. This model is greatly simplified and omits a number

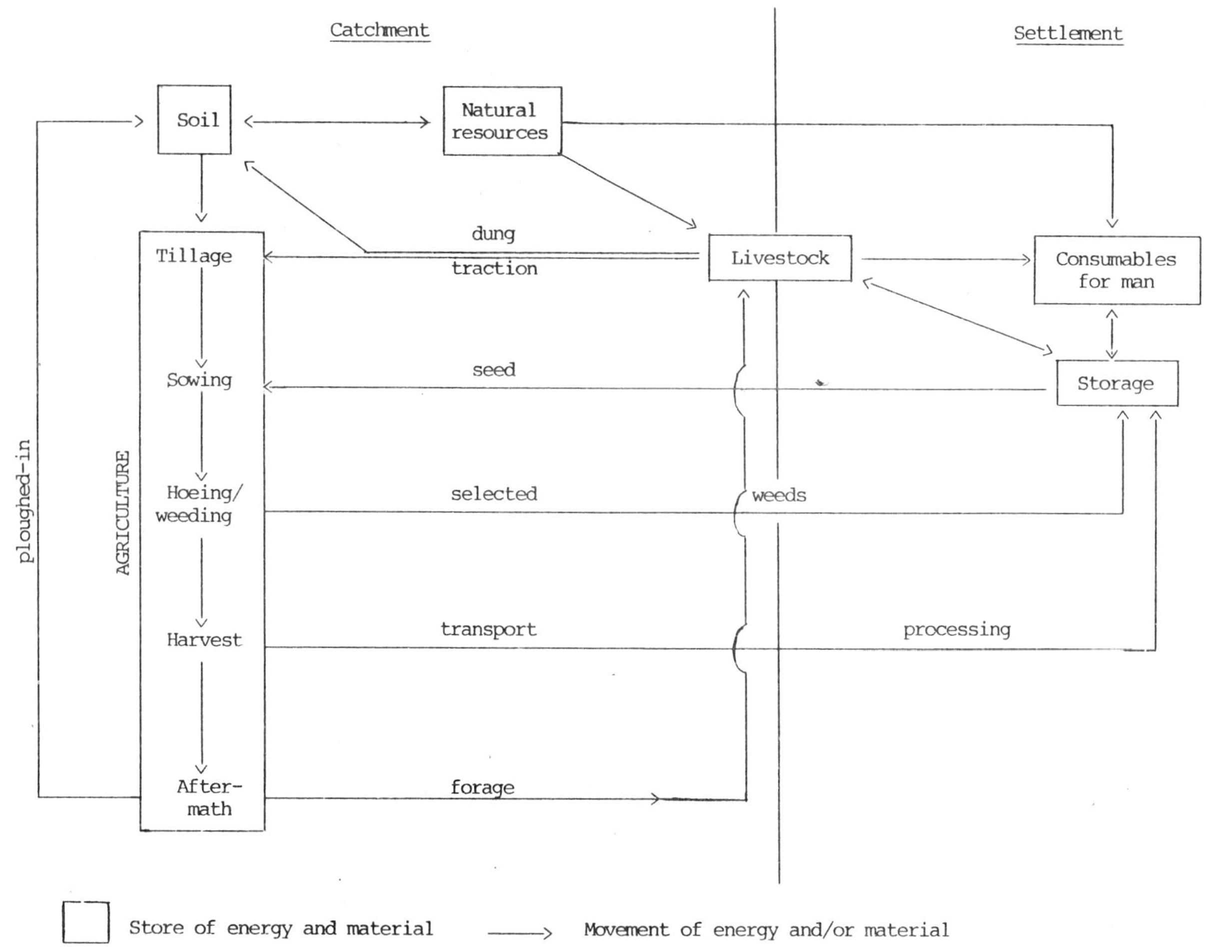

Fig. 13:6 Simple model of an agricultural system based upon a nucleated settlement. Boxes represent storage of energy and materials; arrows show the movements of energy and materials within the system.

of variables, such as the energy inputs from the human population and losses of energy and materials from the system. Some of the transactions of energy and materials could occur through a variety of pathways. For example, the movements of nutrients in dung from livestock to the soil may occur *in situ* (where animals graze on fallow fields, on the aftermath of a crop or on a forage crop) or as a result of the transportation of dung from either the settlement or the rangelands. Overall patterns of movement and utilization of dung may be quite complex and variable (a few pathways are illustrated in Fig 13:7), especially in relation to the availability of alternative fuel resources.

VI AGRICULTURE AND SUBSISTENCE IN THE BANNU BASIN IN THE THIRD MILLENNIUM B.C.

1. Introduction

Limited excavations at Tarakai Qila and Lewan (Fig 4, sites 5 and 13) have yielded evidence of past agricultural activities. Tarakai Qila (mid-late 3rd millennium) was a small village settlement (*circa* 7.5 ha) which persisted for some four or five hundred years. Both animal bones and carbonized plant remains were recovered from the excavations. The remains of plants indicated not only the range of crops grown but, also crop-processing activities on the site (a more detailed analysis of this will be published elsewhere). Lewan (late 4th - late 3rd millennium) is a larger site (*circa* 10 ha) but with very shallow stratigraphy, suggesting numerous short-term occupations on a low natural gravel spread (probably the degraded remnant of an alluvial terrace (Rendell 1981: 223)). Excluding fragments of wood charcoal, no carbonized plant remains were recovered from the excavations at this site. Animal bones were recovered from pits.

Of the other sites of this period, little is known. Islam Chauki (Fig 13:4, site 3) is a small mound of the mid-late 3rd millennium, located on a gently sloping alluvial fan. Deposits of coarse riverine gravels in the stratigraphy of this site indicate that it was overwhelmed by flooding at least twice, the second event apparently causing its final abandonment. Dad Kala Kachkot and Mirz Ali Khan Dheri (Fig 13:4, sites 7 and 19) are small nucleated settlements dated, by the surface pottery, to the mid-late 3rd millennium. Laklargai (Fig 4, site 21) is a large mound, apparently of the same period as these other sites, located in an area where there are no modern settlements. This area around this site is now quite barren (Rendell 1981: 224) and the nearby channels carry only ephemeral torrents. The size of the site would suggest that environments were once favourable for highly productive agricultural activities in this part of the Basin.

2. Crops

The range of species exploited by the inhabitants of Tarakai Qila is quite wide. Domesticated plants include

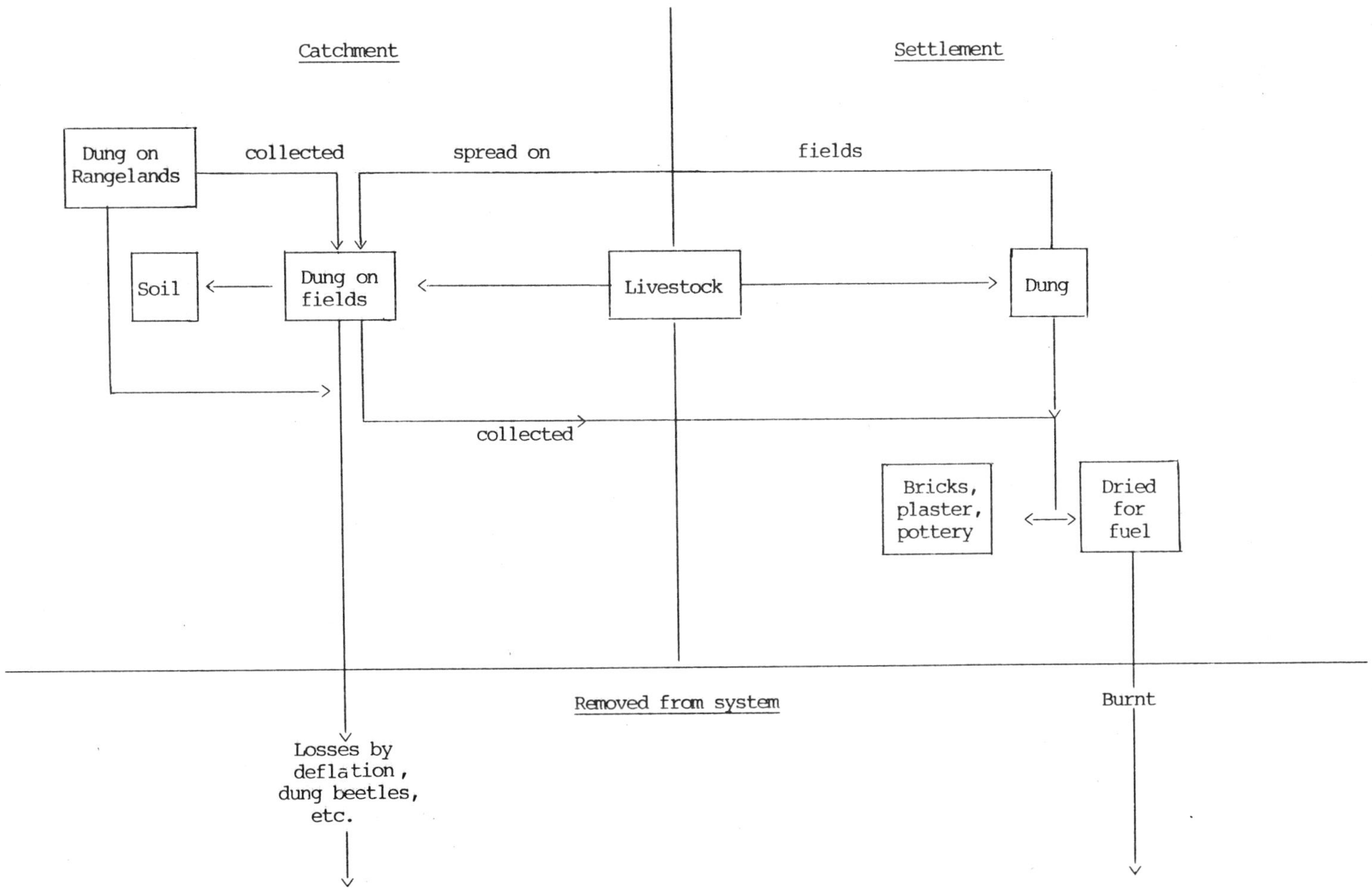

Fig. 13:7 Movements and storage of dung and associated energy and nutrients: a model based upon Figure 13:6 and using the same conventions.

hexaploid free-threshing wheats, 2-row and 6-row barleys (both hulled and naked varieties) and lentils. Other seeds include those of Vicia/Lathyrus, which may have been exploited for food. However, these small seeds are only common in plant assemblages dominated by rachis and awn fragments which, on the basis of models developed by Hillman (1981), are considered to represent the waste from crop-processing activities. Seeds of Malva may represent either exploitation of this plant or its occurrence as a weed. Domesticated animals from Tarakai Qila include cattle (Bos indicus, judging by the terracotta figurines of this animal), sheep, goat and water buffalo (Bubalus bubalis). A similar range of domesticates was recovered at Lewan. Both sites had bones of an equid, probably the wild half-ass Equus hemionus, while Tarakai Qila also yielded the bones of wild pig. Both of these species may have been eaten. Other animals from Tarakai Qila include a wild felid (Felis c.f. caracal) and freshwater fish and molluscs; both these latter categories may represent exploitation for food.

3. Products

Costantini (1981) has pointed out that agricultural systems cannot be identified directly from archaeological data in the form of mere taxonomic catalogues of plant remains; this is true, and the generalization can be extended to include zooarchaeological data. Such information can be used to reconstruct agricultural systems if the uses of such crops are considered (i.e. the products which can be derived from them). The various products which could be obtained from the crops discussed above are listed in Table 13:1. Some of these products are attested directly in the archaeological record, others are inferred with varying degrees of confidence. Crops and products therefore form the basic elements of the agricultural system; the biological requirements of the crops can also suggest the ways in which they were probably exploited and manipulated.

4. Arable activities

The domesticated plants from Tarakai Qila are all characteristic winter/spring crops (rabi); no summer crops (kharif) were recovered. This indicates a marked seasonality of arable activities, probably from November to May (as today in the non-irrigated parts of the Bannu Basin). Barley dominates the charred assemblages from Tarakai Qila: it occurs in almost all the samples and is numerically dominant both in terms of seeds and rachis segments. If this representation reflects the true importance of barley in the ancient arable system, we may suggest that rain-fed (barani) and/or run-off (sailaba) farming were the main activities. Wheat may have been grown in better watered fields bordering the Tochi river, or even within the channel.

Barley is still an important crop in those parts of the basin which are not artificially irrigated. Successful rain-fed and run-off farming in the basin today depends upon a number of

Table 13:1: The main arable, pastoral and wild products utilized at Tarakai Qila; inferred categories are in brackets.

Food for man:

Wheat, Barley, Lentils
(<u>Malva</u>, <u>Vicia/Lathyrus</u>, wild fruits)
Meat of <u>Bos</u>, <u>Bubalus</u>, <u>Ovis</u>, <u>Capra</u>
(Milk of these species)
(Meat of <u>Equus</u>, <u>Sus</u>, freshwater fish, freshwater molluscs)

Feed for animals:

(Forage on the aftermath of harvest)
(Fodder as by-product of crop-processing)
(Forage crops)
(Grazing and browsing wild plants)

Fibres:

(Wool, hair, fur, tendons, intestines)
(Straw)
(From wild plants, such as Gramineae, <u>Calotropis</u>, <u>Crotalaria</u>, etc.)

Leather:

(Skins and hides)

Building materials:

Wood, twigs, grasses
(Dung)

Fuel:

Wood, (charcoal)
(Dung)

Other materials:

Bone, shell, (horn)

Work:

Transport, (traction), (threshing)

factors. Winter rainfall is highly variable in quantity and spatial distribution. As a result, fields tend to be small and dispersed: more fields are sown than are eventually harvested. Barley is planted in the drier soils; although less desirable than wheat, it is regarded as a safer option. Mixed cropping, often of wheat and barley or of barley and pulses, is another strategy to insure against total loss of crops.

Success of crops also depends upon intensive soil husbandry. Tillage of the soil occurs in the summer monsoon, to ensure that a maximum amount of water enters the soil, and before and during the winter rains. Tillage reduces weeds (hence water loss by transpiration), increases infiltration of water, and reduces loss of water and topsoil through run-off. Seeds are sown thinly to reduce competition for water. Fields with crops are regularly hoed and weeded to reduce water-losses. Fields are left fallow to allow recovery of both the water content and organic matter in the soil.

These techniques of soil-water conservation may give us some idea of the range of activities undertaken by the farmers of Tarakai Qila. We cannot say if and how such techniques were applied, but the long-lived nature of the settlement at Tarakai Qila attests to the fact that its inhabitants were successful in their crop-growing efforts. Today, arable activities depend upon the use of animals for traction (ploughing) and transport. Archaeological evidence for this is discussed below (section VI.6.ii).

Harvesting was probably done by hand, as it is today in the Basin. Lunate microliths with 'silica gloss' have been described from Tarakai Qila (Allchin 1981: 234), probably indicating their use in sickles. Today the cereals are cut just above ground level but in the past harvesting may have been of the ears only. This is difficult to establish from the archaeological record as the residues of crop-processing activities are relatively clean of weed species (could this suggest regular weeding of crops, as practised today?), except for seeds of _Vicia/Lathyrus_. If the field were some distance from the settlement, transportation would have been easier if the crops were harvested at the base (although heads of cereals could be carried in baskets or skins). Since the straw would be a valuable by-product of the crop (for thatch, basketry, fodder, bedding, mud-brick, floors etc.) it is unlikely that it was left in the fields.

Crop processing appears to have occurred at the settlement of Tarakai Qila. Some carbonized assemblages are dominated by rachis and awn fragments. Saddle querns and rubbing stones were also found at the site.

No plant remains were recovered from Lewan. It is possible that this settlement was not involved in either arable activities or the primary processing of harvested crops. Silica gloss has not been reported as occurring on the microliths from this site (Allchin 1981), but saddle querns and grinding stones

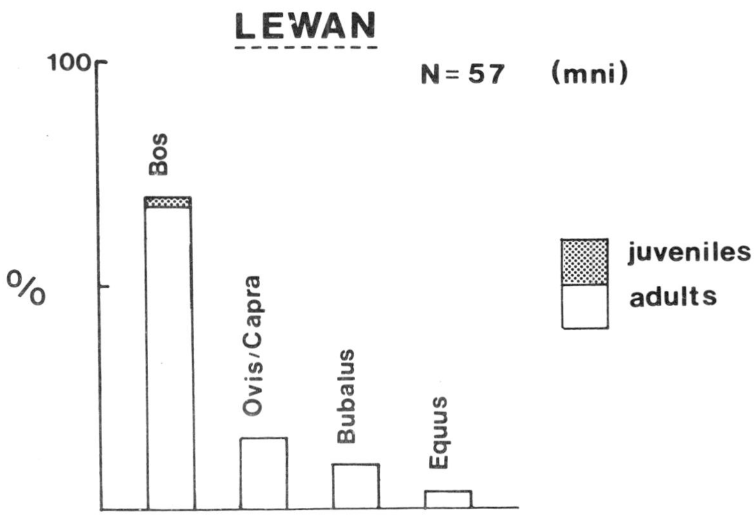

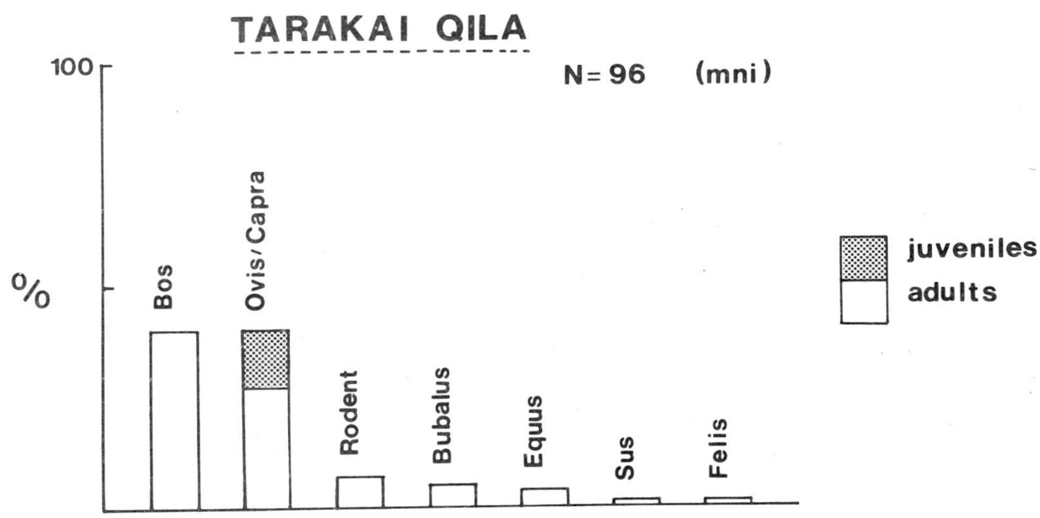

Fig. 13:8 The relative abundance by minimum number of animals from Lewan and Tarakai Qila (mni = minimum number of individuals). Juveniles assessed by both dental and bone-fusion criteria.

were recovered. Such artefacts may have been used for a range of purposes other than the preparation of plant foods (e.g. Alexander 1969), but nonetheless could indicate the exploitation of processed grain brought to the site.

5. Pastoral activities

Fig 13:8 shows the relative abundance, by minimum number of individuals, of the mammals recovered from Tarakai Qila and Lewan. Some caution is required in the interpretation of these data because the bones from Lewan come from rubbish pits only and may be a biased sample. The assemblage is dominated by bones of the larger bovids, particularly Bos indicus. This may be due to taphonomic factors associated with patterns of disposal of food refuse; large bones are more likely to be thrown into rubbish pits than smaller ones and are therefore more likely to persist in the archaeological record. An analagous case has been described by Meadow (1975). Thus, while the species list recorded for Lewan is of interest, the relative proportions of the taxa may not be relevant to the 'animal economy' of the site. The bones from Tarakai Qila, on the other hand, derive from a broader range of archaeological contexts and may suffer less from this bias.

Among the domesticated animals from Tarakai Qila, cattle (Bos) and sheep/goat (Ovis/Capra) are equally abundant while water buffalo (Bubalus) is represented by only a few individuals. Particularly striking is the low frequency of juvenile individuals, except for sheep/goat where some 30 per cent of the individuals were younger than 1.5 years when they died. This probably represents culling for meat; unfortunately, sexing of the bones was not possible so we have no idea if selective culling of young males was practised. The wholly adult representation of cattle and water buffalo may reflect their exploitation for secondary products (see section VI.6.ii, below).

Today, water buffalo are rarely herded with cattle, sheep and goats, or driven out onto rangelands to graze. Usually they are kept close to water. Water buffalo are valued for their rich milk but are also used for transport (rarely for traction). In rural settlements, their numbers tend to be low in comparison with those of the herded bovids; the data from Tarakai Qila resemble this pattern.

Of course, the assemblage of bones from Tarakai Qila need not reflect the structure of the herds. The animals were slaughtered, probably for meat, and the bones have been subject to differential preservation. Cattle may be over-represented, and juveniles of sheep/goat under-represented, in the assemblages. However, other workers (e.g. Meadow 1981) have noted the abundance of remains of cattle in sites of this period. Harris (1966) has suggested that in India large numbers of cattle are needed for the intensive ploughing operations which begin at the onset of the monsoon period. In Bannu District (as we have seen above), large numbers of animals are required for intensive soil husbandry in the weeks before, and during, the winter rains. That cattle were important to the inhabitants of Tarakai Qila is shown by the overwhelming dominance of humped cattle among the non-anthropomorphic terracotta figurines.

As we have seen (section VI.I), Lewan does not appear to represent a nucleated, permanently occupied, settlement. It has been interpreted (Allchin 1981; Allchin and Knox 1981a) as a seasonally-occupied site concerned principally with the manufacture of stone-tools. The inhabitants of the site may have undertaken pastoral activities, but there is no evidence for arable activities. What relationships, if any, existed between Lewan and the nucleated villages in the area? Lewan could represent a wholly nomadic community which traded with the local settled communities. A full discussion of this complex question is beyond the scope of this paper; suffice it to say that the presence of water buffalo in the assemblage of animals from Lewan militates against a nomadic way of life. Lewan may represent seasonally-transhumant exploitation by the nucleated villages (Mirzali Khan Dheri is the nearest such settlement), being a site at which artefacts were made (perhaps for trade) and pastoral activities carried-out. It is possible that other such sites remain to be found in the Basin; even the arid eastern area could have been incorporated into the pastoral-catchment of the nucleated settlements of the north-west. Any such activities would probably have occurred during the summer months, when pastoralism would dominate the human activiity patterns and when there would be a need to extend the area of grazing during this period of low plant productivity.

6. Secondary products

Many of the products listed in Table 13:1 are secondary products arising from both arable and pastoral activities. Archaeological evidence for the exploitation of secondary products is difficult to interpret (see Sherratt (1981) for a review) and some of the products discussed here are inferred, on the basis of likelihood, rather than demonstrated.

(i) Secondary products from arable activities. Most discussions of secondary products (e.g. Sherratt 1981) concentrate wholly on animals, but a number of important products may be derived as by-products of crop-growing. The aftermath of the harvested crop may be a valuable source of grazing for domestic animals; it can also be ploughed into the soil to add organic matter. Straw can be used for animal bedding and feed, as well as for baskets, thatch etc. The use of chaff as a filler in pottery, mudbrick and mudplaster is shown by impressions in these materials. Secondary crops may have been grown, either to provide green forage for animals or green manure for fields.

(ii) Secondary products from pastoral activities. Animal production has often been viewed as oriented towards the provision of meat and other consumables derived from the slaughter of the animals. Sherratt (1981) has suggested that in Eurasia since the late 4th millennium B.C., work, wool and milk may have been the main products obtained from domesticated animals; possibly these were much more important than meat. We should not assume that 'secondary products' were of secondary significance in agricultural systems (see Ryder, this volume).

We have interpreted the animal remains from Tarakai Qila as indicating the exploitation of, at least, cattle and water buffalo for secondary products (section VI.5). These products may include work, fibre, dung and milk.

Work: Sherratt (1981: 266) has noted that "there are many examples of models showing solid-wheeled carts from the Harrapan (*sic*) culture of the mid third millennium B.C. in north-east (*sic*) India". Wheels and 'carts' in terracotta were recovered from the excavations at Tarakai Qila, suggesting the use of animals for transport (perhaps the terracotta figurines of *Bos indicus* were intended for 'use' with these model carts?). There is no evidence from Tarakai Qila for the use of the plough, but ploughmarks have been excavated beneath the pre-Harappan levels at Kalibangan in north-west India (Lal 1971). Mughal has discovered terracotta models of ploughs on Harappan sites in Bahawalpur, Pakistan (Mugal 1981: Plate 17). It seems likely therefore that the plough was used by farming communities of the 3rd millennium B.C. in Bannu District. Lastly, it is possible that cattle were used in threshing, by trampling the harvest to free the grains, although there is no evidence for this either from Tarakai Qila or from other sites of this period in South Asia.

Fibre: the exploitation of wool and hair can only be suggested as a possibility; direct evidence is lacking.

Dung: because wood is so scarce, dung (of cattle and water buffalo) mixed with straw is an important fuel in the Indian subcontinent (Sopher 1980: 191). This may have been so at Tarakai Qila; the only evidence comes from hearths containing friable ashy material which has a "felted" appearance - perhaps the residue of burned dung cakes? Other possible uses of dung (based upon current usage in the area) include: as manure, as a constituent of mud plaster and mudbrick, and as a filler in pottery.

Milk and milk products: the predominantly adult composition of the cattle and water buffalo remains from Tarakai Qila points to the possibility that they were kept to provide milk as well as work. Adult female sheep and goats could also have yielded milk. Unfortunately, we have no reliable data on the ratios of males to females or (for potential work and milk yielding species) the ratio of males to castrates to females. Artefact evidence is very difficult to interpret; perhaps some of the diverse forms of pottery were used in the collection and

Table 13:2

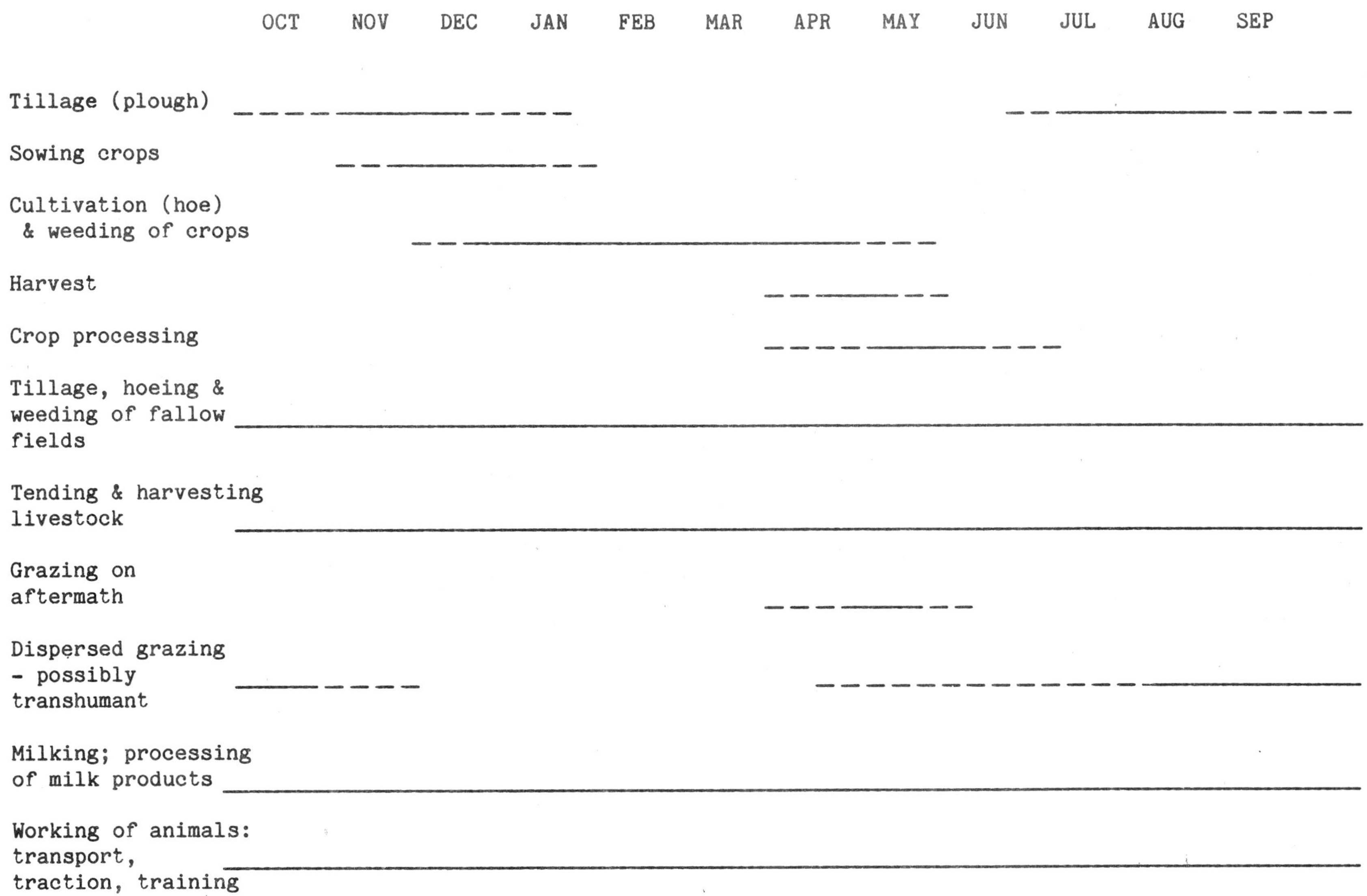
OCT
NOV
DEC
JAN
FEB
MAR
APR
MAY
JUN
JUL
AUG
SEP
Tillage (plough)
Sowing crops
Cultivation (hoe) & weeding of crops
Harvest
Crop processing
Tillage, hoeing & weeding of fallow fields
Tending & harvesting livestock
Grazing on aftermath
Dispersed grazing - possibly transhumant
Milking; processing of milk products
Working of animals: transport, traction, training

processing of milk (c.f. Sherratt 1981: 280-282 and Ryder, this volume).

VII APPROACHES TO THE INTEGRATION OF AGRICULTURAL, SUBSISTENCE AND ECONOMIC SYSTEMS

The elements of the agricultural systems of settlements like Tarakai Qila have either been determined from the archaeological record or inferred. These elements can only be integrated by applying some dynamic to them, involving the movement of energy, materials and people. In constructing a general model which attempts to integrate the diverse components of the system, I have assumed a rational motivation and that movements are therefore purposive (i.e. from places where materials and/or energy are superfluous to requirements to other places where they are deficient). To state it more obviously: from areas of production to areas of consumption or storage (e.g. movements of crops and products to the settlement, movements of wastes from the settlements to manure the fields, etc.).

Some aspects of integration between the settlement and its catchment are shown in Figs 13:6 and 13:7. Fig 13:9 emphasises the importance of secondary products in the integration of the agricultural system, and shows how they link with some components of the rest of the subsistence system (building materials, pottery, spinning activities, and other processing of secondary products). Movements of livestock between settlements and their territories, perhaps on a daily basis, provide an important means of integrating the working space. Such models are useful, but they do not include all possible relationships and feedback loops; neither do they reveal the spatial and temporal variations in the system, such as the distribution of fields with different crops and the distribution of available grazing land.

Table 13:2 presents a model for the annual scheduling of arable and pastoral activities. The time and quantity of winter rainfall was probably a major determinant of activity patterns from October to May. No doubt some of the activities listed here were age or sex specific, as they often are today. Adult males usually undertake tillage activities, males and females may hoe, weed and harvest, while females play a major role in crop processing. The tending of livestock is often the job of either older children or old men. Such roles are fundamentally important in linking the agricultural system into the sociocultural system.

We have noted elsewhere that the annual scheduling of activities depends upon the vagaries of the weather. The relative importance of various products may vary from year-to-year and farmers may have to exercise a range of long-term and short-term options. Fig 13:10 shows some of the short-term options which may arise from a failure of the winter rains. The options are not simple; indeed, each may have important consequences. The reduction of the size of herds could reduce

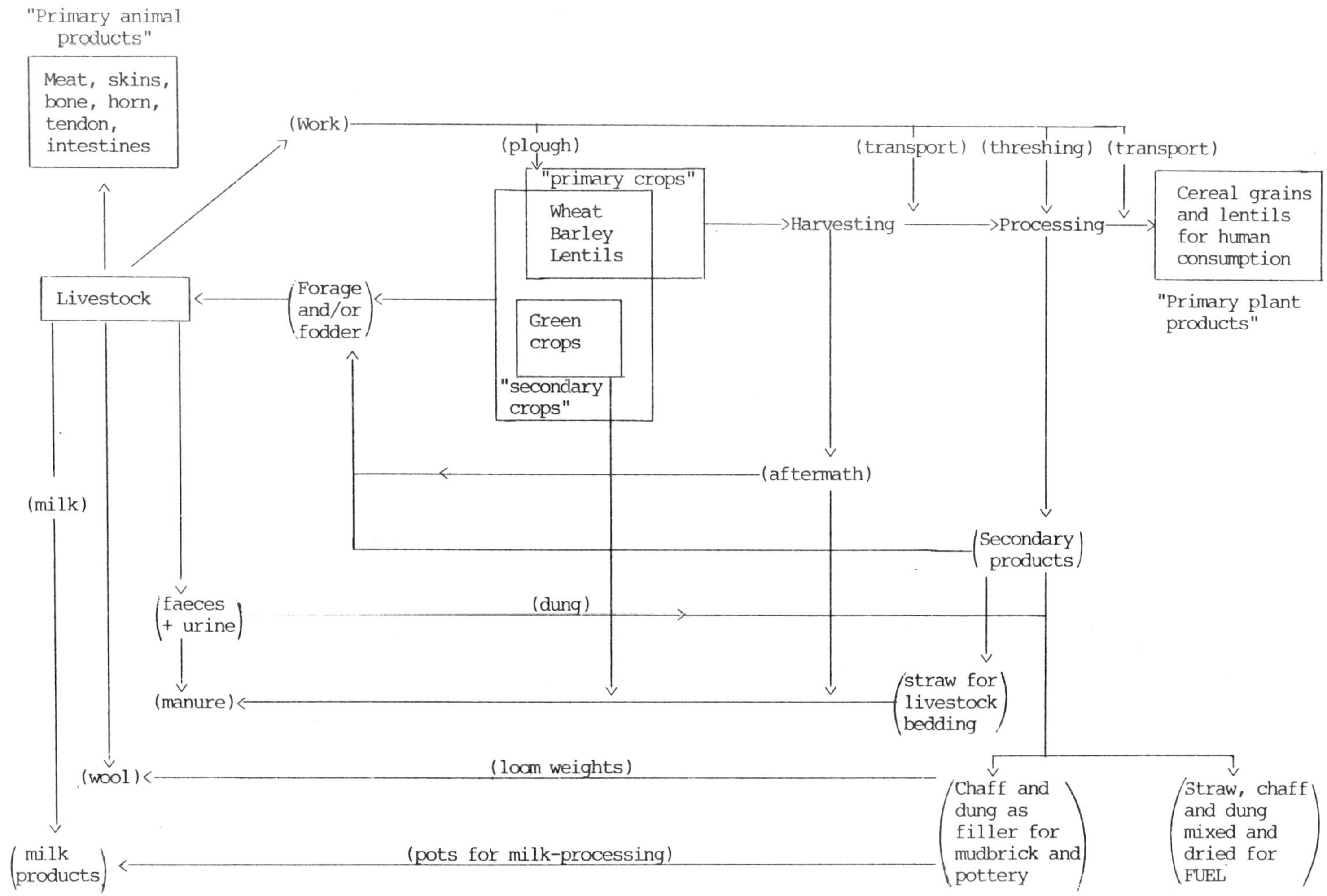

Fig. 13:9 Simplified model of an agricultural ecosystem in which the important role of secondary products in integrating the various components is emphasized. Boxes enclose crops and primary products, brackets enclose various categories of secondary products and arrows show the movement of energy or materials.

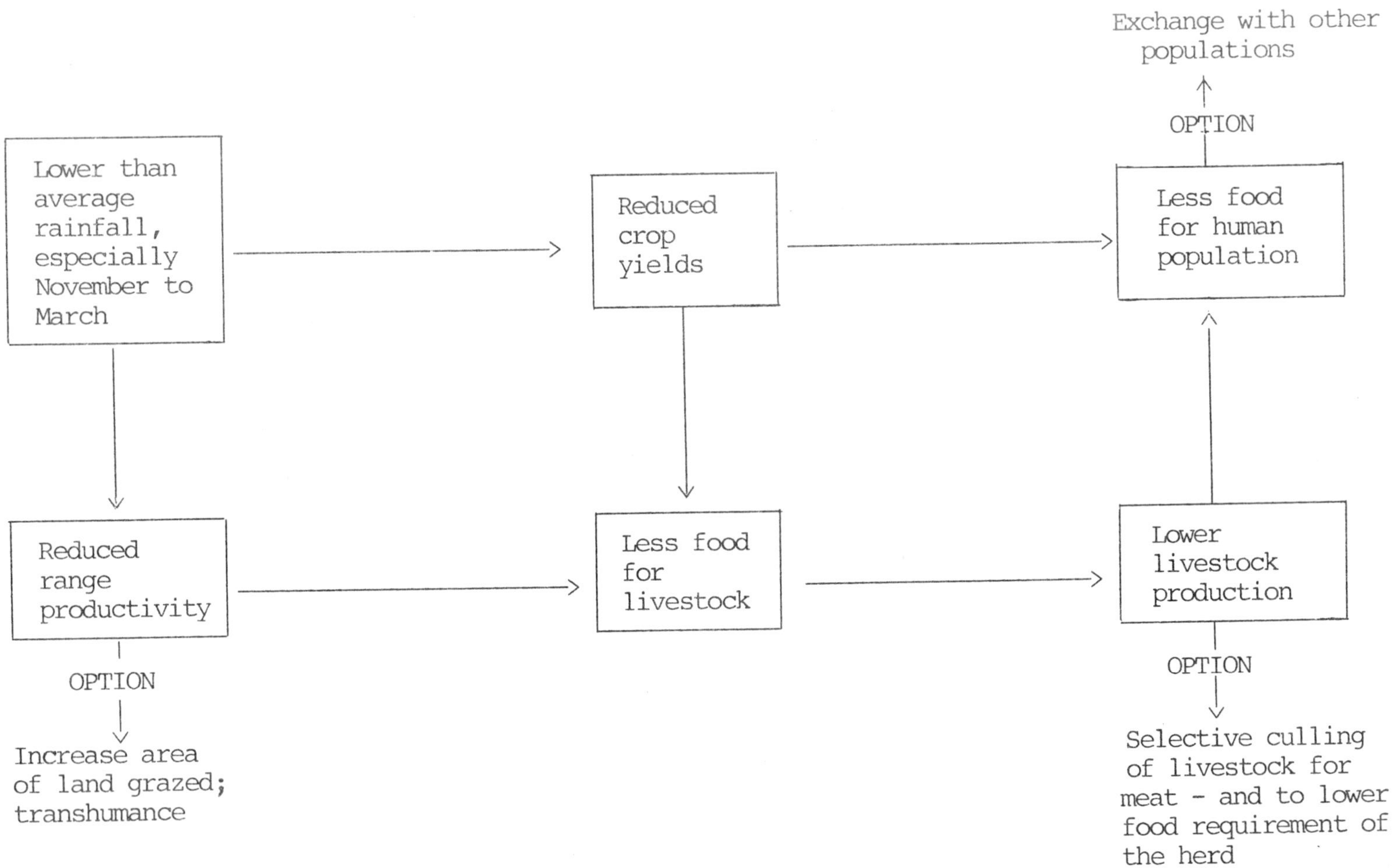

Fig. 13:10 Decision-making: short term options which might be faced as a result of inter annual fluctuations in rainfall (see text).

the output of important secondary products for a number of years afterwards. Trade or exchange may not be possible if other settlements have suffered from the same disaster. Attempts to increase the area of rangeland may lead to conflict with other communities. Longer term strategies to cope with such environmental stresses could include: a diversification of crops; varying the sowing times between different plots; increasing the number and widening the distribution of arable plots; maximizing annual production to ensure adequate storage; growing dependable crops (e.g. barley); mixed-cropping; growing more forage crops to support a larger pastoral element in the system; extending the 'recognized' area of grazing land; encouraging trade and "social storage" (O'Shea, 1981, 169) with other settled communities, or with nomadic groups.

This theme of integration could be extended beyond the level of the individual settlement and its catchment to include the settlement system (see section V.2, above, and Fig 13:3). We have see that Tarakai Qila and Lewan represent quite different modes of exploitation of the landscape in the Bannu Basin. One possible model for this situation is that sites such as Lewan (although we know no other sites of this type) represent summer camps of transhumant pastoralist groups, possibly derived from the nucleated settlements like Tarakai Qila. It is possible that the whole range of habitats in the Bannu Basin could have been integrated by such patterns of seasonal exploitation.

Finally, there is the question of inter-regional contact (Fig 13:3, level V). The cultural assemblages from the sites of the 3rd millennium B.C. in the Bannu Basin are of Kot Dijian facies. Comparable manifestations occur in sites on the Gomal Plain, such as Rehman Dheri (Durrani 1982); the Punjab, such as Sarai Khola (Halim 1972) and the Greater Indus Valley (Khan 1981; Mughal 1981). This argues for at least contact and probably economic and political relations between these areas. Bangles of the chank shell *Xancus pyrum* (L) were found at Lewan and Tarakai Qila, attesting to direct or indirect trade connections with the Indian Ocean (Durante 1979 a, b). Nomadic groups may have provided the trade and communication links between these areas (Possehl (1979) has postulated similar connections between Harappan settlements). Even today, nomadic Powindahs migrate annually between Afghanistan and the Bannu Basin, *via* the passes created by the Kurram and Tochi Rivers (Robinson 1934). No doubt such movements were also possible in the third millennium B.C., although there is no evidence of settlement in south-east Afghanistan during this period (Ball 1982: Fig 58); this may be due to the lack of intensive archaeological survey in this difficult border area.

VIII POSTCRIPT

Much of this paper is speculative and more archaeological fieldwork (both survey and excavation) needs to be done in order to test and refine various hypotheses. My aim has been to incorporate the various elements of the subsistence system into a dynamic model, linking them to one another and to the environment at various levels of analysis (the settlement and its catchment, the inter-site nexus, the settlement system, the region, and with other regions). Analysis at these various levels has, of necessity, been uneven, with most attention focused upon the settlement and its catchment. Tarakai Qila provided the data for this, and was the focus for the model. The relationships within the supposed systems have been inferred by analogy with recent dry-land farming communities, both within the region and elsewhere in north-western South Asia.

Acknowlegements

I wish to thank numerous colleagues for their help and encouragement:

Mr. M. Ishtiaq Khan, Professor F.A. Durrani, Professor Farid Khan, Dr. Abdur Rehman, Dr. Bridget Allchin, Dr. Raymond Allchin, Mr. J. R. Knox and Dr. Helen Rendell for their help in the field; Gill Taylor provided invaluable assistance in the analysis of the animal bones; Mr. Gordon Hillman kindly examined the charred seeds, corrected some of my identifications and confirmed others; Professor D. R. Harris made valuable comments on part of an earlier draft of this paper. Fig. 13:1b was drawn by Yvonne Brown. The fieldwork was supported by research grants from the University of London Hayter Fund.

Bibliography

Alexander, J. (1969) The indirect evidence for domestication. In: The domestication and exploitation of plants and animals. (ed Ucko, P. J. and Dimbleby, G.W.) London: Duckworth pp. 123-129

Allchin, B. (1981) Stone industries of Lewan, Tarakai Qila and Lak Largai in the Bannu Basin. In: South Asian archaeology 1979. (ed Härtel, H.) Berlin: Dietrich Reimer Verlag pp. 233-239

Allchin, B. and Allchin, R. (1982) The rise of civilization in India and Pakistan. Cambridge: University Press

Allchin, F.R. and Knox, R.J. (1981 a) Preliminary report on the excavations at Lewan (1977-78). In: South Asian archaeology 1979. (ed Härtel, H.) Berlin: Dietrich Reimer Verlag pp. 241-244

Allchin, F.R. and Knox, R.J. (1981 b) Preliminary report on the excavation at Tarakai Qila (1978-79). In: South Asian archaeology 1979. (ed Härtel, H.) Berlin: Dietrich Reimer Verlag pp. 245-250

Ball, W. (1982) Archaeological gazetteer of Afghanistan. Paris: Editions Recherche sur les civilisations

Blaikie, P.M. (1971 a) Spatial organization of agriculture in some north Indian villages. Part I. Transactions of the Institute of British Geographers 52 1-40

Blaikie, P.M. (1971 b) Spatial organization of agriculture in some north Indian villages. Part II. Transactions of the Institute of British Geographers 53 15-30

Butzer, K.W. (1982) Archaeology as human ecology. Cambridge: University Press

Clark, C. and Haswell, M. (1970) The economics of subsistence agriculture. 4th edition. London: Macmillan

Costantini, L. (1979) Notes on the palaeoethnobotany of protohistorical Swat. In: South Asian archaeology 1977. (ed Taddei, M.) Naples: Istituto Universitario Orientale Seminario di Studi Asiatici. Series Minor VI pp. 703-708

Costantini, L. (1981) Palaeoethnobotany at Pirak: a contribution to the 2nd millennium B.C. agriculture of the Sibi-Kacchi Plain, Pakistan. In: South Asian archaeology 1979. (ed Härtel, H.) Berlin: Dietrich Reimer Verlag pp. 271-277

Dani, A.H. (1971) Excavations in the Gomal Valley. Ancient Pakistan 5 1-177

Dennell, R.W. (1979) Prehistoric diet and nutrition: some food for thought. World Archaeology 11 121-135

Durante, S. (1979 a) The utilization of Xancus pyrum (L.) at Shahr-i Sokhta. In: South Asian archaeology 1975. (ed van Lohuizen-de Leuw, J.E.) Leiden: E.J. Brill pp. 27-42

Durante, S. (1979 b) Marine shells from Balakot, Shahr-i Sokhta and Tepe Yahya: their significance for trade in ancient Indo-Iran. In South Asian archaeology 1977. (ed Taddei, M.) Naples: Istituto Universitario Orientale Seminario di Studi Asiatici. Series Minor VI pp. 317-344

Durrani, F.A. (1982) Rehman Dheri and the birth of civilization in Pakistan. Bulletin University of London Institute of Archaeology 18 (for 1981) 191-207

Halim, M.A. (1972) Excavations at Sarai Khola. Part I. Pakistan Archaeology 7 (for 1970-71) 23-89

Harris, D.R. (1978) Group territories or territorial groups? Comments on an interdisciplinary problem in cultural ecology. In: Social organization and settlement. Contributions from Anthropology, Archaeology and Geography. (ed Green, D., Haselgrove, C. and Spriggs, M.) Oxford: British Archaeological Reports International Series (Supplementary) No. 47 pp. 187-196

Harris, M. (1966) The cultural ecology of India's sacred cattle. Current Anthropology 7 51-59

Higgs, E.S. and Vita-Finzi, C. (1972) Prehistoric economies: a territorial approach. In: Papers in economic prehistory. (ed Higgs, E.S.) Cambridge: University Press pp. 27-36

Hill, P. (1982) Dry grain farming families: Hausaland (Nigeria) and Karnataka (India) compared. Cambridge: University Press

Hillman, G. (1981) Reconstructing crop husbandry practices from charred remains of crops. In: Farming practice in British prehistory. (ed. Mercer, R.) Edinburgh: University Press pp. 123-162

Hutchinson, J. (Sir) (1976) India: local and introduced crops. Philosophical Transactions of the Royal Society of London Series B 275 129-141

Jarrige, J.-F. and Meadow, R.H. (1980) The antecedents of civilization in the Indus Valley. Scientific American 243 122-133

Khan, F.A. (1981) Kot Diji culture - its greatness. In: Indus Civilization: new perspectives. (ed Dani, A.H.) Islamabad: Centre for the Study of the Civilization of Central Asia, Quaid-i-Azam University pp. 15-24

Lal, B.B. (1971) Perhaps the earliest ploughed field so far excavated anywhere in the world. Puratattva 4 1-3

Meadow, R.H. (1975) Mammal remains from Hajji Firuz: a study in methodology. In: Archaeozoological studies. (ed Clason, A.T.) Amsterdam: North Holland Publishing Company pp. 265-283

Meadow, R.H. (1981) Early animal domestication in South Asia: a first report of the faunal remains from Mehrgarh, Pakistan. In: South Asian archaeology 1979. (ed Härtel, H.) Berlin: Dietrich Reimer Verlag pp. 143-179

Mughal, M.R. (1981) New archaeological evidence from Bahawalpur. In: Indus Civilisation: new perspectives. (ed Dani, A.H.) Islamabad: Centre for the Study of the Civilization of Central Asia, Quaid-i-Azam University pp. 33-41

O'Shea, J. (1981) Coping with scarcity: exchange and social storage. In: Economic archaeology. (ed Sheridan, A. and Bailey, G.) Oxford: British Archaeological Reports International Series No. 96 pp. 167-183

Possehl, G.L. (1979) Pastoral nomadism in the Indus Civilization: an hypothesis. In: South Asian archaeology 1977. (ed Taddei, M.) Naples: Istituto Universitario Orientale Seminario di Studi Asiatici. Series Minor VI pp. 537-551

Raverty, H.G. (1878) Notes on Afghanistan and Beluchistan. London. (Reprinted in Lahore: Sang-e-Meel Publications, 1976)

Rendell, H. (1981) A preliminary investigation of the sedimentary history of the Bannu Basin in the later Holocene. In: South Asian archaeology 1979. (ed Härtel, H.) Berlin: Dietrich Reimer Verlag pp. 219-225

Robinson, J.A. (1934) Notes on nomad tribes of eastern Afghanistan. London. (Reprinted in Quetta: M/S Nisa Traders, 1978)

Schneider, H.K. (1974) Economic man: the anthropology of economics. New York: The Free Press

Sherratt, A. (1981) Plough and pastoralism: aspects of the secondary products revolution. In: Pattern of the past: studies in honour of David Clarke. (ed Hodder, I., Isaac, G. and Hammond, N.) Cambridge: University Press pp. 261-305

Singh, G., Joshi, R.D., Chopra, S.K. and Singh, A.B. (1974) Late Quaternary vegetation and climate of the Rajasthan Desert. Philosophical Transactions of the Royal Society of London Series B 267 467-501

Sopher, D.E. (1980) Indian civilization and the tropical savanna environment. In: Human ecology in savanna environments. (ed Harris, D.R.) London: Academic Press pp. 185-207

Spedding, C.R.W. (1979) An introduction to agricultural systems. London: Applied Science Publishers Ltd.

Stein, A. (Sir) (1928) An archaeological tour in Waziristan and northern Baluchistan. Memoirs of the Archaeological Survey of India 37 29-30

Thomas, K.D. (1981) Palaeoecological studies in the Bannu Basin: the sources of evidence. In: South Asian archaeology 1979. (ed Härtel, H.) Berlin: Dietrich Reimer Verlag pp. 227-232

Thomas, K.D. (in press) Tarakai Qila: site, economy and environment. In: Site, environment and economy. (ed Proudfoot, V.B.) Oxford: British Archaeological Reports International Series (forthcoming)

Vishnu-Mittre and Savithri, R. (1982) Food economy of the Harappans. In: Harappan Civilization: a contemporary perspective. (ed Possehl, G.) Warminster, England: Aris and Phillips Ltd. pp. 205-221

14. ECONOMIC PLANTS IN AMSTERDAM: QUALITATIVE AND QUANTITATIVE ANALYSIS

Norbert Paap

INTRODUCTION

Over the past 25 years archaeological excavations have been carried out in Amsterdam. A catalogue has been made on the finds from the 27 sites from the first 20 years (Baart *et al*. 1977). In this catalogue are also included the remains from the four sites that were botanically investigated (Baart *et al*. 1977: 492-494). Only two of these sites gave information on economic plants in Amsterdam. During the last five years three excavations have taken place. These were also investigated botanically, one for pollen, the other two for seeds. These last two gave abundant information on economic plants. So there are altogether four sites that give information on economic plants in Amsterdam:

1. Remains of a watchtower with part of the wooden town wall, dated 1375-1425. In the ditch outside the wall latrine deposits from the watchtower were found (Pals 1972).

2. House with cesspit "Damrak 49", dated 1580-1590 (Pals in press).

3. "Nieuwendijk" (New Dike). This site covers a considerable period. Until now it represents the earliest phase of Amsterdam (*circa* 1225) and it goes on into the 20th century. At this site the first settlers built along the river Amstel a terp of about 25 m. wide and with a length of probably one kilometer. The weight of the heightening layers caused them to sink down into the peat on which the terp was built. This forced the settlers to heighten their habitation level about every five years. The result is 16 heightening layers, dated 1225-1350. On the top of this came a layer of several meters thick. Foundations made further heightening unnecessary (Baart 1982). It is easily understood that this site offered very good opportunities for finding seeds of economic plants, especially since the 15th century when cesspits appear.

4. "Waterlooplein" (Waterloo Square). During the 16th century, the position of this site was occupied by a great bend in the River Amstel. The bend was rapidly filled in at the end of the 16th century and a map of 1625 shows the area covered with houses. The area originally had a mixed population, but with time an increasing number of Jews, especially from Spain and Portugal,

settled the site. In the beginning of the 18th century between 60 and 80% of the occupants were jews, this figure based on the find of kosher tokens for ritually slaughtered animals (Baart 1982). Taking into consideration the (illegal) slaughtering at home, this figure could be increased to 85% (Elize van Dienst: personal communication). A reason for the concentration of the Jews in this area could be the installation of a synagogue in the neighbourhood (Vaz Dias 1938). Whatever the reason, the result was that rich and poor lived together in one area, though in different streets. This situation offered the possibility, apart from qualitative analysis, to compare the food remains to see if evidence could be found of different social status.

METHODS

While the older excavations were mainly treated as separate accidental finds, with the "Nieuwendijk" excavation the town archaeologist Jan Baart started a period of systematic investigation to produce an archaeological framework into which the results of each subsequent excavation could be fitted. As this framework was also meant for the economic plants, systematic sampling became necessary. At the "Nieuwendijk" all the layers that fell within one ceramic period were sampled at the most promising spot. From each sample one litre was fully investigated. The results of these samples made up the basis of a diagram that shows the introduction and period of use of a wide range of fruits and food plants, from 1225 until the 19th century (figures 14:1 & 14:2). The results from the other sites gave additional information that was incorporated in the diagram.

Before analysis the samples were sieved through sieves with mesh 2 mm, 1 mm, 0.5 mm and 0.25 mm. In general only 10% by volume of the finest fraction was examined, except where the total sample volume fell below 100 ml, when 10 ml. was examined. While analysing the samples of the cesspits it appeared that one litre sediment was too little to produce quantifiable amounts of the larger remains such as the stones of Prunus avium/cerasus (sweet/sour cherry), Prunus insititia (damson), Prunus domestica (plum) and to a certain extent Vitis vinifera (grape). At the "Waterlooplein" nearly all the samples would come from cesspits, it was necessary to modify sampling tactics accordingly. Apart from qualitative results the main purpose of the botanical investigation was here to obtain quantifiable amounts to see if social differences could be traced. Therefore 5 litre samples were used of which all but a half litre sub-sample was sieved at 2 mm, and the sub-sample was passed through a stack of finer sieves. A trial analysis had made it clear that in most cases half a litre was sufficient for the smaller fractions; greater volumes produced only greater amounts, not more species. The whole method takes no more time than a full investigation of one litre, but gives better results for the food plants.

RESULTS

Qualitative analysis. Figures 14:1 & 14:2 show the results of the qualitative analysis. The Roman figures at the top denote centuries. A solid line indicates that the species has been found, a broken line indicates that the occurrence in that period is very probable, but remains have not been found.

Fig.14:1

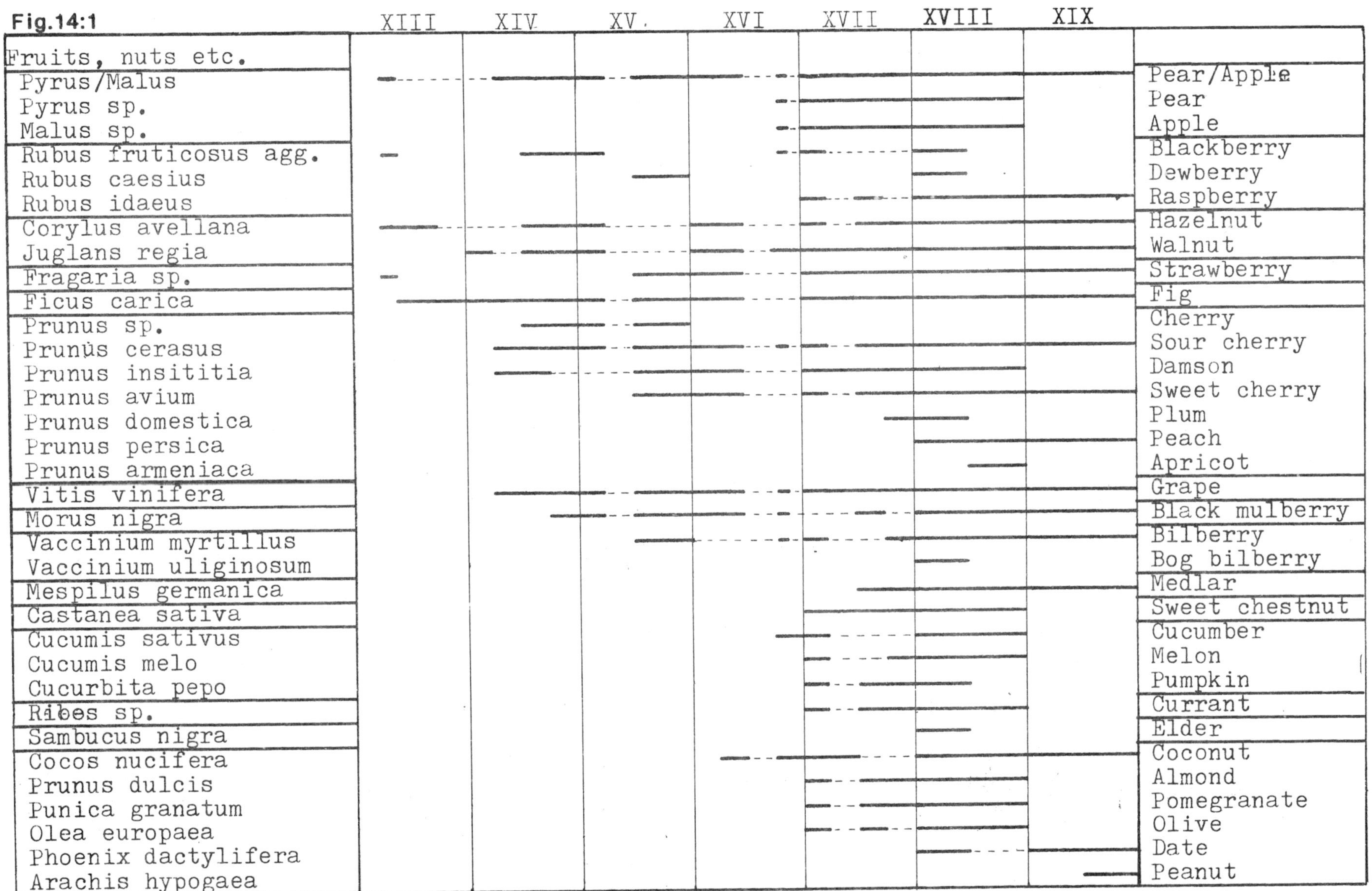

XIII
XIV
XV
XVI
XVII
XVIII
XIX
Fruits, nuts etc.
Pyrus/Malus
Pyrus sp.
Malus sp.
Rubus fruticosus agg.
Rubus caesius
Rubus idaeus
Corylus avellana
Juglans regia
Fragaria sp.
Ficus carica
Prunus sp.
Prunus cerasus
Prunus insititia
Prunus avium
Prunus domestica
Prunus persica
Prunus armeniaca
Vitis vinifera
Morus nigra
Vaccinium myrtillus
Vaccinium uliginosum
Mespilus germanica
Castanea sativa
Cucumis sativus
Cucumis melo
Cucurbita pepo
Ribes sp.
Sambucus nigra
Cocos nucifera
Prunus dulcis
Punica granatum
Olea europaea
Phoenix dactylifera
Arachis hypogaea
Pear/Apple
Pear
Apple
Blackberry
Dewberry
Raspberry
Hazelnut
Walnut
Strawberry
Fig
Cherry
Sour cherry
Damson
Sweet cherry
Plum
Peach
Apricot
Grape
Black mulberry
Bilberry
Bog bilberry
Medlar
Sweet chestnut
Cucumber
Melon
Pumpkin
Currant
Elder
Coconut
Almond
Pomegranate
Olive
Date
Peanut

Fig.14:2

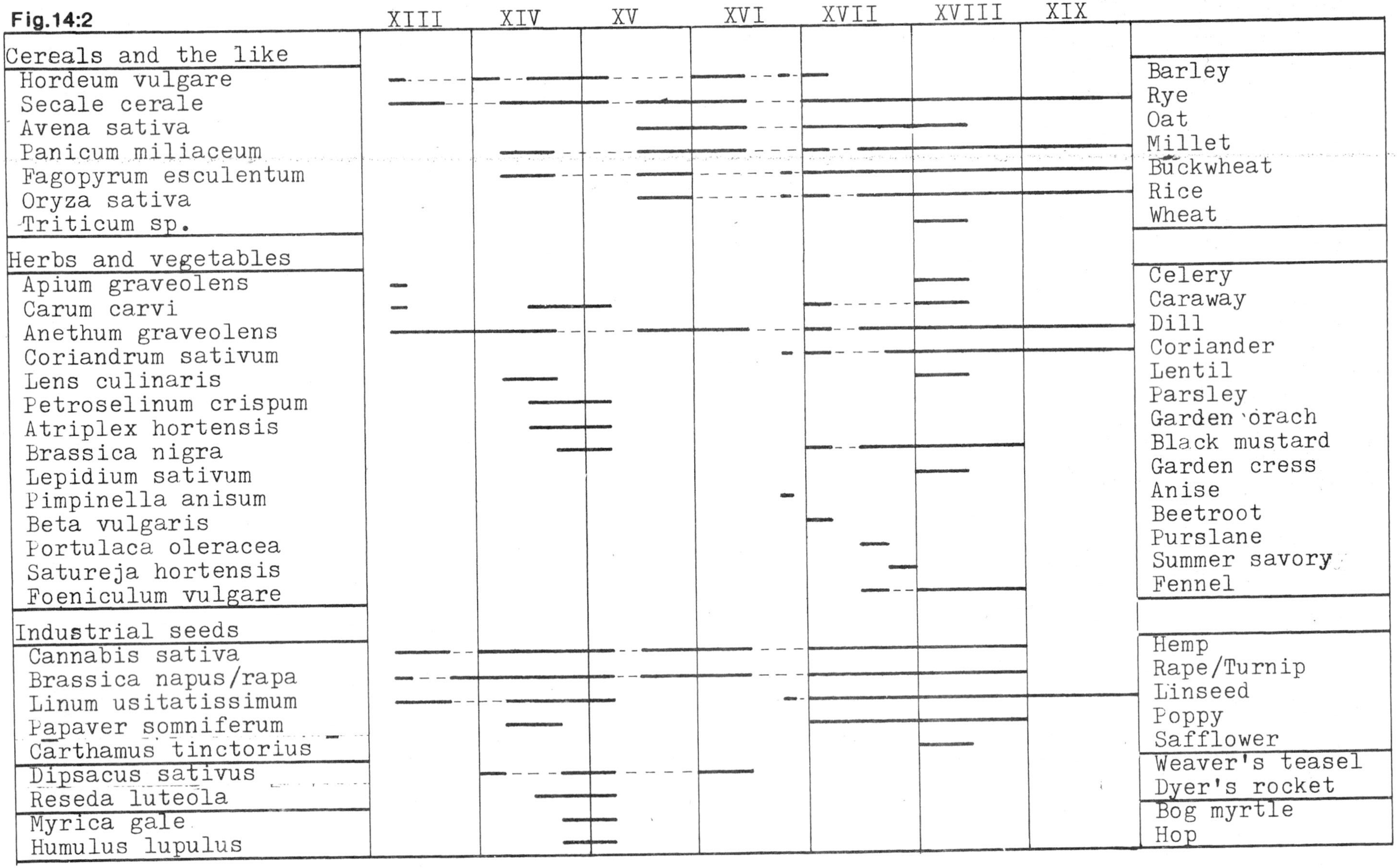
XIII
XIV
XV
XVI
XVII
XVIII
XIX
Cereals and the like
Hordeum vulgare
Secale cerale
Avena sativa
Panicum miliaceum
Fagopyrum esculentum
Oryza sativa
Triticum sp.
Herbs and vegetables
Apium graveolens
Carum carvi
Anethum graveolens
Coriandrum sativum
Lens culinaris
Petroselinum crispum
Atriplex hortensis
Brassica nigra
Lepidium sativum
Pimpinella anisum
Beta vulgaris
Portulaca oleracea
Satureja hortensis
Foeniculum vulgare
Industrial seeds
Cannabis sativa
Brassica napus/rapa
Linum usitatissimum
Papaver somniferum
Carthamus tinctorius
Dipsacus sativus
Reseda luteola
Myrica gale
Humulus lupulus
Barley
Rye
Oat
Millet
Buckwheat
Rice
Wheat
Celery
Caraway
Dill
Coriander
Lentil
Parsley
Garden orach
Black mustard
Garden cress
Anise
Beetroot
Purslane
Summer savory
Fennel
Hemp
Rape/Turnip
Linseed
Poppy
Safflower
Weaver's teasel
Dyer's rocket
Bog myrtle
Hop

The species have been ordered in four main groups:

1. Fruits, nuts etc.

2. Cereals and the like.

3. Herbs, spices and vegetables.

4. Industrial seeds.

The first and last groups are again divided in clusters so that more or less comparable species appear side by side, such as Pyrus sp. (pear) and Malus sp. (apple). The order of these clusters is chronological, according to the first occurence of the earliest representative of a cluster. Within the cluster the order is also chronological.

The scarcity of species in the earlier centuries has to do with the nature of the samples. They originate from the habitation levels of the heightening layers at the "Nieuwendijk". In later periods they come also from cesspits, which are not generally found in Amsterdam before the 16th century.

The lines for Pyrus/Malus and Prunus sp. correspond to seeds that were too damaged to allow closer identification.

1. Fruits, nuts etc.

When looking at the diagram, almost immediately the questions arise: what was local, what was imported, when did the importation of food plants start? To answer these questions in detail, a thorough study of historical sources should be undertaken. As long as this is not done, one has to use the available literature. For Amsterdam a good piece of work is Sangers (1952). Many of the interpretive remarks below originate from this work.

Corylus avellana (hazelnut) and Juglans regia (walnut) are nearly always found, although in small amounts. They may be partly of native origin, but walnuts were already being imported from Germany in the 14th century, and hazelnuts have arrived here in great quantities from the Baltic area since the 16th century. Fragaria sp. (strawberry) may have been picked only in the earlier centuries, but when importation started it is hard to discover by seed evidence alone. There seems to be no morphological difference between pips of wild and cultivated strawberries. It is certain however that since the 17th century Beverwijk, a little village about 20 kilometers north-east of Amsterdam was known as centre of cultivation of strawberries and exported them almost exclusively to Amsterdam. Vitis vinifera (grape) would be normally imported from abroad, but around 1800 Hoorn, a town about 30 kilometers north of Amsterdam, was growing grapes solely for Amsterdam. A highly speculative origin of the grapes could be the home growing of them by jews of the "Waterlooplein". The reason would be that they were not allowed to drink wine produced by non-Jews (Elize van Dienst:

personal communication). The last cluster of highly exotic fruits does not include citrus-fruits. These fruits are never listed among the food residues from archaeological contexts, and it seems probable that their remains very easily and completely decompose.

2. Cereals and the like.

Within this group Oryza sativa (rice) is an interesting component. It occurs in nearly every cesspit at the "Waterlooplein". The way of life among Portuguese and Spanish Jews is said to have had some oriental affinities, and rice was a common staple food (Encyclopadia Judaica: Food). The origin of the rice may have been Italy. Indications for this are the find of Scirpus mucronatus, a weed from mediterranean ricefields, and the known intensive trade contacts of Amsterdam with Italy.

The species Hordeum vulgare L. emend (barley), Secale cereale L. (rye), Avena sativa L. (oat) and Triticum sp. (wheat) are represented by occasional chaff fragments or accidently mineralised grains which are in most cesspits. The amounts are too small to draw conclusions about their occurrence.

3. Herbs, spices and vegetables

The last comment on the cereals is equally valid for this group. Apart from a few species of which the seeds were used, like Brassica nigra (black mustard) and Pimpinella anisum (anise), most species belonging to this group are grown for their leaves. It is therefore accidental that seeds of these species are found in a cesspit. The only possible way for this to happen seems to be the gathering and drying of the plants in autumn (Greig 1982: 51). It is clear however that this group is underrepresented, in species as well as in periods of occurrence.

4. Industrial seeds

This group starts with a cluster of oilseeds. Of great interest here is the species Carthamus tinctorius (safflower), nowadays used as a source of vegetable oil for margarine production. Papaver somniferum (poppy) is grouped here with the oil seeds, but at the "Waterlooplein" its main use was different. The Jewish sabbath bread "challe" was sprinkled abundantly with poppy seeds, while the Haman's ears, small pies eaten at Purim, were stuffed with them. (Encyclopaedia Judaica: Food).

The finds of Dipsacus sativus (weaver's teasel) and Reseda luteola (dyer's rocket) indicate the presence of someone who was employed in the finishing process of textile manufacture. The dried-out inflourescences of the teasel were used to roughen fabrics. This plant was cultivated for this purpose in Germany and France (Chomel 1778: 1370). The great quantities of Myrica gale (bog myrtle) and Humulus lupulus (hop), also found at the "Nieuwendijk", suggest the presence of a brewer in this area. Although this conclusion is often drawn to eagerly, even based on only a few hop-seeds, it seems justified here. One sample consists of myrtle branches, leaves and seeds, while a sample from the near vicinity and the same date contains about 1.5% hop seeds.

Quantitative analysis. As mentioned above, the "Waterlooplein" excavation offered good possibilities to see whether social differences could be traced in the food remains.

The excavated area was known to consist of a well-to-do to very rich part and a lower class to very poor part (Baart 1982). For the present analysis five samples were randomly chosen each dated to the first half of the 18th century from each part. To test whether the observed difference in pottery between the two parts (Jan. Baart: personal communication) was really a difference on all levels, discriminant analysis was thought most appropriate. Its application consists of finding a combination of discriminating variables that is responsible for the greatest difference between the groups. Species were used as discriminating variables that occured in both groups and in at least three samples per group. All samples had the same volume, so that absolute amounts could be transformed to proportions indicating the relative importance of the species within a sample.

As discriminant analysis assumes that the discriminating variables have a normal distribution, the arcsine transformation (Sokal and Rohlf 1969: 386-387) was used on the proportions calculated to produce a distribution that a least approximated to the normal distribution.

The version of discriminant analysis used was the one in SPSS (Klecka 1975). The criterion used for finding the discriminating species was stepwise Rao's V. V is the measure for distance between the groups for each variable. The stepwise method selects first the variable that discriminates most between the groups, next the variable that discriminates most in combination with the first one etc. Before a variable is actually accepted it must pass a tolerance test to see if it will really have an influence upon the result.

The results are shown in table 14:2. An interesting division into two groups is seen. It is based on a few discriminating species, which are certainly not the species that were generally expected (with the possible exception of *Mespilus germanica*). The groups are each about 40 standard deviations from 0, which is the grand mean of all the classified cases. The probability that a case is really that far from the group centroid is given below (maximum probability is 1.0000):

sample number	probability	group
8	.7579	1
109	.8183	1
118	.1272	1
191	.5932	1
195	.0325	1
12	.5873	2
17	.5582	2
94	.8771	2
161	.9325	2
163	.9748	2

Table 14:1

	Group 1: richer part					Group 2: poorer part				
Sample numbers	8	109	118	191	195	12	17	94	161	163
Pyrus sp.	215	91	59	235	58	87	199	151	105	151
Malus sp.	130	69	100	463	60	222	-	50	168	70
Rubus fruticosus agg.	2	-	-	-	-	-	9	2	4	-
Rubus caesius	3	-	-	-	-	-	-	-	-	-
Rubus idaeus	27	-	16	65	1	2	9	54	81	-
Corylus avellana	30	1	12	1	25	14	7	2	5	-
Juglans regia	-	1	1	1	-	1	-	2	2	1
Fragaria sp.	4080	-	918	3703	1101	914	174	1634	1169	540
Ficus carica	248	855	2465	2441	1669	579	2496	719	3504	1958
Prunus cerasus	123	6	41	21	8	8	30	34	35	30
Prunus insititia	47	3	83	35	13	14	16	15	14	5
Prunus avium	34	8	65	86	26	176	79	37	105	16
Prunus domestica	9	1	12	-	-	4	-	1	-	-
Vitis vinifera	762	448	1673	892	984	726	2044	716	1462	503
Morus nigra	67	-	-	-	2	2	-	-	117	-
Vaccinium myrtillus	5	1	2	61	-	-	131	150	-	-
Mespilus germanica	41	3	99	3	56	4	39	24	-	7
Castanea sativa	6	-	-	1	-	1	-	2	1	1
Cucumis sativus	-	-	-	-	16	-	-	-	8	6
Cucumis melo	20	-	-	8	-	-	-	3	-	-
Cucurbita pepo	-	2	-	-	-	-	-	-	-	-
Ribes rubrum	1460	36	1931	1233	1020	4338	3942	3171	2546	2296
Ribes uva-crispa	150	-	-	72	-	284	92	36	44	62
Ribes nigrum	-	-	-	22	-	23	-	46	-	-
Sambucus nigra	1	-	-	-	-	-	1	3	-	-
Cocos nucifera	1	-	-	-	-	-	-	-	-	-
Prunus dulcis	1	1	5	-	-	-	-	-	-	-
Punica granatum	38	15	-	-	-	-	-	-	-	-
Olea europaea	11	2	-	10	3	8	7	2	2	2
Phoenix dactylifera	-	-	-	-	1	-	-	-	-	-
Secale cerale	-	-	-	1	-	1	-	2	179	20
Avena sativa	-	-	-	-	-	-	-	-	-	-
Panicum miliaceum	1	16	-	1	-	-	-	3	7	-
Fagopyrum esculentum	4	2	1	8	5	4	-	-	8	1
Oryza sativa	3	2	1	6	5	6	34	1	1	8
Triticum sp.	-	-	-	-	-	1	-	-	-	-
Apium graveolens	1	-	-	-	-	-	-	-	-	-
Carum carvi	-	-	-	7	-	46	-	-	-	-
Anethum graveolens	-	-	-	-	-	-	-	3	-	-
Coriandrum sativum	22	4	-	2	-	-	7	3	4	-
Lens culinaris	-	-	3	-	2	-	-	-	-	2
Brassica nigra	175	57	-	45	-	25	59	57	7	-
Lepidium sativum	-	-	-	1	-	-	-	-	-	-
Foeniculum vulgare	-	-	1	-	-	-	-	-	-	-
Cannabis sativa	3	1	3	-	-	-	34	-	2	-
Brassica napus/rapa	146	1	-	-	-	8	136	-	3	-
Linum usitatissimum	1	-	-	-	-	-	1	-	6	5
Papaver somniferum	-	-	12	30	20	10	1	10	1	155
Carthamus tinctorius	-	1	-	-	-	-	-	-	-	-
indet.	-	14	3	-	-	-	-	-	17	-

Table 14:2: HISTOGRAM OF THE GROUPS

```
                   4 +                                                                                 +
                     .                                                                                 .
              F      .                                                                                 .
              R      .                                                                                 .
              E    3 +                    2                                       1                    +
              Q      .                    2                                       1                    .
              U      .                    2                                       1                    .
              E      .                    2                                       1                    .
              N    2 +                   22                                       1                    +
              C      .                   22                                       1                    .
              Y      .                   22                                       1                    .
                     .                   22                                       1                    .
                   1 +                   22                                     111                    +
                     .                   22                                     111                    .
                     .                   22                                     111                    .
                     .                   22                                     111                    .
                    OUT.........+.........+.........+.........+.........+.........+.........+..........OUT
                               +60       -40       -20        0         20        40        60
CLASSIFICATION        22222222222222222222222222222222222222221111111111111111111111111111111111111111
GROUP CENTROIDS                           2                                        1
```

Discriminating species:
- for group 1 (richer): Fragaria sp., Malus sp., Pyrus sp., Mespilus germanica, Oryza sativa
- for group 2 (poorer): Ribes rubrum, Prunus avium

It is clear that the samples from group 2 have a greater probability of belonging to their group than those from group 1. The probability of the samples 118 and 195 in particular is very low.

This first trial to see if social differences in food remains could be traced seems rather successful. It has, however, no general value. The selected species have only discriminating value for this subset of 10 samples. Only when more subsets are treated in the same way, could it be shown that certain species are constantly discriminating. These species would have then general value as indicators of difference between samples.

FINAL REMARKS

Only the qualitative analysis that has led to a result that is complete in itself. As for the historical interpretation a lot of work remains to be done. A thorough study of historical sources should be made to find out from where imported species came, when their use started and/or finished, what their (relative) price was etc. This could only be done by a historian and should not be undertaken by someone who has only the possibility to collect minor bits and pieces. The acquisition of detailed information about the people living at the "Waterlooplein" would also be useful. What does it mean when you say someone there is poor, how rich were the rich and so on. The quantitative analysis is highly preliminary too. As mentioned before, many subsets will have to be analysed before the results can be of general value.

Acknowledgements

Apart from all the people who have helped me with this work I would thank especially Drs. B. Voorrips, Instituut voor Prae en Protohistorie van de Universiteit van Amsterdam, for his help with statistics and computer use, and the Association of Environmental Archaeologists in the person of Martin Jones, Department of Archaeology of the University of Durham, for offering me the possibility of publishing this paper.

Bibliography

Baart, J. (1982) Amsterdam 1. *Holland* 14 214-217

Baart, J. (1982) Opgravingen bij Nieuwendijk tonen het oudste Amsterdam. *Ons Amsterdam* 34 86-90

Baart, J., Krook, W., Lagerweij, A., Ockers, N., Van Regteren Altena, H. Stam, T., Stoepker, H. Stouthart, G., Van der Zwan, M. (1977) *Opgravingen in Amsterdam, 20 jaar stadskernonderzoek*. Amsterdam: Fibula-Van Dishoeck

Chomel, N. (1778) *Algemeen huishoudelijk-, natuur-, zedekundig- en konst- Woordenboek*. Leeuwarden: J.A. de Chalmot pp. 1369-1370: Kaardendistel

Encyclopaedia Judaica (1971). Jerusalem pp. 1414-1427: Food

Greig, J. (1982) Garderobes, Sewers, Cesspits and Latrines. *Current Archaeology* 85 49-52

Klecka, W.R. (1975) Discriminant Analysis. In: *SPSS: statistical package for the social sciences* (ed Nie, N.H.) New York: McGraw-Hill pp. 434-467

Pals, J.P. (1972) Vruchten en zaden uit de stadsgracht. In: *Vondsten onder de St. Olofskapel* (ed Van Regteren Altena, H.H.) Amsterdam: Instituut voor Prae- en Protohistorie pp. 39-40

Pals, J.P. (in press) Plantaardigheden onder Damrak 49

Sangers, W.J. (1952) *De ontwikkeling van de nederlandse tuinbouw*. Zwolle: Tjeenk Willink

Sokal, R.R. and Rohlf, F.J. (1969) *Biometry*. San Francisco: W.H. Freeman and Company

Vaz Dias, A.M. (1938) Een verzoek om de joden in Amsterdam een bepaalde woonplaats aan te wijzen. *Vijf en dertigste jaarboek van het genootschap Amstelodamum* 181-202

15. FEEDING LINCOLN IN THE 11TH CENTURY - A SPECULATION

T. P. O'Connor

The use of the word "speculation" in the title of this paper is deliberate. I propose to consider the meat production sector of the subsistence economy of Lincoln for the years around the Norman Conquest, the mid 11th century, with the wider intention of providing justification for considering the medieval economy of England not on a site, or even regional, basis, but as one coherent economic structure consisting of many interdependent micro-economies. In order to undertake such a reconstruction, it will be necessary to guesstimate numerous items of data, and to take a very simple view of the meat production, distribution, and consumption processes. Accordingly, what follows must not be seen as an attempt to generate a model of the economy of the city; it is a series of informed guesses about some aspects of what might have been happening.

The first point to establish is the approximate population, or at least the meat-consuming population, of Lincoln during the period in question. At the time of the Domesday survey, Lincoln is reckoned to have comprised 1270 households (Hill 1948: 54). If we assume that each household was just a large family, with unusually large households being balanced out by solitary maiden aunts, and then deduct a factor to allow for small children eating a negligible amount of meat, 1270 households could represent a meat-eating population of around 4000 people. How much meat did each of these 4000 people consume in a single, typical year? The present national average is 63.5 kg, compared with 45.5 kg around the end of the last century (Melvyn Howe 1972: 54). Medieval accounts of diet are not helpful: it is seldom apparent which stratum of society is being described. For the purposes of this speculation, I wish to take a medieval average of 25 kg per person per annum to represent consumption of beef, mutton and pork. Fish and fowl were undoubtedly important elements in the diet, but in order to simplify the calculations, they are assumed not to have been important as bulk elements. Multiplying this estimate of per capita consumption by the estimate of population gives a total annual meat consumption of 100,000 kg. This can probably be regarded as a minimum estimate: the values for population and consumption are probably both rather low.

In order to convert this total annual meat consumption estimate into numbers of livestock, we need to estimate what proportion of the meat supply was provided by each of cattle, sheep, and pigs. The evidence on which the following estimates are based is contained in O'Connor (1982: 12), and only summarised here. Taking relative abundance estimates based on minimum numbers of individuals calculated for ten samples of bones from the period under consideration, and then weighting the numbers of individuals by factors derived from the estimated meat yield per individual for each of the three taxa, we arrive at figures of 80% beef, 13% pork, and 7% mutton. This is piling estimate upon estimate, and these figures can only be regarded as very approximate. By using figures for meat yield per individual based on data from extant primitive breeds, the percentages can be converted to figures of 500 cattle, 700 sheep, and 400 pigs. To this should be added a considerable number of hens and geese, fowls of the air, and fishes of the deep. To summarise, during a typical year in the mid 11th century, the citizens of Lincoln were munching their way through a considerable number of animals.

The full socioeconomic impact of this appetite can best be illustrated if we consider the smallest element, the 7% represented by sheep. Examination of sheep mandibles recovered from excavated features in the city dated to 1040-1070 shows a concentration on the slaughter of individuals aged 3-4 years, (O'Connor 1982: 23-4). Thus the total meat producing flock would have required adult breeding stock, and 1- and 2-year hoggs being fattened. Allowing for this, and assuming a modest rate of lamb production, it is difficult to see the annual 700 sheep being produced from a total flock of much less than 5000 head. Assuming a system of extensive, rather than intensive, grazing, we can perhaps expect a grazing intensity similar to that recorded from the Sussex Downs before the 18th century Improvements movement. Taking figures from Fussell (Fussell 1951: 80), we can calculate that a mere 7% of the meat supply was being raised on something like 2000 to 4000 hectares of grazing. To provide comparative figures for a different commodity, even with low medieval grain yields, the same area of land (although obviously not the same land) could have yielded between 20 million and 40 million litres of grain.

Obviously, this direct comparison of meat with grain is spurious, as the estimated 5000 head of sheep would have been producing wool and milk as well. Indeed, meat might well only have been a by-product.

The purpose of all these numbers has been to give some illustration to the general statement that medieval cities would have had an all-pervading influence on the economies of their hinterland. Providing 500 cattle, 700 sheep, and 400 pigs for slaughter each year would have engaged the output of an appreciable proportion of rural Lincolnshire, quite apart from the meat requirements of the rural population. And there are wider ranging effects. The bones recovered from 11th century deposits include species of wildfowl which are typically abundant in estuarine and coastal areas during the winter months, implying a seasonal trade in hunted birds over distances of 60 km or more. In addition to these birds, bones of marine fish were also recovered, indicating regular trade with the coast (O'Connor 1982: 44-6).

It is quite obvious that the subsistence economy of this particular city was inextricably tied up with that of most of Lincolnshire, and probably beyond. Thus any events which affected the urban subsistence economy would have had some effect in distant and supposedly rural settlements.

The excavation at Flaxengate, in the lower walled area of Lincoln, provided a useful example of just such an economic event. Immediately after his success at Hastings, William the Conqueror set about pacifying the North of England by garrisoning castles in major towns. One such castle was constructed in Lincoln. In the years immediately following the Conquest, this castle was garrisoned, bringing several hundred more meat-eaters into the city, and this appears to have had a fairly striking effect on the urban subsistence economy. Roughly contemporaneous with the garrisoning of the castle, samples of sheep and pig mandibles show an increased slaughter of young individuals, with large numbers of 1- and 2-year sheep appearing in domestic debris for the first time. By about 1080, the economy seems to have adjusted, and the majority of sheep were slaughtered at 3 to 4 years old, as before. What appears to have happened is that the garrisoning of the castle put sudden pressure on the meat supply, and the increased demand was met by slaughtering the sheep earlier and not waiting for them to have produced two annual clips of wool. Interestingly, the earlier slaughter age for pigs is maintained from the Conquest right through the medieval period, (O'Connor 1982: 49).

The effect which garrisoning the castle had on the urban economy can thus be identified and explained. What is less clear, but nonetheless important, is what effect these events would have had on small, ostensibly rural, communities living perhaps 10 to 20 km away from the city, where the sheep were being raised.

The moral of the story is this. Roman and post-Roman Britain, or at least the southern half of it, was economically dominated by its cities. When examining economic evidence from a rural site, the possible effect that historically recorded events in the cities may have been having in the countryside must be taken into account. When considering subsistence economies, we cannot afford to work in isolation, or to consider any one site on its own. It seems quite likely that in areas where urban catchments overlap, the economic information derived from one city might be directly relevant to the interpretation of another.

During the discussion which followed the presentation of this paper, a number of points were raised which emphasised the difficulties inherent in any attempt to model a subsistence economy. These included such imponderables as the degree of adherence to fast days, the role of non-bone producing animal foods such as offal and milk products, and the additional estimates which have to be made for the consumption of the food-producing rural population. Possibly the speculative figures contained in this paper will initiate attempts (preferably by somebody else) to construct such a model. Either way, it is hoped that this paper has drawn attention to the need to view the economic evidence from each site as just one part of a complex of inter-related macro- and micro-economies.

Bibliography

Fussell, G.E. (1951) Four centuries of farming systems in Sussex. Sussex Archaeological Collections 90 60-95

Hill, J.W.F. (1948) Medieval Lincoln Cambridge: University Press

Melvyn Howe, G. (1972) Man, Environment and Disease in Britain Newton Abbot: David & Charles

O'Connor, T.P. (1982) Animal Bones from Flaxengate, Lincoln c 870-1500. The Archaeology of Lincoln XVIII-I London: C.B.A.